Lecture Notes in Computer Science   1998
Edited by G. Goos, J. Hartmanis and J. van Leeuwen

W0106547

Springer
Berlin
Heidelberg
New York
Barcelona
Hong Kong
London
Milan
Paris
Singapore
Tokyo

Reinhard Klette   Shmuel Peleg
Gerald Sommer (Eds.)

Robot Vision

International Workshop RobVis 2001
Auckland, New Zealand, February 16-18, 2001
Proceedings

Springer

Series Editors

Gerhard Goos, Karlsruhe University, Germany
Juris Hartmanis, Cornell University, NY, USA
Jan van Leeuwen, Utrecht University, The Netherlands

Volume Editors

Reinhard Klette
The University of Auckland
Center for Image Technology and Robotics (CITR Tamaki)
Tamaki Campus, Building 731
1005 Auckland, New Zealand
E-mail: r.klette@auckland.ac.nz

Shmuel Peleg
The Hebrew University of Jerusalem, Department of Computer Science
Givat Ram, Ross Building
91904 Jerusalem, Israel
E-mail: peleg@cs.huji.ac.il

Gerald Sommer
Universitäet Kiel, Institut für Informatik
Preusserstr. 1-9, 24105 Kiel, Germany
E-mail: gs@ks.informatik.uni-kiel.de

Cataloging-in-Publication Data applied for

Die Deutsche Bibliothek - CIP-Einheitsaufnahme

International Workshop RobVis <2001, Auckland>:
Robot vision : proceedings / International Workshop RobVis 2001,
Auckland, New Zealand, February 16 - 18, 2001. Reinhard Klette . . .
(ed.). - Berlin ; Heidelberg ; New York ; Barcelona ; Hong Kong ;
London ; Milan ; Paris ; Singapore ; Tokyo : Springer, 2001
  (Lecture notes in computer science ; Vol. 1998)
  ISBN 3-540-41694-3

CR Subject Classification (1998): I.4, I.2.9, I.5

ISSN 0302-9743
ISBN 3-540-41694-3 Springer-Verlag Berlin Heidelberg New York

Springer-Verlag Berlin Heidelberg New York
a member of BertelsmannSpringer Science+Business Media GmbH
© Springer-Verlag Berlin Heidelberg 2001
Printed in Germany

Typesetting: Camera-ready by author, data coversion by DA-TeX Gerd Blumenstein
Printed on acid-free paper   SPIN 10782191   06/3142   5 4 3 2 1 0

Preface

On behalf of the organizers we would like to welcome all participants to the "Robot Vision 2001" workshop. Our objective has been to bring together researchers in robot vision, and to promote interaction and debate. Participants of the workshop come from Europe, US, the Middle East, the Far East, and of course from New Zealand.

Fifty-two papers were submitted to the workshop, and each paper was thoroughly reviewed by at least three reviewers. Seventeen papers were selected for oral presentation, and seventeen papers were selected for poster presentation. There were no invited technical papers, to give all participants the sense of equal opportunity.

The technical scope of the workshop is very wide, and includes presentations on motion analysis, 3D measurements, calibration, navigation, object recognition, and more. The schedule of the workshop was therefore prepared to allow, in addition to the technical presentation, ample time for discussions and interaction. We hope that interaction among researchers of such different areas, yet all part of robot vision, will result in better understanding and research of the robot vision area.

February 2001   Reinhard Klette, Shmuel Peleg, and Gerald Sommer

Organization

RobVis 2001 was organized by the Center for Image Technology and Robotics (CITR), Tamaki campus, The University of Auckland.

Co-chairs

Reinhard Klette (Auckland, New Zealand)
Shmuel Peleg (Jerusalem, Israel)
Gerald Sommer (Kiel, Germany)

Program Committee

Jacky Baltes (Auckland, NZ), Thomas Bräunl (Nedlands, AUS), Ross Clarke (Hamilton, NZ), Georgy Gimel'farb (Auckland, NZ), Atsusuhi Imiya (Chiba, J), Reinhard Klette (Auckland, NZ), Claus-E. Liedtke (Hanover, D), Bruce MacDonald (Auckland, NZ), Takashi Matsuyama (Kyoto, Japan), Allan McIvor (Auckland, NZ), Josef Pauli (Kiel, D), Shmuel Peleg (Jerusalem, IL), Moshe Porat (Haifa, IL), Gerald Sommer (Kiel, D), Bill Trigs (Grenoble, F), and Friedrich Wahl (Braunschweig, D).

Local Organizing Committee

Jacky Baltes, Georgy Gimel'farb, Ulrich Günther, James Harper, Reinhard Klette, Cecilia Lourdes, Bruce MacDonald, S. Manoharan, Sudhir Reddy, and Sharon Walker.

Sponsors

The International Association for Pattern Recognition
MORST/DFG
Institute of Electrical and Electronic Engineers (IEEE) - NZ North Section
The University of Auckland, Tamaki campus
Visual Impact Auckland Ltd.

Table of Contents

Poster Session 2: Robotics & Video

Computational Stereo

Robotic Vision

Image Acquisition

Visual Cues for a Fixating Active Agent

Mårten Björkman and Jan-Olof Eklundh

Computational Vision and Active Perception Laboratory (CVAP)
Department of Numerical Analysis and Computing Science
Royal Institute of Technology (KTH), S-100 44 Stockholm, Sweden
{celle,joe}@nada.kth.se

Abstract. In order for an active visual agent to act in a dynamic environment, it needs the ability to fixate onto objects that might be of interest. In this article we will discuss issues concerning the design of a binocular system with such capabilities. The problems range from gaze shifting and saccading, to epipolar geometry and ego-motion estimation. In the end of the paper it will be shown how scene parts of independent motion, that will be used to trigger saccades, can be efficiently detected.

1 Introduction

A person that moves around in the world, while looking at various locations and things in his way, will experience that objects will enter and leave his field of view, due to his ego-motion or the motion of the objects. He will sometimes shift his gaze to such objects, to find out e.g. if their trajectories cross his path or to determine what kind of things they are. In fact, he may be looking for specific types or instances of objects, for example they may be obstacles to avoid or something he needs.

These activities form part of what an intelligent agent, acting and existing in the world, use visual perception for. In our research we are engaged in a long-term effort to study principles and methods for developing such agents and implement them in terms of mobile robots, capable of a set of behaviors, including also grasping and manipulating things. These systems will be engaged in various foreground and background tasks, defined by their interests and drives. In this paper we will discuss work on the early visual mechanisms of such an active and purposive agent, and present how it can derive a spatial understanding of its environment that can serve higher level processes, such as recognition, and also motor behaviors.

2 Gaze Shifting and Saccading

Depending on the task at hand, an active visual agent moves its gaze differently from occasion to occasion. An agent that has found an object of interest, fixates onto the object and tracks it as it moves, in order to gather as much information about the object as possible. If the agent is looking for something special, it keeps

R. Klette, S. Peleg, G. Sommer (Eds.): Robot Vision 2001, LNCS 1998, pp. 1–9, 2001.
© Springer-Verlag Berlin Heidelberg 2001

moving its cameras between different locations, extracting enough information to judge whether an object is of interest has been found. It could also be the case that the agent is not at all interested in what is going on in the neighborhoods and moves its cameras only to stabilize the image data. To detect objects that might enter the scene, it is necessary to have a background process, that reacts on radical changes in the images. If the visual data is not stabilized, such changes might not be properly detected.

There can be a number of reasons why saccades are triggered in a system like ours. If the agent loses interest in an object that is being tracked, the cameras may be moved towards a more interesting part of the scene, which means that saccades are triggered by the robot itself in a top-down fashion. A triggering might also originate from low-level processes that pass information about possible interesting objects entering the field of view bottom-up, so that the visual system can react accordingly. Consequently, it is necessary for an artificial system to include low-level vision processes that are capable of detecting areas, that might be of interest, over the whole visual field.

A difficult question is what is supposed to be considered as interesting. Objects that directly affect the performance of the agent, will always be necessary to identify. There are a number of visual cues that might be important for the agent to know where to look next and since they all might be useful, they have to be considered in parallel. In this study we have been concentrating on the ability to identify areas of independent motion. The work is much in the spirit of [12], but rather than using image motion, figure-ground segmentation is performed using motion in 3D. In cases of translations, it might otherwise be hard to separate moving objects from the background.

3 Epipolar Geometry Estimation

In order to identify image regions as objects in 3D space, an agent would benefit from the use of binocular disparities. Disparities have been used extensively in active vision and robotics, but the applications have often been limited to behaviors such as navigation and obstacle avoidance [16,9] or simply binocular tracking [4]. In robotics, cameras have typically been mounted in parallel, so as to simplify the procedure of calculating the actual disparities. However, if the binocular camera system is to be used in for example manipulation, parallel cameras might lead to objects only being visible in one of the two cameras. In order to maximize the applicability and flexibility of the system, it is necessary for the system to be able to verge dynamically.

The reason why verged cameras are seldom used in practise, is because it is hard to keep track of the epipolar geometry, that is the relative position and orientation of the cameras. Knowing the epipolar geometry is essential if image features are to be matched between the two cameras and distances measured in metric space. Most stereo head systems have counters on their motors, which can be used to estimate the epipolar geometry. However, all systems include delays and if the cameras are in continuous motion it is often hard to know the true relation between two newly grabbed images. The problem gets even harder when the computational load of the system is varying as the agent goes from one task to another. In our study, we have concluded that motor counters are preferably used to get a rough estimate of the epipolar geometry, but if you like to calculate a dense disparity map, the epipolar geometry better be estimated using information available in the images themselves.

3.1 The Essential Matrix

Typically the essential matrix $\mathbf{E}$ is used to describe the relationship between the projections of 3D points onto the left and right image planes [11]. For a system such as the one used in this study, an image point $\mathbf{x_l}$ in the left image is constrained to a line, defined by the corresponding projection in the right image $\mathbf{x_r}$, according to the equation:

$$\mathbf{x_l^T E x_r} = \begin{pmatrix} x_l \\ y_l \\ 1 \end{pmatrix}^T \begin{pmatrix} 0 & -\sin(\alpha_l) & 0 \\ -\sin(\alpha_r) & 0 & \cos(\alpha_r) \\ 0 & -\cos(\alpha_l) & 0 \end{pmatrix} \begin{pmatrix} x_r \\ y_r \\ 1 \end{pmatrix} = 0. \qquad (1)$$

In the equation the projections, $\mathbf{x_l}$ and $\mathbf{x_r}$, are in homogeneous coordinates. The pan angles, α_l and α_r, are the unknown parameters to be searched by the epipolar estimation process. The stereo head is constrained such that it only involves two degrees of freedom, even if a typical binocular stereo system might have as many as six [3]. However, rotations around the optical axes do not change the visual data, only its orientation, and will not be needed here. Since the system always will be in fixation, there is no relative tilt between the cameras. A joint tilt of both cameras does not change the nature of the problem.

In our work, the angles α_l and α_r are iteratively searched in a least square framework. In order to minimize the influence of outliers, we use random sampling. Random sets of six points each are generated and for each set the angles are estimated. Unlike RANSAC [6,15], where a winner is selected among the resulting estimates, we simply calculate the mean of all values that can be considered as feasible. It turns out that sets including outliers result in slower convergence and can easily be eliminated from the final result. Further details and evaluations can be found in [2].

3.2 An Optical Flow Model

The essential matrix has got one major disadvantage. If the cameras are located almost parallel, the least square problem described in Section 3.1 will be illconditioned and the results can not be trusted. The problem originates from the fact the flow induced by a small rotation can not be separated from a translation along the baseline, since the depths have been eliminated in the essential matrix and the magnitude of the translational flow can not be used.

In the presented system, we use an alternative model based on optical flow [10], that is typically used in structure from motion algorithms. The disparities, that is the difference in image position between the two cameras, can be described as follows:

$$\begin{pmatrix} dx \\ dy \end{pmatrix} = \begin{pmatrix} 1 + x^2 \\ xy \end{pmatrix} \beta + \frac{1}{Z} \begin{pmatrix} \cos(\alpha_l) - x\sin(\alpha_l) \\ -y\sin(\alpha_l) \end{pmatrix}. \tag{2}$$

The vergence angle β is the sum of the two pan angles, α_l and α_r, but unlike the essential matrix, the depth Z has not been eliminated from the equation and has to be estimated as well. The reason why this approximate model is rarely used in stereo vision, is because it collapses if the difference in position and orientation between the cameras is too large. However, our study show that the model is in fact appropriate for a stereo head system under typical working conditions.

The unknown parameters are solved iteratively, with one pass estimating the angles and another pass determining the depths. Considering the fact that horizontal disparities never change ordering as the cameras are verging, it is possible to get a rough estimate of the vergence angle, that can be used to initialize the procedure. We match a guessed median depth of points in 3D space to the median disparity of extracted feature points and calculate the corresponding vergence angle. Simulations show that the procedure will converge, even if the initial depth is relatively far away from the truth. If the guess is within a factor of two from the true value, the procedure rarely diverges, but if the error is as large as 400%, the iterations often diverge for large vergence angles.

3.3 Simulations

Both methods were tested through a number of simulations based on randomly generated 3D points, spread around an area 10 to 30 baselines in front of the cameras. Gaussian noise, with a standard deviation of one pixel, was added to the points that were projected onto two 360×288 pixel image planes. Each simulation included about 500 feature correspondences, out of which 20% were outliers. These outliers, that were modeled with an additional noise source of 30 pixels, represent features that have been wrongly matched, in the sense that the two projections do not originate from the same point in 3D space.

The results can be divided into two components, a rotational component and a translational one, that is the vergence angle and the position of right camera in the coordinate frame of the left. Figure 1 shows the translational and rotational errors for a number of cases. The rotational error is about $0.2°$ for the optical flow method and slightly larger for the essential matrix. For near parallel systems, the translation is considerably harder to estimate and for vergence angles of $2°$, this error can be as much as ten times larger. Asymmetric systems are slightly harder, then symmetric ones. The major difference between the methods is in the convergence. As the essential matrix method rarely converges for parallel systems, the second method have problems with vergence angles larger than $15°$, especially if the initial median depth is far from the truth. The optical flow method is about twice as fast as the first method and requires about 27 ms on a 195 MHz MIPS R10K processor.

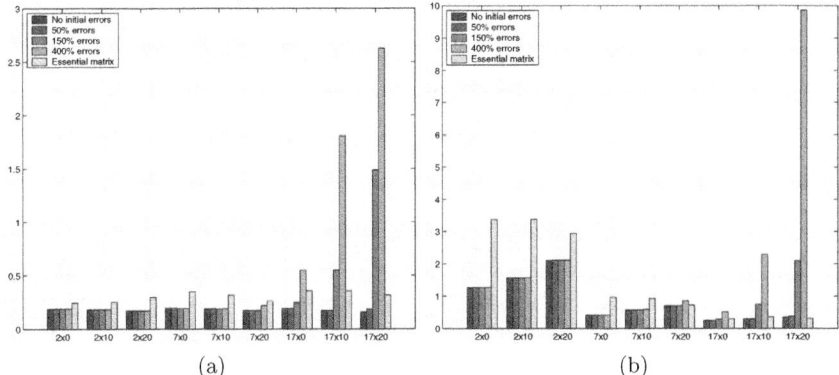

(a)   (b)

Fig. 1. The standard deviation in degrees of the rotational (a) and translational (b) errors for various combinations of true rotations and translations. The results of the optical flow based method are shown in the first four bars of each group, for different errors in the initial median depth. The last bar in each group represents the results based on the essential matrix

4 Ego-Motion Estimation

Knowing the motion of the agent itself, the ego-motion, is essential in order to stabilize the image data and find image regions of independent motion. Just like the case of epipolar geometry estimation using motor feedback, odometry can be used to get an idea of the ego-motion, but the problem is somewhat harder than that. If any conclusions are to be drawn from the images based on ego-motion, the estimated ego-motion has to be relative to the cameras, not to the agent itself. As the agent is moving, the neck is rotating and the cameras are constantly changing orientation, its terribly hard to know the exact direction and position of the cameras at every instance of time, especially since all systems involve delays that may be hard to predict. It becomes a necessity for the system to estimate the ego-motion based on extracted image data, even if an initial prediction based on odometry might still be of good use.

In a monocular system estimating ego-motion is a very difficult problem, if the cameras are in translation. There are a number of alternative methods that can be used [8,14], but so far none of these reach the requirements on robustness and speed needed for a system like ours. Fortunately, the problem is greatly simplified in the binocular case. In our system we use reconstructed 3D features, that are easily found once the epipolar geometry is known, and try to minimize the equation

$$\sum_{i=1}^{N} \|\mathbf{y_i} - (\mathbf{R}\mathbf{x_i} + \mathbf{t})\|^2, \tag{3}$$

where $\mathbf{R}$ is a rotation matrix, $\mathbf{t}$ a translation and, $\mathbf{x_i}$ and $\mathbf{y_i}$ are the reconstructed feature points of two consecutive time frames. Since it can be expected that the clouds of 3D points have the same shape in both instances of time, the translation can be found as the difference in position between the centers of the two clouds, that is $\mathbf{t} = \bar{\mathbf{y}} - \bar{\mathbf{x}}$. Subtracting the centers from the 3D points, we get two new sets of points $\acute{\mathbf{x}}_i = \mathbf{x}_i - \bar{\mathbf{x}}$ and $\acute{\mathbf{y}}_i = \mathbf{y}_i - \bar{\mathbf{y}}$.

We use an approach based on singular value decomposition (SVD), as proposed by Arun et al [1]. It can be shown that minimizing Equation 3 is equivalent to maximizing

$$f_2(\mathbf{R}) = \sum_{i=1}^{N} \acute{\mathbf{y}}_i^{\mathbf{T}} \mathbf{R} \acute{\mathbf{x}}_i = \text{Trace}(\mathbf{RH}), \tag{4}$$

where $\mathbf{H} = \sum_{i=1}^{N} \acute{\mathbf{x}}_i \acute{\mathbf{y}}_i^{\mathbf{T}}$. The rotational component $\hat{\mathbf{R}}$ that maximizes $f_2(\mathbf{R})$ can be found by first performing a SVD of $\mathbf{H}$. The decomposition $\mathbf{H} = \mathbf{UDV^T}$ consists of two orthogonal matrices $\mathbf{U}$ and $\mathbf{V}$, and a diagonal matrix $\mathbf{D}$ of non-negative elements. The minimizer of Equation 4 is then given by $\hat{\mathbf{R}} = \mathbf{VU^T}$, which will be used as our estimated rotation. In practise we perform these operations twice and use weights, to minimize the influence of reconstructed points with large errors.

4.1 Simulations

Simulations show that the rotational errors are rarely more than about 0.1°, which only happens in cases of large rotations and small translations parallel to the image plane. In such cases the translational error might be as large as 6° in direction and 0.03 baselines in speed, whereas typical errors are 0.02° in rotation, 2° in translational direction and 0.01 baselines in speed. The feature points involved in the simulations were generated such as described in Section 3.3. Since features have to be visible in both cameras at two different instances of time, only 100 reconstructed points were used, with 20% being outliers representing areas of independent motion. A systematic error of 0.3° was added to the vergence angle before reconstruction, modeling the inaccuracy of the epipolar geometry estimation. The computational cost of the method is as low as 3 ms, excluding the matching of corner features. Since it relies on reconstructed 3D points, the features have to be matched in stereo, as well as in motion, such as shown in Fig. 2. The feature extraction itself costs as much as about 15 ms per image, using Harris [7] corner detector. Further details about the matching processes and how outliers are identified, can be found in [2].

5 Independent Motion

The benefit of knowing the ego-motion of the cameras is the fact that images of future time frames can be predicted. Changes in the scene can then easily be detected through simple image subtraction, even if the agent itself is in continuous motion. In the system presented here, disparities are calculated using a method based on correlations and dynamic programming [13,5]. These calculations are

(a)   (b)   (c)

Fig. 2. The left image (a), the right image including stereo feature correspondences (b) and features matched in motion (c)

done after images have been rectified, using the results from the epipolar estimation. Using the disparities, such as the example in Figure 3(a), and the estimated ego-motion, a warping is then performed from the current image to the next one. This predicted image is then subtracted from the true next image. Results after thresholding can be seen in Fig. 3(b). Points above the threshold are finally put into a three-dimensional histogram, the largest peak of the histogram is found and a segmentation, such as the one shown in Fig. 3(c), can be done.

(a)   (b)   (c)

Fig. 3. The disparity map (a), the result after warping and subtraction (b) and image regions of independent motion (c)

The methods presented in previous sections were implemented on a Nomad 200 platform, powered by a 450 MHz Pentium III. The complete system runs in about 6 Hz, including feature extraction, matching, epipolar and ego-motion estimation, as well as the calculation of dense disparity maps and segmentation.

6   Discussion

In the presented paper we have been dealing with the problem of designing a system capable of dynamic fixation. The technical part of the paper included a presentation of two different methods for estimating the epipolar geometry of a binocular system. It was shown that this is in fact feasible in terms of speed, as well as robustness. Once feature points have been reconstructed in 3D space, the ego-motion was estimated at a very low computational cost. In the end of the article we show how knowledge about depth and ego-motion can be used to find image region of independent motion. An issue that has not been covered is how to exploit the available 3D information during tracking and gradually improve the speed as new information becomes available. How information about objects in 3D space is to be stored and updated is another important and difficult question.

References

1. K. S. Arun, T. S. Huang, S. D. Blostein: Least-squares fitting of two 3-D point sets. *IEEE Trans. Pattern Analysis and Machine Intelligence*, **9** (1987) 698–700. 6

2. M. Björkman, J-O. Eklundh: Real-time epipolar geometry estimation and disparity. Tech. Report ISRN KTH/NA/P-00/09-SE, NADA, Royal Institute of Technology (March 2000). 3, 6

3. M. J. Brooks, L. de Agapito, D. Q. Huynh, L. Baumela: Direct methods for self-calibration of a moving stereo head. In: *Proc. of the 4th European Conf. on Computer Vision*, (1996) 415–426. 3

4. D. Coombs, C. M. Brown: Real-time binocular smooth-pursuit. *Int. Journal of Computer Vision*, **11** (1993) 147–165. 2

5. I. J. Cox, S. L. Hingorani, S. B. Rao, B. M. Maggs: A maximum likelihood stereo algorithm. *Computer Vision and Image Understanding*, **63** (1996) 542–567. 6

6. M. A. Fischler, R. C. Bolles: Random sampling consensus: a paradigm for model fitting with applications to image analysis and automated cartography. *DARPA Image Understanding Workshop*, (1980) 71–88. 3

7. C. Harris, M. Stephens: A combined corner and edge detector. In: *Proc. of the 4th Alvey Conf.*, (1988) 189–192. 6

8. D. J. Heeger, A. D. Jepson: Subspace methods for recovering rigid motion I: algorithm and implementation. *Int. Journal of Computer Vision*, **7** (1992) 95–117. 5

9. K. Konolige: Small vision systems: hardware and implementation. *8th Int. Symposium on Robotics Research*, Hayama, Japan (1997). 2

10. H. C. Longuet-Higgins, K. Prazdny: The interpretation of a moving retinal image. *Proc. of Royal Society of London*, **B-208** (1980) 385–397. 3

11. H. Longuet-Higgins: A computer algorithm for reconstructing a scene from two projections. *Nature*, **293** (1981) 133–135. 3

12. D. W. Murray, K. J. Bradshaw, P. F. McLauchlan, I. D. Reid, P. M. Sharkey: Driving saccade to pursuit using image motion. *Int. Journal of Computer Vision*, **16** (1995) 205–228. 2

13. Y. Ohta, T. Kanade: Stereo by intra- and inter-scanline search using dynamic programming. *IEEE Trans. on Pattern Analysis and Machine Intelligence*, **7** (1985) 139–154. 6

14. J. Oliensis: Computing the camera heading from multiple frames. In: *Proc. of IEEE Computer Vision and Pattern Recognition*, (1998) 203–210. 5

15. P. H. S. Torr, D. W. Murray: The development and comparison of robust methods for estimating the fundamental matrix. *Int. Journal of Computer Vision*, **24** (1997) 271–300. 3

16. V. Tucakov, M. Sahota, D. Murray, A. Mackworth, J. Little, S. Kingdon, C. Jennings, R. Barman: Spinoza: a stereoscopic visually guided mobile robot. In: *Proc. the 13th Annual Hawaii Int. Conf. on System Sciences*, (Jan 1997) 188–197.   2

Tracking with a Novel Pose Estimation Algorithm

Bodo Rosenhahn, Norbert Krüger, Torge Rabsch, and Gerald Sommer

Institut für Informatik und Praktische Mathematik
Christian-Albrechts-Universität zu Kiel
Preußerstrasse 1-9, 24105 Kiel, Germany
{bro,nkr,tr,gs}@ks.informatik.uni-kiel.de

Abstract. In this paper we apply a novel pose estimation algorithm to the tracking problem. We make use of error measures of the algorithm which enable us to characterize the quality of an estimated pose. The key idea of the tracking algorithm is random start local search. The principle of the heuristic relies upon a combination of iterative improvement and random sampling. While in many approaches a manually designed object representation is assumed, we overcome this condition by using accumulated object representations and combine these successfully with the tracking algorithm.

1 Introduction

In this work we apply a novel 2D-3D pose estimation algorithm [12] to the tracking problem. This algorithm shows some interesting characteristics which makes it especially useful for this purpose. Beside features such as stability in the presence of noise and online–capabilities its main advantage in the tracking context is that it can unify different kinds of correspondences within one algebraic framework. which were

To apply the pose estimation algorithm to the tracking problem we intend to solve two problems which were avoided in [12] but are important for further applications like robot navigation or object recognition:

1. **Correspondences:** Correspondences between model data and image data have been defined manually.
2. **Object Representation:** A manually designed representation of the object to be tracked has been presupposed.

In this paper we describe an automatic procedure to find correspondences between an object model and its image projection which makes use of features of the pose estimation algorithm [12] and of the specific tracking condition. We suppose a 3D model of the object consisting of 3D points and 3D lines and we extract lines in the image sequence by a Hough transformation combined with a new algorithm to extract lines from the Hough array. We find correspondences between 3D lines and 2D lines by a local search. The essential attribute is that

R. Klette, S. Peleg, G. Sommer (Eds.): Robot Vision 2001, LNCS 1998, pp. 9–18, 2001
© Springer-Verlag Berlin Heidelberg 2001

a discrete local neighborhood of states is defined with respect to the current
state, in this context the Hamming distance n–neighborhood [11]. Further, we
allow correspondences only for entities with small distance. This assumption is
justified by the specific tracking situation. The pose estimation algorithm is able
to use correspondences as 3D point to 2D point, 3D point to 2D line and 3D line
to 2D line to estimate the rotation and translation between two frames. In this
paper only line correspondences are used. Note that this kind of correspondence
allows to avoid the so called aperture problem, i.e. the impossibility to define
correspondences between a point on a line in two frames.

To avoid a manually designed object representation we also applied the tracking algorithm with an accumulated object representation consisting of local 3D
line segments. The object accumulation is based on a scheme which accumulates
confidences for entities representing the object and which allows to extract representations in even quite complicated environments [4]. We could show, that
with such a representation tracking is possible and therefore both assumptions
of manual intervention in [12] can be substituted by automatic procedures.

2   Description of the Tracking

In this context tracking means to minimize a matching error by solving two
problems:

1. The correspondence problem: Determine the mapping between model elements (here 3D model lines) and image features (extracted Hough lines).
2. The spatial fitting problem (pose estimation): For each correspondence determine the best parameters (here rotation R and translation t), so that the
   spatial fit error of the model lines to image lines is minimized.

In the following sections we describe the automatic extraction of lines (section 2.1), the pose estimation algorithm (section 2.2), the automatic finding of
correspondences (section 2.3), and the accumulating of object representations
(section 2.4).

2.1   Hough Transformation

To extract lines in an image we apply the well known Hough transformation [3].
The robustness of the Hough transformation can be increased by using not only
information about the presence of edges but by also checking the agreement
of lines and local orientation, i.e. by applying the orientation selective Hough
transformation [9]. The Hough transformation results in an accumulator array
(see Fig. 1) from which the representative lines show up as peaks. These are
easily detectable for 'simple' images such as the one in Fig. 1 but difficult to
extract in more complex situations.

To avoid the extraction of additional lines caused by locally neighbored peaks
in the accumulator array (often occurring in the presence of noise in the image
data) usually some kind of metric on the accumulator array is defined to allow

Fig. 1. Standard-Hough-transformation and Orientation selective Hough-transformation

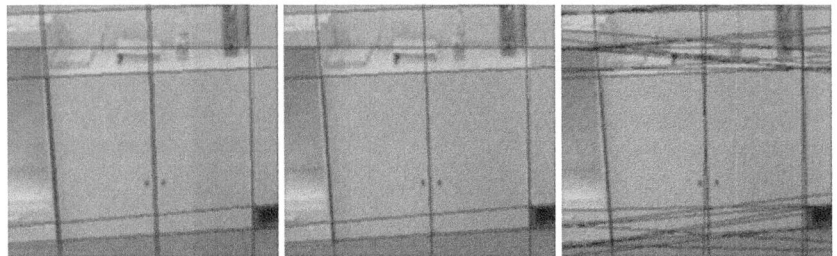

Fig. 2. Representative Hough lines extracted by different methods

only lines corresponding to peaks with certain distance. A problem of these methods is that important lines may have small distance in the Hough space (see e.g., narrow parallel lines in Fig. 2). To extract the significant lines we also use information about the areas which do support lines, i.e. we evaluate also image information. This allows us to extract lines with small distance in the accumulator array which are usually not extractable by other methods (for details see [1]).

Figure 2 shows extracted Hough lines using different kind of metrics. In the left image our method has been used, in the middle image for each selected peak a neighborhood in the accumulator array is set to zero (as, e.g. in [8]), while in the right image connected areas which occur after thresholding the accumulator array are treated as one line (as e.g., in [6]). Note that the narrow parallel lines could only be extracted by our method. The procedure used in the middle image extracts the most significant lines but not the narrow parallel lines because the corresponding peaks are too close in the accumulator array. The procedure used for the right image has great difficulties with locally neighbored peaks which are above threshold.

2.2 Pose Estimation

The problem of pose estimation means to estimate the transformation (the rigid body motion) between the two coordinate frames of measured data and model data. In [12,10] the problem of 2D-3D pose estimation is described in the algebraic language of kinematics. The key idea is that the observed 2D entities together with their corresponding 3D entities are constraint to lie on other, higher order entities which result from the perspective projection. The observed 2D entities in this context are extracted Hough lines.

To be more detailed, in the scenario of figure 3 we describe the following situation: We assume 3D points Y_i, and lines S_i of an object or reference model. Further, we extract line subspaces l_i in an image of a calibrated camera and match them with the model. Three constraints can be depicted:

1. **3D point 2D point correspondence:** A transformed point, e.g. X_1, of the model point Y_1 must lie on the projection ray L_{b1}, given by c and the corresponding image point b_1.
2. **3D point 2D line correspondence:** A transformed point, e.g. X_1, of the model point Y_1 must lie on the projection plane P_{12}, given by c and the corresponding image line l_1.
3. **3D line 2D line correspondence:** A transformed line, e.g. L_1, of the model line S_1 must lie on the projection plane P_{12}, given by c and the the corresponding image line l_1.

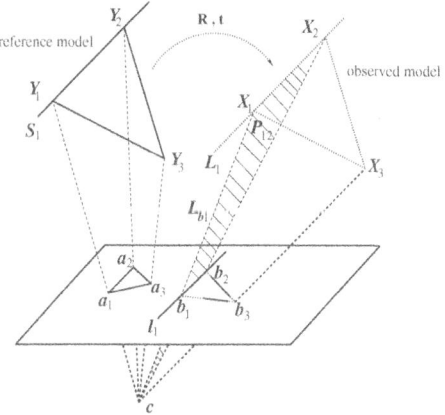

Fig. 3. The scenario. The solid lines at the left describe the assumptions: the camera model, the model of the object and the initially extracted lines on the image plane. The dashed lines at the right describe the actual pose of the model, which leads to the best fit of the object with the actual extracted lines

The use of the motor algebra [2] allows to subsume the pose estimation problem by compact constraint equations since the entities, the transformation of the entities and the constraints can be described economically in one unifying language. Furthermore the constraint equations express a natural distance measure, in this case the Hesse distance between the entities, which is also explained in [12]. This property is important for the robustness of our algorithms since we work with digital images and noisy data. To solve these constraint equations a special extended motor Kalman filter was developed [13].

2.3 Testing of Correspondences

It is well known, that for $l = m \times n$ potential pairs, there are $S = 2^{|l|}$ correspondences. This means, the search space is in general very large and not practicable for applications. The tracking assumption allows to use local crite-

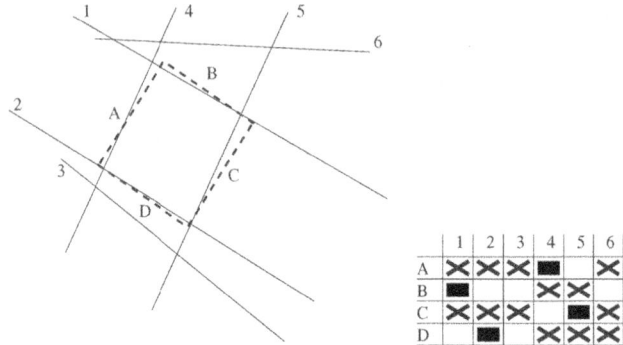

Fig. 4. Match example for a rectangle. The model lines are labeled with letters and the extracted image lines are labeled with numbers. The table indicates the correspondence space with the allowed possibilities (white/black), the impossible matches (cross) and the current match (black)

ria like distances and angles to reduce the search space significantly, depending on the error boundaries. In this context the correspondence space for m model lines and n image lines is represented by a $m \times n$ fit-matrix. In this matrix flags represent the needed information for a match, mismatch or potential match, Fig. 4 shows an example. In this example the model lines are labeled with letters and the extracted image lines are labeled with numbers. The table indicates the correspondence space with the allowed matches (white/black), the impossible matches (cross) and the current match (black). See also [11] for further information.

Random start local search [11] is the basis for our algorithm, which is summarized in Fig. 5. The principle of the heuristic relies upon a combination of iterative improvement and random sampling. Iterative improvement refers to a repeated generate-and-test principle by which the algorithm moves from an initial state to its local optimum. So the algorithm consists of two main steps: First find an initial state for a minimum of correspondences and then refine the result by the other correspondences. For the first step we choose five random model

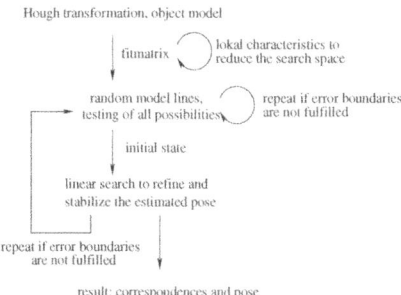

Hough transformation, object model

fitmatrix   lokal characteristics to
                 reduce the search space

random model lines,   repeat if error boundaries
testing of all possibilities   are not fulfilled

initial state

linear search to refine and
stabilize the estimated pose

repeat if error boundaries
are not fulfilled

result: correspondences and pose

Fig. 5. A scheme of the tracking algorithm

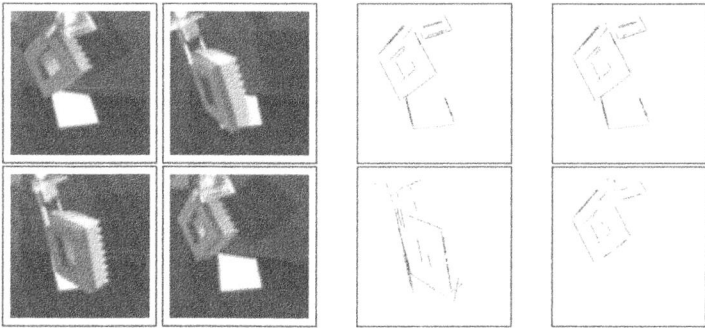

Fig. 6. Accumulation of an object representation (first and fifth iteration). The robot has physical control over the object. Line segments corresponding to the background vanish after a few iterations. Left: the stereo images of left and right camera. Middle: Representation extracted from one stereo image pair. Right: Accumulated representation

lines and try every combination of the object lines to the allowed image lines to estimate an optimal pose and use the error function to characterize the quality of the pose. This is possible since the error measure corresponds directly to the Hesse distance and leads to a suitable error measure. Once the initial pose is estimated, in the second step an additional model line will be tried to match an allowed image line to stabilize and refine the result. Note, that this part of the algorithm is linear, since the use of the Kalman filter leads to recognizable peaks for the detection of mismatches [14]. So the first assumption of [12], i.e. the knowledge of the correspondences can be solved by the algorithm, which is summarized in Fig. 5.

2.4   Object Accumulation

The second assumption, i.e., a manually designed object model can be avoided by applying the methods described above with a model extracted from a stereo

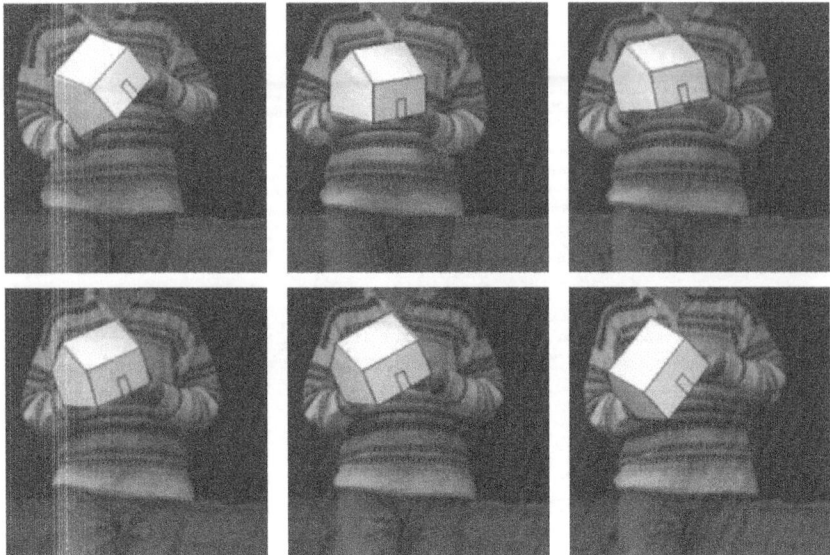

Fig. 7. Tracking with a manually designed object representation

image sequence. The key idea of the algorithm (described more precisely in [4] and [5]) is to accumulate evidences for entities used to represent an object over time. In our case the object was manipulated by a robot (see Fig. 6). This allows us to solve the correspondence problem during accumulation since the knowledge of the parameters of motion could be used in the accumulation scheme. Here the entities used to represent an object are local 3D line segments. However, the accumulation scheme can be applied for a wide range of visual entities. After forty iterations the object model was good enough to be applied in our tracking algorithm.

3   Experiments

In our first experimental scenario we use a manually designed model of a house for tracking. Figure 7 shows some results of the sequence with the superimposed model of the house. The slight displacements between the model and the house on the image in some of the frames emerge from calibration errors, extraction errors and match errors.

In our second experimental scenario we accumulate an object representation of a model house by the algorithm described in section 2.4. Our accumulated object model consists of 130 line segments.[1] Though the accumulated represen-

---

[1]   Our object representation consists of a large number of statistically very dependent entities. For matching it would be advantageous if these entities become connected

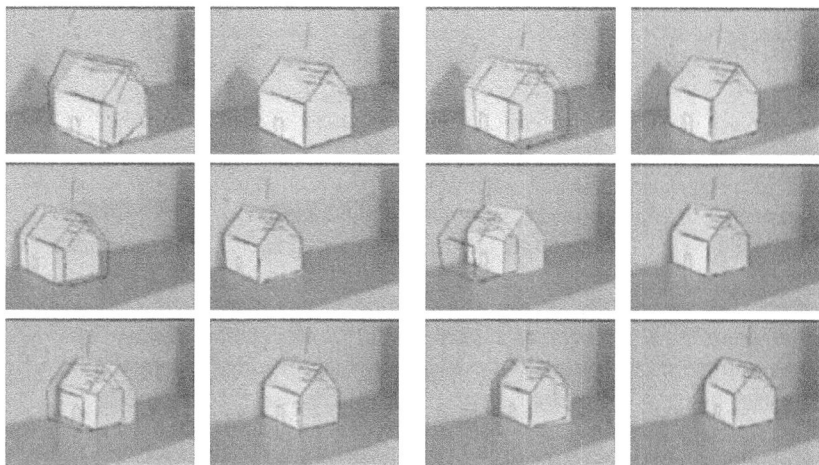

Fig. 8. Tracking with an accumulated object model. In this sequence we show the results before tracking and after tracking for each image to visualize the movements

tation also consists of noisy line segments, which do not belong to the house, the algorithm is able to estimate the transformations, which are necessary to get a good fit of the object model with the image lines. Since our algorithm is also able to neglect object lines, our algorithm is able to deal with hidden or not extracted object features in the image, or noisy line segments of the object model. Some results of the required and estimated movements are visualized in Fig. 8. The performance of our algorithm is not optimized yet and the main steps, the Hough transformation and the testing of correspondences are not in real time. The Hough transformation itself needs about two seconds, and the testing of the correspondences needs about 5 seconds to 15 seconds in artificial designed objects and 3 to 5 minutes with the accumulated object (because of its 130 line segments). But still the algorithm is heuristic and we also had cases where it never converged. The time performance is also dependent on the parameters of the tracking assumptions and the parameters of the Hough transformation.

4   Conclusion and Outlook

We applied the novel pose estimation algorithm described in [12] to the tracking problem. For tracking we could automatically find correspondences between model data and Hough lines by a local search algorithm. Furthermore, we could

by some kind of grouping process to achieve a representation with a smaller set of more complex features to speed up matching. The formalization of such grouping processes is part of our research.

demonstrate that tracking is even possible with an accumulated object representation.

In this paper we only used 2D line to 3D line correspondences. However, with a more elaborated object representation consisting of point features (such as corners) as well as line features, other kind of correspondences could be applied for tracking as well. The possibility to deal with these different entities within one framework as in the pose estimation algorithm in [12] would be an interesting extension of the tracking algorithm introduced here.

In this paper tracking and accumulation are distinct competences, for object accumulation it was necessary to have physical control over the object by a robot to solve the correspondence problem. With the tracking algorithm introduced here we aim to replace the need of physical control. The pose estimation gives us the parameters of object motion which are needed in our accumulation scheme and which were granted by the knowledge of the motor commands of the robot. Therefore, by combining tracking and accumulation we might achieve learning while doing object tracking.

All algorithms introduced here were implemented in the C++-software library KiViGraP [7] which allows us to combine competences as the one introduced in this paper into one system. In [5] a framework of such a system is discussed in which basic competences can be combined to more complex behavior patterns.

Acknowledgment

We would like to thank Kord Ehmcke and Oliver Granert for performing the experiments necessary to accumulate the object representation applied here. Furthermore we thank Yiwen Zhang for his implementation of the Kalman filter within KiViGraP.

References

1. M. Ackermann: Akkumulieren von Objektrepräsentationen im Wahrnehmungs–Handlungs Zyklus. Diplomarbeit, Lehrstuhl für Kognitive Systeme der Christian-Albrechts-Universität zu Kiel, (2000). 11
2. E. Bayro-Corrochano: The geometry and algebra of kinematics. In: *Geometric Computing with Clifford Algebra* (Sommer G., ed.), Springer, Berlin (2000). 12
3. P. V. C. Hough: Methods and means for recognizing complex patterns. U. S. Patent 3,069,654, (Dec. 18, 1962). 10
4. Norbert Krüger, Marcus Ackermann, Gerald Sommer: Accumulation of object representations utilizing interaction of robot action and perception. *DAGM 2000* (2000) 365–372. 10, 15
5. Norbert Krüger, Daniel Wendorff, Gerald Sommer: Two models of a vision–based robotic system: visual haptic attention and accumulation of object representations. In: *Proceed. Robot Vision 2001*, Auckland (2001). 15, 17
6. S. Kunze: Ein Hand-Auge-System zur visuell basierten Lokalisierung und Identifikation von Objekten. Diplomarbeit, Lehrstuhl für Kognitive Systeme der Christian-Albrechts-Universität zu Kiel, (2000). 11

7. KiViGraP (Homepage of the Kieler Vision and Grasping Project): http://www.ks.informatik.uni-kiel.de/~kivi/kivi.html. 17

8. J. Pauli: Geometric/photometric consensus and regular shape quasi-invariants for object localization and boundary extraction. Technical Report 9805, Christian-Albrechts-Universität zu Kiel, Institut für Informatik und Praktische Mathematik, (1998). 11

9. J. Princen, J. Illingworth, J. Kittler: An optimizing line finder using a Hough transform algorithm. *Computer Vision, Graphics, and Image Processing* **52** (1990) 57–77. 10

10. B. Rosenhahn, Y. Zhang, G. Sommer: Pose estimation in the language of kinematics. 2nd Internat. Workshop *Algebraic Frames for the Perception-Action Cycle*, AFPAC 2000, **LNCS 1888** (2000) 284–293. 11

11. J. R. Beveridge: Local search algorithms for geometric object recognition: Optimal correspondence and pose. Technical Report CS 93-5, University of Massachusetts, (1993). 10, 13

12. G. Sommer, B. Rosenhahn, Y. Zang: Pose estimation using geometric constraints. Technical Report 2003, Christian-Albrechts-Universität zu Kiel, Institut für Informatik und Praktische Mathematik, (2000). 9, 10, 11, 13, 14, 16, 17

13. Y. Zhang, B. Rosenhahn, G. Sommer: Extended Kalman filter design for motion estimation by point and line observations. 2nd Internat. Workshop *Algebraic Frames for the Perception-Action Cycle*, AFPAC 2000, **LNCS 1888** (2000) 339–348. 13

14. Y. Zhang, G. Sommer: 3D rigid object tracking in computer vision. Seminar presentation, Oberseminar, Institut für Informatik und Praktische Mathematik, Christian-Albrechts-Universität zu Kiel (July 2000). 14

Real-Time Tracking of Articulated Human Models Using a 3D Shape-from-Silhouette Method

Jason Luck[1], Dan Small[2], and Charles Q. Little[3]

[1] Colorado School of Mines, Dept. of Engineering
jluck@mines.edu
[2] Sandia National Labs, Intelligent Systems
University of New Mexico, Dept. of Computer Science
[3] Sandia National Labs, Intelligent Systems

Abstract. This paper describes a system, which acquires 3D data and tracks an eleven degree of freedom human model in real-time. Using four cameras we create a time-varying volumetric image (a visual hull) of anything moving in the space observed by all four cameras. The sensor is currently operating in a volume of approximately 500,000 voxels (1.5 inch cubes) at a rate of 25 Hz. The system is able to track the upper body dynamics of a human (x,y position of the body, a torso rotation, and four rotations per arm). Both data acquisition and tracking occur on one computer at a rate of 16 Hz. We also developed a calibration procedure, which allows the system to be moved and be recalibrated quickly. Furthermore we display in real-time, either the data overlaid with the joint locations or a human avatar. Lastly our system has been implemented to perform crane gesture recognition.

1 Introduction

Due to the enormous number of applications involving human-computer interaction, real-time 3D human motion-tracking has become a highly valued goal. Applications such as virtual reality, telepresence, smart rooms, human robot interaction, surveillance, gesture analysis, movement analysis for sports, and many others all have a need for real-time human motion-tracking. Accordingly there has been a lot of work done in this field. Most of the work has been done off-line, where images are acquired in real-time and analyzed at a later time [1,2]. However, this does not allow for any human-computer interaction. Therefore we must move to a real-time system. The approaches that work in real-time can be divided into two categories. The first category works in the 2D domain and attempts to get 3D information either from a single view or through combining 2D information from multiple cameras [3,4,5]. The problem with this type of approach is that modeling 3D articulated objects from 2D data is often an ill-posed problem due to occlusion/self-occlusion. The second category works directly in the 3D domain. At the time of this paper only one other team has attempted to

R. Klette, S. Peleg, G. Sommer (Eds.): Robot Vision 2001, LNCS 1998, pp. 19–26, 2001.

work directly in the 3D domain [6]. In their work ellipsoids are fit to body segments that provide a coarse model of the body. Our system uses a similar data acquisition approach, but instead fits a stick model of the human so that joint angles can be recovered. In this way a detailed analysis of the human motion can be performed for a wide variety of applications. Specifically, our system is recovering the shoulder and elbow joint angles, which allows us to use a much richer set of pose-analysis functions.

The system developed in this paper acquires data of an area composed of 500,000 3D volume elements (voxels) of dimension 1.5 inches at a rate of 25 Hz. The system then uses the data to track an eleven degree-of-freedom (DOF) humanoid stick figure model at 16 Hz, which is fast enough to track very rapid movements. Currently our upper body model incorporates the (x,y) horizontal location of the torso, a rotation of the torso about the vertical (z) axis, and 4 rotations for each arm (3 in the shoulder and 1 at the elbow). However, testing to allow 6 DOF in the torso and include leg dynamics has already begun. When the user is in the workspace real-time visual feedback is provided to the user in terms of viewing the voxels overlaid with the joint locations or viewing a human avatar driven from the tracking results. Our system has also been implemented for a crane gesture analysis system and achieved excellent results.

Our system can be divided into three main components. The first acquires our 3D data using 4 cameras and a technique called shape-from-silhouette. The second component uses the data to track our model. The last component uses the tracked model to provide feedback to the user and to perform gesture recognition.

2   The 3D Video Motion Detector

The system we are using to acquire 3D data is a real-time shape-from-silhouette sensor that we call the 3D Video Motion Detector, or 3DVMD. This sensor uses a combination of industry standard components including a high-end PC, four RS170 monochrome video cameras, and four PCI-bus frame grabber cards. Using this hardware we create a time-varying volumetric image of the visual hull of whatever object is moving in the space observed by all four cameras.

2.1   Algorithm Description

The algorithm for performing the shape from silhouette involves extracting silhouettes from the four images using an adaptive background subtraction and thresholding technique, much like the algorithm described in detail in [7]. This algorithm indicates which pixels have changed from the background in each of the cameras. The calibration procedure creates look up tables that store voxel-pixel associations, which relate each voxel to a pixel in each camera. By traversing the voxels and examining the appropriate image-pixels we can tell which voxels are occupied. Voxels that are occupied will have the appropriate pixel in each of the cameras active. To speed the process of traversing through the voxels we are defining a subvolume around the region in which the person is moving, thereby minimizing the search. This results in a very fast, low-latency system that is appropriate for our tracking work, as well as for development of uninstrumented 3D user interfaces. Figure 1 shows an example of our data.

Fig. 1. An example of 3DVMD data

2.2   Calibration of the 3DVMD Sensor

The 3DVMD system relies on an accurate calibration of each camera's intrinsic and extrinsic parameters. We estimate parameters for each camera's intrinsic characteristics by recording six images of a flat 8 by 10 inch checkerboard at a variety of angles between the target plane and the camera. From this information we can determine the focal length, center pixel location, and the coefficients of radial distortion[8]. The next step is to estimate the camera's world-frame position. We place targets on the floor of the space we are observing and determine their image-plane locations with subpixel accuracy. We then use the same planar calibration methods, but instead of solving for the intrinsic parameters, we now solve for the extrinsic parameters. Because all of the cameras observe the targets in the same positions, we are able to obtain the relative positions of each camera in a global coordinate frame. By using this analytical calibration method, we can now set these systems up much more rapidly than previous manual calibration methods.

3   Tracking the Humanoid Model

3.1   Initialization

To simplify tracking we require that the user performs an initialization pose upon entering the workspace. The pose is a simple cross formation with the user facing along the y axis with his arms extending parallel to the floor and straight out to the sides of the body (along the x axis). Once in this pose, the system measures the body parameters required for tracking: body radius, shoulder height, and arm length. Currently the user must always begin by initializing the system. However, the model parameters can easily be saved to a file and read in before the user enters the workspace. Our system already uses voice communication, therefore it would be easy to incorporate an option to simply tell the system your name while entering the workspace and skip the initialization phase.

3.2   Tracking

The tracking procedure it broken into two phases. The current algorithm first finds the (x,y) location and heading (the direction the person is pointed) of the torso. We can then predict the location of the shoulders, and assume that their locations remain constant for a particular torso orientation. These locations are then used to anchor the arms. In this way we only need to solve for the angles at which the upper arms extend from the shoulders, and then for which the lower arms extend from the elbows.

Currently we only compute the x, y location of the torso and a rotation about z; accordingly we are assuming the user is standing straight up. The x, y location can easily be computed using the median value of all the voxels. The heading is computed by fitting an ellipsoid to all the data within the body radius of the x, y location. This is done by performing an eigen-decomposition on the moment matrix M. The eigenvectors of this matrix correspond to the principal axes of the ellipsoid. The principal axis closest in orientation to the last known heading is taken as the new heading of the person.

$$M = 4 \begin{pmatrix} M_{xx} & M_{xy} & M_{xz} \\ M_{yz} & M_{yy} & M_{yz} \\ M_{xz} & M_{zy} & M_{zz} \end{pmatrix} \quad \text{where } M_{ab} = \frac{1}{n} \sum_{i=1,...,n} a_i b_i \qquad (1)$$

To compute the angles for a particular arm segment, we simply compute the angles from all voxels that are not within the body radius to the anchor point. The result is used as a pulling force from the current orientation to a new orientation that passes through this voxel, as shown in Fig. 2. In this way each voxel can have an effect on all of the arm segments, which overcomes the problem of having to decide which arm segment a voxel belongs to. However, we weight each pull by the distance, d, to each arm segment, so that a voxel exerts a stronger pull on closer arm segments than on those further away. For each voxel, let the minimum distance to any arm segment be denoted as d_m. The weight for each segment is given by $(d_m / d)3$. Accordingly as the model is pulled into the correct orientation, the forces exerted from a particular voxel should be almost entirely on the segment closest to the voxel. This weighting strategy works extremely well for small adjustments, which is all it should have to make since our tracking rate is extremely high. In addition we employ zero weighting to any arm segment further than a certain distance from the voxel (d_t, computed from a maximum arm velocity divided by our tracking rate). Also, if a voxel is within a very small distance of an arm segment (less than d_s), then we assume that it belongs to only that segment, and assign a zero weight to all other segments. The weighting assignment is shown below:

$$Weight = \begin{cases} 1 & \text{if} & d = d_m \\ (\frac{d_{min}}{d})^3 & \text{if} & d_m < d < d_t \\ 0 & \text{if} & d > d_t \\ 0 & \text{if } d > d_m \text{ and } d_m < d_s \end{cases} \qquad (2)$$

Once the adjustment is computed for all voxels outside of the body radius, an adjustment is made to each of the arm segment angles and the loop repeats. The process stops for each arm segment when either the adjustment is too small or the move was bad (fewer points are close to new orientation of the segment).

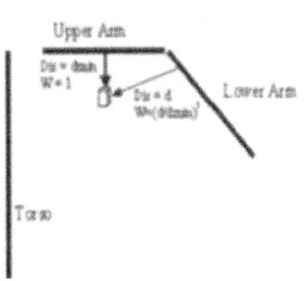

Fig. 2. Example of alignment forces (the weight to the upper arm is 1 while the weight to the lower arm, which is further away, is scaled down)

Currently our system stops tracking an arm when the arm lies along the body. If we hope to track the arms when they are extremely close to the body we must move to a much more precise torso model (work is in progress). In addition we will need data which has far less noise from shadows. Employing color cameras and using hue to distinguish shadows as done by Cheung and Kanade [6] might solve this problem.

The process also has a recovery algorithm in case it "loses" an arm. If too few voxels are close to one of the arm segments, the process assumes that the optimization has failed and attempts to grow the arm instead. Since we know the shoulder position, we start a growing algorithm from this point. Growing can include any neighboring point outside of the body radius, and continues until no new points are found. In this way the last point found should again be the hand and the elbow should be somewhere in the middle. Accordingly we compute the elbow to be at the voxel that was 2/3 of the way down the arm. Angles are calculated using these locations. The growing algorithm may not be able to find the arm if we have missing data (data acquisition has failed to find the arm voxels) or if the arm lies within the body radius. In this case our algorithm will leave the arm where it was last located and wait for the next set of data.

4   Results

The current system is able to collect data and track our model at 16 frames per second on one computer, while a second computer provides visual feedback to the user as shown in Fig. 3. Currently we are not able to indicate precisely how accurate our tracking is because we do not know the ground truth for the users movement within the workspace (see future work). However, the fit looks good when comparing the movement of the user and the avatar side by side (http://egweb.mines.edu/cardi/3dvmd.htm). In addition we were able to use the joint angles to distinguish several gestures to control a crane as will be discussed in the application section.

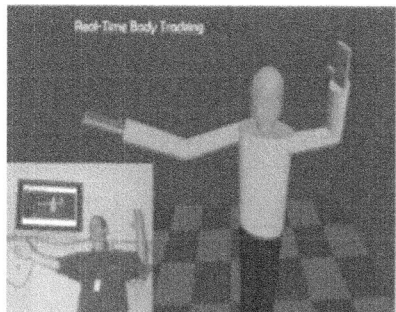 

Fig. 3. The two figures show the two display options. On the left is our human 3D avatar. On the right is a display of the voxels with large colored spheres representing joint positions (the body voxels have been removed)

In general the system worked extremely well when the arms were moved at normal velocities, and were held away from the body and from each other. However, when the arms were positioned close to one another the arm segments from one arm would be pulled towards the other. A similar problem occurred when the elbow was bent well past ninety degrees. In this configuration the pull from the upper arm voxels on the lower arm was often large enough to dominate. Accordingly when the actual arm was straighten the tracking algorithm would leave the elbow bent. Lastly when an arm was moved extremely fast the algorithm can fall behind and eventually loose the arm. In general our routine was able to detect that it had lost the arm in these situations, and would then revert to the growing algorithm to find the arm.

5   Application

We employed our system to perform crane gesture analysis. A separate computer, with a voice recognition system, was utilized to read the joint locations, use them to interpret gestures, and control the crane.

The process follows this pattern. When a user enters the workspace the computer says "Hello" and asks the user to verbally identify himself (eventually this will be used to skip the initialization step). The user is then asked to stand in the initialization pose and the system verbally communicates when initialization is done and gesturing can begin. Once in this state, the system continuously inspects the joint angles to see if a gesture is occurring. Because joint angles are being used for gesture interpretation the size of the user has no impact, and the user can perform a gesture facing any direction and even while moving. Currently the system is able to reliably interpret all of the gestures shown in Fig. 4. A movie can be seen at http://egweb.mines.edu/cardi/3dvmd.htm.

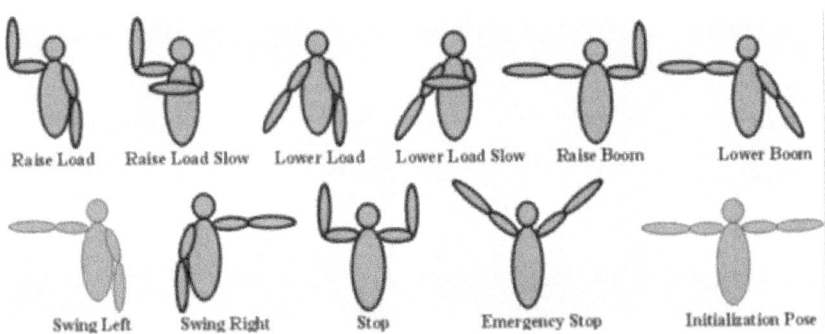

Fig. 4. The 17 gestures depicted were successfully recognized by our system (gestures on the top row can be performed with the arms switched)

6   Conclusions and Future Work

The system we have developed is able to collect high-resolution 3D voxel data at 25 Hz. We have also developed a calibration procedure that allows the system to be rearranged or even moved and to be quickly recalibrated. We are currently able to use the hardware and software that we have developed to track a low-dimensional human avatar figure at 16 Hz. The avatar we are using has a total of eleven degrees of freedom: four in each arm, and XY-theta for the position of the body. The system provides visual feedback in real-time with either human avatar display or through displaying the voxels overlaid with the joint positions. Lastly we were able to employ our system to perform crane gesture interpretation and successfully interacted with a system using a robotic arm to simulate crane movement.

By working directly in the 3D domain we eliminate problems caused by occlusion/self-occlusion, thereby allowing complicated motions to be accurately captured and tracked. By tracking a human stick figure model we are able to directly calculate human joint angles, which greatly simplifies motion analysis.

In addition, because our system works at a high cycle rate, we are able to track extremely fast motions.

This project is currently undergoing many improvements. We are currently testing a tracking algorithm that will use a tighter torso model and allow for six DOF movement. The new torso model will contain both shoulder and hip locations, hence our current tracking algorithm can be implemented to track the legs. Once this is complete we plan to further extend our algorithm to track the head and possibly even some hand orientations. We also plan to test the accuracy of our system by obtaining a ground truth to compare to the joint angles obtained through tracking. This will be accomplished by comparing our tracking results to values obtained with a traditional electromagnetic tracking system (Ascension's Flock-of-Birds). In addition, we plan to implement our system in several applications. For instance, the system will allow us to interact in real-time with another human avatar driven at a remote location (a 3D conference-call). Lastly we would like to extend our algorithms to allow for multiple people to interact within the workspace of a single sensor.

Acknowledgements

This work was funded by Sandia National Laboratories under contract PO 12727. Special thanks go to the Intelligent Systems and Robotics Center, in particular: Dan Schmidt, Brian Rigdon, Jeff Carlson, and Carl Diegert.

References

1. C. Bregler, J. Malik: Tracking people with twists and exponential maps. Proc. IEEE CVPR, (1998).
2. D. M. Gravrila: The visual analysis of human movement: a survey. *Computer Vision and Image Understanding*, **73** (1999) 82–98.
3. C. Wren, A. Azarbayejani, T. Darrel, A. Pentland: Pfinder: real-time tracking of the human body. *IEEE Transactions PAMI*, **19** (1997) 780–785.
4. I. A. Kakadiaris, D. Metaxas: 3D human body model acquisition from multiple views. *Proc. 5th Int. Conf. on Computer Vision*, Boston, MA (June 1995) 618–623.
5. D. Gravrila, L. Davis: 3-D model-based tracking of humans in action: a multi-view approach. In: *Proc. IEEE CVPR*, San Francisco (1996).
6. K. M. Cheung, T. Kanade, J. Y. Bouguet, M. Holler: A real time system for robust 3D voxel reconstruction of human motions. In: *Proc. IEEE CVPR*, (2000).
7. D. Snow, P. Viola, R. Zabih: Exact voxel occupancy with graph cuts. In: *Proc. IEEE CVPR*, (2000).
8. Open Source Computer Vision Library: http://www.intel.com/research/mrl/ research/cvlib/overview.html.

Hierarchical 3D Pose Estimation for Articulated Human Body Models from a Sequence of Volume Data

Sebastian Weik and C.-E. Liedtke

Institut für Theoretische Nachrichtentechnik und Informationsverarbeitung
University of Hanover, Germany
{weik,liedtke}@tnt.uni-hannover.de

Abstract. This contribution describes a camera-based approach to fully automatically extract the 3D motion parameters of persons using a model based strategy. In a first step a 3D body model of the person to be tracked is constructed automatically using a calibrated setup of sixteen digital cameras and a monochromatic background. From the silhouette images the 3D shape of the person is determined using the shape-from-silhouette approach. This model is segmented into rigid body parts and a dynamic skeleton structure is fit. In the second step the resulting movable, personalized body template is exploited to estimate the 3D motion parameters of the person in arbitrary poses. Using the same camera setup and the shape-from-silhouette approach a sequence of volume data is captured to which the movable body template is fit. Using a modified ICP algorithm the fitting is performed in a hierarchical manner along the the kinematic chains of the body model. The resulting sequence of motion parameters for the articulated body model can be used for gesture recognition, control of virtual characters or robot manipulators.

1 Introduction

In recent time emphasis has been put on the extraction of human body shape and motion parameters from videosequences. Application areas appear in the TV and film production where virtual actors have to be taught to exhibit human behavior like human facial expressions and human gestures. Another area of application is the control of remote systems from the passive observation of body motions. Examples are remote control of avatars in multi-player games or the remote control of robots which may act in hazardous and dangerous environments. The creation of those models consists mainly of two parts: firstly the extraction of the shape and texture of the real person and secondly the automatic adaptation and fitting of an interior skeleton structure to extract motion.

In this paper an approach for motion estimation using a hierarchic ICP algorithm is presented. This is illustrated in Fig. 1. From a real person an initial 3D surface model is obtained. The motion of the person and its model can be

R. Klette, S. Peleg, G. Sommer (Eds.): Robot Vision 2001, LNCS 1998, pp. 27–34, 2001.
© Springer-Verlag Berlin Heidelberg 2001

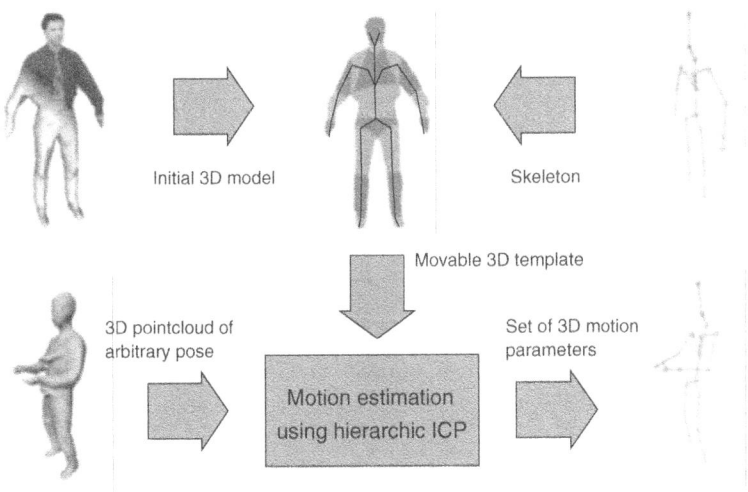

Fig. 1. System Overview for model based 3D estimation of arbitrary human poses

described by the motion of a skeleton, which is fitted to and connected to the model surface. When an arbitrary pose of the same person represented again as surface model is obtained the parameters of this pose can be determined by a model based approach. As model serves the movable 3D template which has been obtained in the first step. The result of the analysis is a set of 3D motion parameters, which describes for a sequence of poses the motion of the gesture.

2   Body Modeling

As shown in Fig. 1 the model based motion estimation requires two steps: the creation of a 3D segmented movable model of the person and the fitting of that model to a 3D measurement of the same person in an arbitrary pose. The first step is performed with a camera based passive full body scanner.

2.1   Shape from Silhouette

The *shape from silhouettes* or "method of occluding contours" approach is a well known technique for the automatic reconstruction of 3D objects from multiple camera views [4]. In this section the reconstruction technique is described briefly.

To capture the body model and to extract the sequence of volume data for motion estimation a special setup of sixteen digital cameras has been constructed which is suitable for using the shape-from-silhouette approach (Fig. 2). The person is situated in front of a monochromatic coated background which is used

Fig. 2. Principal measurement setup (left) and input image and segmented foreground (center and right)

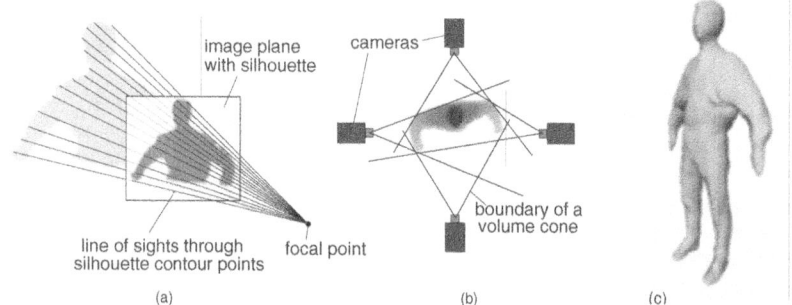

Fig. 3. Volume reconstruction: (a) Construction of a volumetric cone, (b) Topview of the cone intersection, (c) 3D modeling result

later on for silhouette extraction. The combination of background and camera positions need to fulfil mainly two important constraints: firstly, *all* cameras must see the complete person in front of the monochromatic background and secondly no camera should be visible from any other camera.

The principle of the silhouette-based volumetric reconstruction can be divided into three steps. In the first step, the silhouette of the real object must be extracted from the input images as shown in Fig. 2. In the proposed environment the segmentation of the person against the background is facilitated by using the monochromatic background ("blue screen technique").

In the second step, a volumetric cone is constructed using the focal point of the camera and the silhouette as shown in Fig. 3a. The convex hull of the cone is formed by the lines of sight from the camera focal point through all contour points of the object silhouette. For each view point such a volumetric cone is constructed, and each cone can be seen as a first approximation of the volume model.

In the last step, the volumetric cones from different view points are intersected in 3D and form the final approximation of the volume model. This is performed with the knowledge of the camera parameters, which give the information of the geometrical relation between the volumetric cones. In Fig. 3b a two dimensional top view of the intersection of the cones is shown. In Fig. 3c a triangulated 3D point cloud representing the volume model surface is shown.

After the reconstruction of the geometry the model can automatically be textured using the original camera images giving a highly realistic impression [1].

2.2   Skeleton Fitting

To extract the 3D motion parameters of a moving person a template based approach has been used. Therefore an internal skeleton structure is needed which controls the model movements. In order to find the correct set of motion parameters the body model has to be adapted to the specific person that is to be tracked later. Normally this requires a tedious manual positioning of the joint positions within the model. In order to reduce the costs of model creation it is desirable to automate this process. As opposed to other algorithms that use the thinning of 3D data[2][3] we propose to find the skeleton as a multi-step process based on re-projected images of the voxel model of the person.

In a first step a principal axis analysis is performed to transform the model into a defined position and orientation. Using a virtual camera – not to be confound with one of the real cameras – a synthetic silhouette from a frontal viewpoint is calculated. The outer contour of this image is used to extract certain feature points like the bounding box, the position of the neck, the hands and so on as can be seen in Fig. 4 on the left. In the last step the 2D joint positions of the desired skeleton are derived directly from the detected feature points using certain ratios (Fig. 4, left). Using the real model and the virtual camera these 2D joint positions are extended

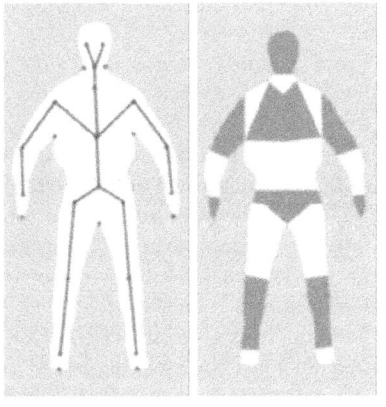

Fig. 4. Extracted features and calculated skeleton (left) - Automatic segmentation in body parts (right)

to their real 3D positions. In Fig. 4 on the right the skeleton has been used to segment the model into different body parts.

The resulting reference model describes the relation between the elements of the skeleton and the surface points of the model and it contains parameters like the bone length which are assumed to remain constant during the following motion analysis. The reference model has been derived from a special pose, which exhibits most distinctly the elements of the model skeleton, like the neck, head, the torso, the limbs and their parts. During pose analysis this reference model

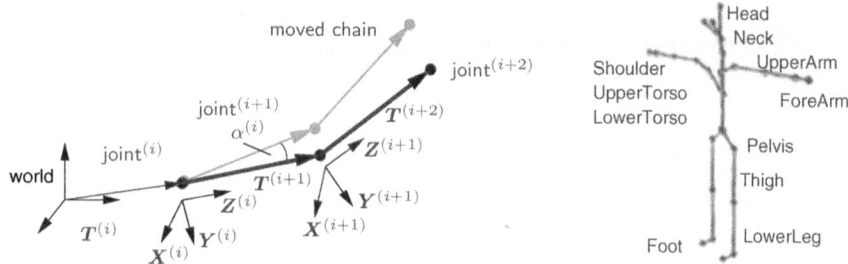

Fig. 5. The used coordinate systems within the kinematic chain(left) and the skeleton structure in neutral position (right)

serves as a movable 3D template in order to estimate the free parameters of the internal skeleton from the observation of the recorded surface points of the particular pose under investigation.

3   3D Motion Analysis

From the body scanner mentioned above a cloud of 3D surface points is obtained for each pose from a real person. The task of the motion analysis is to estimate for each pose the free parameters of the underlying skeleton, i.e. the location of the skeleton and the angular positions. An overview in the area of visual analysis of human movement can be found in [7].

The known approaches for this motion estimation problem can be divided into two different types. The first kind, which is often based on optical flow, tries to register the differential motion of an object between subsequent frames like for instance in [8]. This approach lacks the possibility to find an appropriate motion for sequences of arbitrary length because estimation errors from frame to frame add up until the tracking is lost. The second approach which is proposed here normally requires some kind of a (3D) model. It dispenses with the information yielded from prior processing stages and thus avoids the mentioned problem. Because of the larger movements between the initial pose of the model and the pose to be estimated finding the motion is more difficult. The proposed approach tries to eliminate manifolds in the solution by exploiting the motion hierarchy of the model.

3.1   Skeleton Hierarchy

The internal skeleton structure as shown in Fig. 5 on the right is organized in form of a kinematic chain as shown in Fig. 5 on the left. Each body part is described as a bone of a certain length and is connected to a parent and child part respectively through a joint. Each joint is equipped with a set of two local coordinate systems. The first gives the transformation of the parent to the child in its neutral position and the second coordinate system describes the actual movement around a joint. This approach has been chosen to be able to control maximum rotation angles around the axes of the fixed coordinate system. In addition the fixed local coordinate systems are oriented within the skeleton such that the Z-axis always runs along the longitudinal orientation of the body part and the X-axis is oriented along the viewing direction of the person. The Y-axis follows from a right handed coordinate system. This makes sure that for instance the maximum twist of a body part can always be controlled employing a minimum and maximum rotation angle around the Z-axis.

To transform the coordinates of a locally given point or into the global world coordinate system the following operation has to be performed:

$$P_G = \prod_{i=1}^{n} \mathbf{M}_R^{(i)} \cdot \mathbf{M}_B^{(i)}(\alpha, \beta, \gamma) \cdot P_L, \tag{1}$$

where the homogenous matrix $\mathbf{M}_R^{(i)}$ is constructed from the directions of the coordinate axes X, Y, Z and the position T of the fixed coordinate system:

$$\mathbf{M}_R^{(i)} = \begin{pmatrix} X^{(i)} & Y^{(i)} & Z^{(i)} & T^{(i)} \\ 0 & 0 & 0 & 1 \end{pmatrix}. \tag{2}$$

The homogeneous matrix $\mathbf{M}_B^{(i)}(\alpha, \beta, \gamma)$ is a rotation matrix

$$\mathbf{M}_B^{(i)}(\alpha, \beta, \gamma) = \begin{pmatrix} & & & 0 \\ & \mathbf{R}_{X,Y,Z}^{(i)}(\alpha, \beta, \gamma) & 0 \\ & & & 0 \\ 0 & 0 & 0 & 1 \end{pmatrix}, \tag{3}$$

whose coefficients contain the product of the rotations around the X-, Y- and Z-axis with the values α, β and γ respectively.

Each joint carries two coordinate systems that are responsible for the overall motion. $\mathbf{M}_R^{(i)}$ is the transformation of the fixed coordinate system and gives the neutral position as shown in Fig. 5 on the right whereas $\mathbf{M}_B^{(i)}$ gives the actual motion depending on the angles α, β and γ respectively.

The task of motion estimation is to find the matrices $\mathbf{M}_B^{(i)}$ such that the deformed template fits into the 3D measured point cloud of an arbitrary pose. From the matrices the values α, β and γ are derived and can be used to animate computer graphic models or robot manipulators.

3.2   Hierarchical ICP

The body part which exhibits within several poses of a gesture the smallest motion is the lower torso. Therefore the lower torso serves as root of the motion hierarchy in Fig.6 and its motion is investigated first. The ICP (Iterative Closest Point)-algorithm[5] is used to calculate the translation and rotation parameters of the "closest points" from the measured surface data. The

translational and rota-
tional parameters of the
lower torso represent the
body position and orien-
tation of the pose under
investigation. In the next
step the rotation of the
lower torso of the refer-
ence model is adapted ac-
cording to the previous
measurements. The mo-

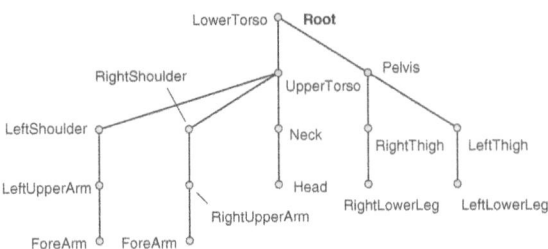

Fig. 6. Estimation hierarchy

tion parameters of the child node in Fig.6 (left), here the upper torso, are calculated using a modified version of the ICP algorithm which only calculates the
rotational parameters. The center point for this calculation is given by the joint
position of that particular body part which has been determined already through
the motion parameters of the hierarchically higher body part (in this case the
lower torso). The five independent kinematic chains from Fig. 6 are calculated
in the described hierarchic manner from the root (lower torso) to the respective
end effectors. Measured pose points, which have served for the previous adaptation are eliminated from the data set in order to prevent manifold assignments
of points for the following estimation steps of the hierarchy.

In order to consider the differing degrees of freedom for the different joints
along the kinematic chains from the root to the end effector a post processing
step follows which shifts additional degrees of freedom (DOF) between joints.
E.g. the additional DOFs in the elbow joint (the algorithm estimates 3 rotational
DOFs) are shifted to the motion in the shoulder joint which must be responsible
for the additional motion since the elbow is only equipped with a single DOF.

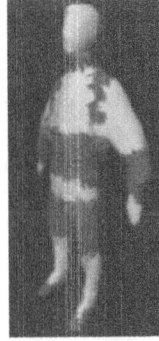

Fig. 7. 3D segmented body model
(left) and initial pose of skeleton
within the pose "carry"

4   Results

Figure 7 shows the pose under investigation overlaid by the skeleton of the reference model. The differences in the arm
and leg positions are obvious. Fig. 8 shows
the results of an automated pose estimation as described in this paper. Since the
pose estimation is done in three dimensions, the fitting of the skeleton to the
cloud of surface points in the pose under
investigation is illustrated by views from
four different spatial positions. It can be
seen, that in this case the skeleton has
been adapted to the pose almost perfectly.
Right now each pose is treated indepen-

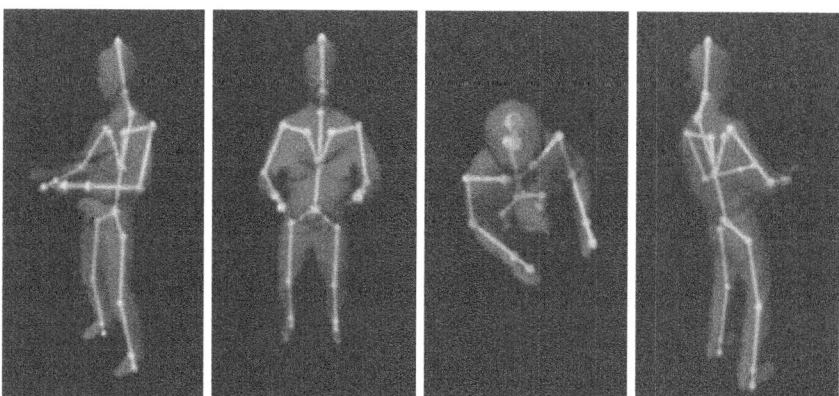

Fig. 8. Estimation results: The initial body skeleton has been fit to the pose "carry"

dently from the others by match against the same, locally adapted reference model. For future applications where body parts merge, for instance in a pose, where an arm is pressed against the torso, it might be necessary to process poses in their natural sequence in order to track the skeleton. However, this endangers the accumulation of estimation errors, which is prevented by our present approach.

References

1. W. Niem, H. Broszio: Mapping texture from multiple camera views onto 3D-object models for computer animation. In: Proceed. Internat. Workshop on *Stereoscopic and Three Dimensional Imaging*, Santorini, Greece, (1995). 30
2. C. Pedney: Distance-ordered homotopic thinning: a skeletonization algorithm for 3D digital images. *Comput. Vis. Image Understanding* **72** (1998) 404–413. 30
3. L. Dekker, I. Douros, B. F. Buston, P. Treleaven: Building symbolic information for 3D human body modeling from range data. 2nd Internat. Conf. on *3D Digital Imaging and Modeling*, Ottawa, Ont., Canada, (4-8 Oct. 1999). 30
4. S. Weik, J. Wingbermuehle, W. Niem: Creation of flexible anthropomorphic models for 3D videoconferencing using shape from silhouettes. *J. of Visualization and Computer Animation* **11** (2000) 145–154. 28
5. P. J. Besl, N. D. McKay: A method for registration of 3D shapes. *IEEE Trans. on Pattern Analysis and Machine Intelligence*, **14** (1992) No.2. 32
6. D. A. Simon, M. Hebert, T. Kanade: Real-time 3D pose estimation using a high-speed range sensor. In: Proceed. IEEE Internat. Conf. on *Robotics and Automation*, Vol. 3 (1994) 2235–2240.
7. D. M. Gavrila: The visual analysis of human movement: a survey. *Computer Vision and Image Understanding*, **73** (1999) 82–98. 31
8. S. Weik, O. Niemeyer: Three-dimensional motion estimation for articulated human templates using a sequence of stereoscopic image pairs. In: SPIE Proceed. of *Visual Communications and Image Processing*, VCIP99, **SPIE-3653** (1999). 31

Vision-Based Robot Localization
Using Sporadic Features

Stefan Enderle, Heiko Folkerts, Marcus Ritter, Stefan Sablatnög,
Gerhard Kraetzschmar, and Günther Palm

Dept. of Neural Information Processing
University of Ulm, D-89069 Ulm, Germany
{steve,folkerts,ritter,stefan,gkk,palm}@neuro.informatik.uni-ulm.de

Abstract. Knowing its position in an environment is an essential capability for any useful mobile robot. Monte-Carlo Localization (MCL) has become a popular framework for solving the self-localization problem in mobile robots. The known methods exploit sensor data obtained from laser range finders or sonar rings to estimate robot positions and are quite reliable and robust against noise. An open question is whether comparable localization performance can be achieved using only camera images, especially if the camera images are used both for localization and object recognition. In this paper, we discuss the problems arising from these characteristics and show experimentally that MCL nevertheless works very well under these conditions.

1   Introduction

In the recent past, Monte-Carlo localization has become a very popular framework for solving the self-localization problem in mobile robots [4,5,6]. This method is very reliable and robust against noise, especially if the robots are equipped with laser range finders or sonar sensors. In some environments, however, for example in the popular RoboCup domain [7], providing a laser scanner for each robot is difficult or impossible and sonar data is extremely noisy due to the highly dynamic environment. Thus, enhancing the existing localization methods such that they can use other sensory channels, like uni- or omni-directional vision systems, is a state-of-the-art problem in robotics. In this work, we present a vision-based MCL approach using visual features which are extracted from the robot's unidirectional camera and matched to a known model of the RoboCup environment. Additionally, we try to use the same feature detectors for a completely different environment – in an office.

2   Monte Carlo Localization

Monte Carlo localization (MCL) [5] is an efficient implementation of the general Markov localization approach (see e.g. [4]). Here, the continuous probability distribution $Bel(l)$ expressing the robots' belief in being at location l is represented

R. Klette, S. Peleg, G. Sommer (Eds.): Robot Vision 2001, LNCS 1998, pp. 35–42, 2001.

by a set of N samples $S = \{s_1, \ldots, s_N\}$. Each sample $s_i = \langle l_i, p_i \rangle$ consists of a robot *location* l_i and *weight* p_i. As the weights are interpreted as probabilities, we assume $\sum_{i=1}^{N} p_i = 1$.

The algorithm for Monte Carlo localization is adopted from the general Markov localization framework. Initially, a set of samples reflecting initial knowledge about the robot's position is generated. During robot operation, the following two kinds of update steps are iteratively executed:

Sample Projection across Robot Motion: As in the general Markov algorithm, a motion model $P(l|l', m)$ is used to update the probability distribution $Bel(l)$. In MCL, a new sample set S is generated from a previous set S' by applying the motion model as follows: For each sample $\langle l', p' \rangle \in S'$ a new sample $\langle l, p' \rangle$ is added to S, where l is randomly drawn from the density $P(l|l', m)$.

Observation Update and Weighted Resampling: Sensor inputs are used to update the robot's beliefs about its position. All samples are re-weighted by incorporating the sensor data o and applying the observation model $P(o|l')$. Given a sample $\langle l', p' \rangle$, the new weight p for this sample is given by

$$p = \alpha \, P(o|l') \, p' \tag{1}$$

where α is a normalization factor which ensures that all beliefs sum up to 1. These new weights for the samples in S' provide a probability distribution, which is then used to construct a new sample set S. This is done by randomly drawing samples from S' using the distribution given by the weights.

3   Vision-Based Localization

There are mainly two cases where MCL based on distance sensor readings cannot be applied: (i) If distance sensors like laser range finders or sonars are not available. (ii) If the readings obtained by these sensors are too unreliable, e.g. in a highly dynamic environment. In these cases, other sensory information must be used for localization. A natural candidate is the visual channel, because many robots include cameras as standard equipment. An example for using visual information for MCL has been provided by Dellaert et al. [1].

An interesting and open question is whether the MCL approach still works when the number of observations is significantly reduced and when particular observations can be made only intermittently. In the following, we show how to adapt the MCL approach in order to overcome these problems.

3.1   Feature-Based Modeling

As described in Equ. 1, the sensor update mechanism needs a sensor model $P(o|l)$ which describes how probable a sensor reading o is at a given robot location l.

Fig. 1. Detection of post, corner and edge features in the RoboCup domain

This probability is often computed by estimating the sensor reading $\tilde{o}$ at location l and determine some distance $dist(o, \tilde{o})$ between the given measurement o and the estimation $\tilde{o}$.

As it is not possible to efficiently estimate complete camera images and then compute image distances for hundreds of samples, we use a feature-based approach. In the RoboCup domain, we use the following features (see [3]): 1) goal posts of blue and yellow goal, 2) corners, and 3) distances to field edges. In the office environment, we only use the distance features in an initial try to show the feasibility of the approach.

Feature Detection: The feature detection process for the RoboCup domain works as follows: In a first step the camera image is segmented into color regions. Based on the segmented image, filters are applied to detect color discontinuities.

The *goal post detector* detects a vertical *white-blue* or a *white-yellow* transition for the blue or yellow goal post, respectively (see left image in Fig. 1). The *corner detector* searches for vertical *green-white-green* transitions in the image (middle image in Fig. 1). The *distance estimator* estimates the distance to the field edges based on detected horizontal *green-anything* transitions in the image. Currently, we select four specific columns in the image for detecting the field edges (right image in Fig. 1).

In Fig. 2, a corridor inside our office building can be seen. The left image shows the original camera image while the right image shows the detected edges. The grey lines starting from the bottom of the image visualize the estimated distances to the walls. Note that there are several errors mainly in the door areas.

In Fig. 3, the estimated distances are projected to a horizontal plane. The brighter dots show the estimated distances by the visual feature detector, while

Fig. 2. Detection of distance features in the office environment

Fig. 3. Distance estimator. The bigger dots are estimated distances, the smaller ones are laser readings for comparison

the smaller darker dots show the laser readings in the same situation. Note, that in the left image a situation is shown where a lot of visual distances can be estimated, while in the right image only four distances are estimated at all. Due to the simplicity of the applied feature detector often only few distance features are detected and therefore the situation is comparable to the RoboCup environment.

Weight Update Let the sensor data o be a vector of n features $f_1 \ldots f_n$. If we assume that the detection of features depends solely on the robot's position and does not depend on the detectability of other features, then the features are independent and we can conclude:

$$P(o|l') = P(f_1 \ldots f_n|l')$$
$$= P(f_1|l') \ldots P(f_n|l') \tag{2}$$

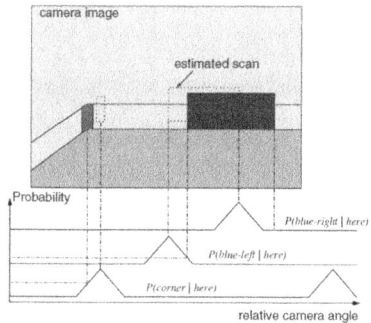

Fig. 4. Heuristics for estimating the probabilities $p(f_i|l)$ from the current camera view

The sensor model $P(f_i|l)$ describes how likely it is to detect a particular feature f_i given a robot location l. In our implementation, this sensor model is computed by comparing the detected feature with an estimated feature given a geometrical world model. The distance measurement between the features is mapped to a probability estimate by applying a heuristic function as illustrated in Fig. 4.

The application of these heuristics to the examples used in Fig. 1 are illustrated in Fig. 5. Probabilistic combination of evidence for several features yields significantly better results, as convincingly demonstrated by the rightmost image in Fig. 5.

The figure illustrates various properties of our approach. The shape of the function causes all samples with comparatively high angular error to be drastically down-valued and successively being sorted out. Thus, detecting a single goal post will reshape the distribution of the sample set such that mostly locations that make it likely to see a goal post in a certain direction will survive in the sample set. Secondly, there is only a single heuristic function that captures all of the ambiguous corner features. Each actually detected corner is successively given a probability estimate. If the corner detector misses corners that we expected to see, this does not do any harm. If the detector returns more corners than actually expected, the behavior depends on the particular situation: detecting an extra corner close to where we actually expected one, has a small negative influence on the weight of this sample, while detecting a corner where none was expected at all has a much stronger negative effect.

Figure 6 shows a similar probability distribution inside a corridor. The distribution is computed for the estimated scan in Fig. 3. As can be seen, the locations along a straight line in the middle of the corridor are the most probable. One can easily imagine, that an additional door detector would strongly reduce the number of possible locations.

4   Experiments

The described method was implemented and evaluated on a Sparrow-99 robot, a custom-built soccer platform equipped with an uni-directional camera [2].

Fig. 5. Probability distributions $P(f_i|l)$ for all field positions l (with fixed orientation) given the detection of a particular feature f_i or all features, respectively. From left to right: goal posts, corners, distance measurements, all three combined

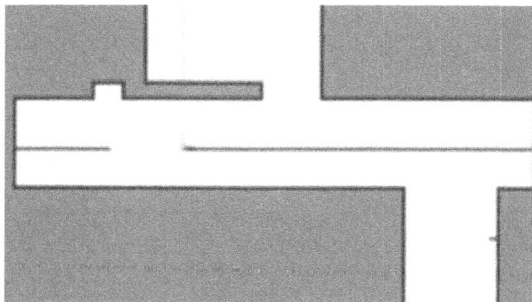

Fig. 6. Probability distributions $P(f_i|l)$ for all positions in the corridor using visually estimated distances

Additionally, we performed an experiment with the B21 robot "Stanislav" for investigating the feasibility of the feature-based localization approach.

Experiment 1: Number of Features In this experiment, we show that the robot can localize robustly and that the accuracy can be improved by adding more features. A Sparrow-99 robot was placed in one corner of the field (the right top circle of Fig. 7). In order to have an accurate reference path, we moved the robot by hand along a rectangular trajectory indicated by the dots.

The first image in Fig. 7 displays the odometry data when moving four rounds and shows the drift error that occurs. The second image displays the corrected trajectory which does not drift away. The third image displays the first round corrected by the localization algorithm using only the goal posts as features. In the fourth image we can see a more accurate path found using all three feature types.

Experiment 2: Number of Samples Implementing sample-based localization, it is important to have an idea of how many samples you need. Obviously, a small number of samples is preferred, since the computational effort increases with the number of samples. On the other side, an appropriate number of samples is

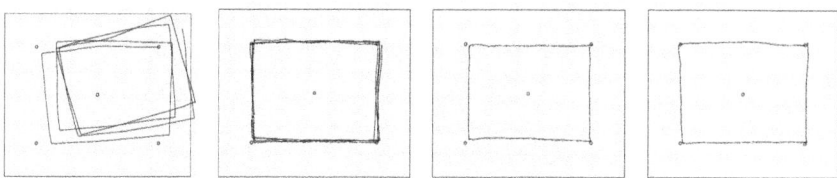

Fig. 7. A single round of the Sparrow-99 being pushed along a rectangular path in the RoboCup field.(Odometrie path, corrected path using goal post features, corrected path using all features)

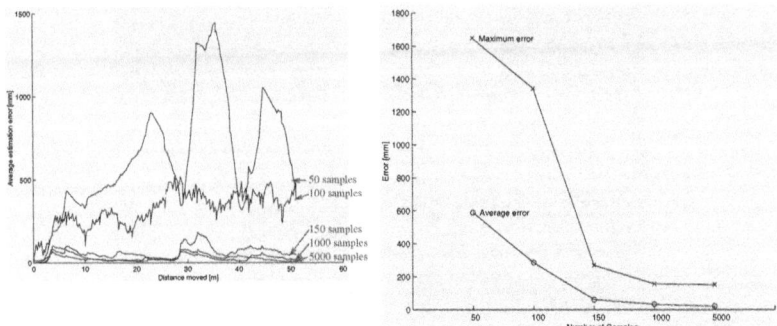

Fig. 8. Left: Average localization errors of the different sample numbers. Right: Average error and maximum error of the same five runs (50,100,150,1000 and 5000 samples used)

needed in order to achieve the desired accuracy. In this experiment, we moved four rounds exactly as in the previous experiment. This time, we used five different numbers of samples: 50, 100, 150, 1000, 5000

In the left image of Fig. 8, the average localization errors of the different sample numbers are shown. One can see that the error decreases when more samples are used. On the other hand, the difference in accuracy does not increase very much from 150 samples over 1000 to 5000. The right image of Fig. 8 shows both the average error and the maximum error of the same five runs (50, 100, 150, 1000 and 5000 samples). Again, you can see that above 150 samples, the accuracy hardly increases.

Experiment 3: Feasibility for Office Environments With this experiment, we wanted to see whether our approach chosen for the RoboCup environment was also feasible for more complex environments, like regular corridors and rooms. Since the first feature detector we implement was the *distance estimator*, we performed the following qualitative experiment with distance features only.

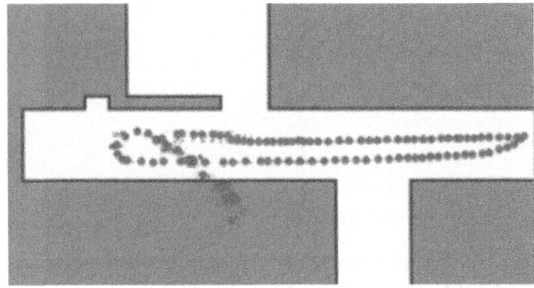

Fig. 9. Robot "Stanislav"

Our B21 robot "Stanislav" was initially located at the beginning (left side of Fig. 9) of a corridor of about 20 m length and 2.2 m width. We used 1000 samples which were initialized at the approximating robot position with a diameter of 1 meter and an angular uncertainty of 20 degrees.

You can see that the robot was able to track the path along the corridor up to the right side and the way back to the starting point. Then, while rotating 180 degrees, to many visual artifacts were detected and the samples drifted away.

5 Conclusions

In this paper, we used the Monte-Carlo approach for vision-based localization of a soccer robot on the RoboCup soccer field. Unlike many previous applications, the robot could not use distance sensors like laser scanners or sonars. Also, special camera setups were not available. Instead, the onboard camera was used for localization purposes in addition to object recognition tasks. As a consequence, sensor input to update the robot's belief about its position was low-dimensional and sporadic. Nevertheless, the experimental evaluation demonstrated that Monte Carlo localization works well even under these restrictive conditions.

We could also show that even with a small number of detected features leading to sporadic observation updates, the localization results are usable. However, by increasing the number of visual features the accuracy enhances dramatically.

The lifting of the presented approach from the RoboCup environment to an office delivery robot seems promising. However, in this area it is still more important to combine different kinds of feature in order to yield a robust system behavior.

References

1. F. Dellaert, W. Burgard, D. Fox, S. Thrun: Using the condensation algorithm for robust, vision-based mobile robot localization. In: Proceed. IEEE Conf. on *Computer Vision and Pattern Recognition*, (1999). 36
2. S. Enderle: The sparrow 99 robot. Technical report, University of Ulm, (1999). 39
3. S. Enderle, M. Ritter, D. Fox, S. Sablatnög, G. Kraetzschmar, G. Palm: Soccer-robot locatization using sporadic visual features. Proceed. of the IAS-6 Internat. Conf. on *Intelligent Autonomous Systems*, (2000). 37
4. D. Fox: Markov localization: a probabilistic framework for mobile robot localization and navigation. PhD thesis, University of Bonn, Bonn, Germany, (December 1998). 35
5. D. Fox, W. Burgard, F. Dellaert, S. Thrun: Monte carlo localization: efficient position estimation for mobile robots. In: Proceed. of the Conf. on *Artificial Intelligence*, AAAI, (1999). 35
6. J.-S. Gutmann, W. Burgard, D. Fox, K. Konolige: An experimental comparison of localization methods. In: Proceed. of the Internat. Conf. on *Intelligent Robots and Systems*, IROS'98, Victoria, Canada, (October 1998). 35
7. H. Kitano, M. Asada, Y. Kuniyoshi, I. Noda, E. Osawa, H. Matsubara: Robocup a challenge problem for AI. *AI magazine*, **18** (1997) 73–85. 35

A Comparison of Feature Measurements
for Kinetic Studies on Human Bodies

Nikki Austin[2], Yen Chen[1], Reinhard Klette[1], Robert Marshall[2],
Yuan-sheng Tsai[1], and Yongbao Zhang[1]

[1] Center for Image Technology and Robotics
[2] Department of Sport and Exercise Sciences
Tamaki Campus, The University of Auckland, Auckland, New Zealand
{r.klette,r.marshall}@auckland.ac.nz

Abstract. The paper reports about a performance comparison within a joint project of computer vision, and sport and exercise sciences. The project is directed on the understanding of human motion based on shape features and kinetic studies. Three shape recovery techniques, a traditional technique as used in sport and exercise sciences (manual measurement based on an elliptical zone assumption) and two computer vision techniques (based on a small number of occluding contours, and a new combination of photometric stereo and shape from boundaries), are compared using a mannequin as test object. The computer vision techniques have been designed to go towards dynamic shape recovery (humans in motion). The paper reports about these three techniques and their measurement accuracies.

1 Introduction

In sport science, biomedical engineering or ergonomics, accurate measurements of segmental anthropometry are often required for the estimation of resultant moments and forces acting on body segments. As it is not possible to directly measure the forces and moments acting on the human body, it is necessary to calculate them from measures of the external forces, the kinematic characteristics of the segment (linear and angular position, velocity and acceleration) and estimates of the segment's inertial parameters (center of mass location, segment length and segment moment of inertia).

We specify features of interest [11]. We assume an XYZ-cartesian world coordinate system with the Z-axis being the axis of gravitation, i.e. the weight acts in Z-direction. Consider a human body segment s and its local xyz-coordinate system with its origin at the center of mass of s, and with the z-axis aligned with the main ("long") axis of the segment. We assume and use uniform density estimates for human body segments [2,3]. The forces acting on the center of mass are in world coordinates

$$\mathbf{F}_{j_1,s} + \mathbf{F}_{j_2,s} + (0,0,w_s) = m_s \cdot \mathbf{a}_s ,$$

where j_1, j_2 are the joints of the given segment s (note: there is only one joint

R. Klette, S. Peleg, G. Sommer (Eds.): Robot Vision 2001, LNCS 1998, pp. 43–51, 2001.
© Springer-Verlag Berlin Heidelberg 2001

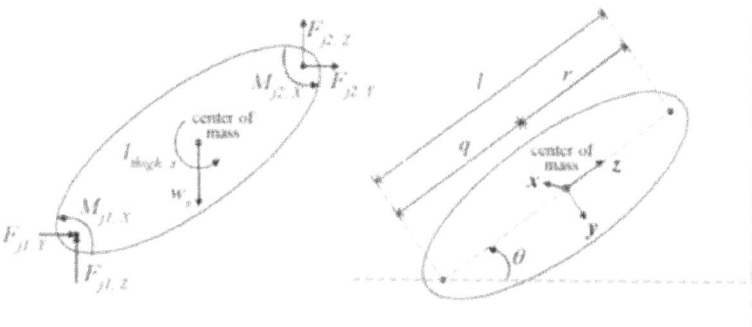

Fig. 1. Left: forces acting on a cross section of a thigh. Right: general scheme of the segment

for some segments), $\mathbf{F}_{j,s} = (F_X, F_Y, F_Z)$ are the forces acting on the segment s at joint j, m_s is the mass of the segment s, and $\mathbf{a}_s = (a_X, a_Y, a_Z)$ are the accelerations of the segmental center of mass, all expressed with respect to the 3D world coordinate system. Figure 1 shows the forces acting on a two-dimensional cross section of the thigh, with $j_1 = $ knee, $j_2 = $ hip.

Kinematic studies require estimates of the *center of mass* (which requires shape data), of *accelerations* $\mathbf{a}_s$ (which requires video sequence analysis) and of m_s, i.e. of the *volume* of s using uniform segmental density estimates for deriving m_s. The location of the center of mass (in 2D specified by a value of r, see Fig. 1) is estimated based on shape analysis results in general, and an often used simplification is to assume that it is on the line of slope θ from joint to joint. The motion of the segment is assumed to be in the YZ-plane.

Furthermore, the moment of force in X-direction acting on the center of mass is [12]

$$M_{j_1,X} + M_{j_2,X} + q \cdot \sin\theta \cdot F_{j_1,Y} - q \cdot \cos\theta \cdot F_{j_1,Z} - r \cdot \sin\theta \cdot F_{j_2,Y}$$
$$+ r \cdot \cos\theta \cdot F_{j_2,Z} = I_x \cdot \alpha_x ,$$

where $q = l - r$, $\mathbf{M}_j = (M_{j,X}, M_{j,Y}, M_{j,Z})$ are the moments of force acting at joint j (we are not using $M_{j,Y}, M_{j,Z}$), I_x, I_y, I_z are the values in the segmental moment of inertia tensor (I_y, I_z are not of interest), and $\alpha_s = (\alpha_x, \alpha_y, \alpha_z)$ is the angular acceleration of the segment. The *base length* of the segment is given by the distance l between both joints, and r represents the distance from the center of mass to one of these joints, θ is the angle made by the line connecting the joints, and the horizontal axis. Kinematic studies require estimates of the *segmental moment of inertia* I_x and of the *segmental angular acceleration* α_x.

Accurate estimations of the listed parameters are essential to accurate calculations of forces and moments of forces. In this paper we only discuss estimates of the static parameters (center of mass, of the volume, and of the moments of inertia). The dynamic parameters (acceleration, angular acceleration) will be discussed in another publication. Traditional techniques for estimating these static parameters as used in sport and exercise sciences include proportional estimates from cadaver measures, regression equations, mathematical modelling and scanning/imaging (MRI) techniques.

We also make use of a standard simplification of the model assuming that segments are only considered in a 2D YZ-coordinate system as shown in Fig. 1, and moments of force are considered with respect to rotation about the x-axis only.

Automatic computer-vision based whole body reconstruction systems that are currently available include a whole body color 3D scanner WB4, developed by Cyberware [4], a phase based body measurement system, developed by TC^2 [13], and a PC cluster system, under development at Kyoto university [14].

Cyberware's system uses structured lighting with four scanning instruments mounted on two vertical towers. The scanning instruments start scanning from the person's head and continue down to capture the shape and color of the human body. Recently such a scan requires about 17 seconds. The system is priced at US$410,000 each, excluding the graphical workstations required to run the system. TC^2's system uses a phase based approach to recover surface information. Six stationary sensors capture different light patterns that are projected onto the body. Recently a scan requires 8 seconds and the system is priced at US$100,000. The cited PC cluster system uses the shape from contours approach for shape reconstruction and is intended to achieve the reconstruction of a human body in real time. The cluster system includes 9 cameras and 10 PCs with dual Pentium-III 600 MHz CPUs and 256MB memories, connected by a high speed network.

The paper is structured as follows. Section 2 describes two methods for measuring the static parameters. Section 3 compares the results obtained by both methods.

2   Methods

For our performance test we choose the elliptical zone method [7] which is still in use for manual shape recovery, shape from contours (with a limitation to just a small number of contours, say 9), and a new computer-vision based method which combines photometric stereo based 2.5D shape data and an even smaller number of captured contours (say 3 or 4) to recover a full 3D shape. The motivation is that these methods allow dynamic 3D shape estimation due to possible frequencies. However, we have to answer the question up to what accuracy. We compare these techniques with the traditional elliptical zones method. True volume data may be obtained by water displacement measurement. A mannequin is used as a test object which is positioned on a turntable for both computer vision approaches (allowing that only one camera has to be used).

2.1   Elliptical Zones

The elliptical zone technique [7] is a traditional technique in sport and exercise sciences. It considers each segment to be composed of a sequence of right *elliptical cylinders* e that follow the shape fluctuations of the segment. High resolution digital images from the front and side of a body are taken. For the segment under consideration assume that the side view corresponds to a projection into the YZ-plane.

An interactive program allows to divide a segment s into elliptical cylinders e. The two axes of such an elliptical cylinder, x_e, y_e, are measured in x- and y-direction of the segment, and the height h_e (which is typically about 2 cm) in z-direction. The volume of an elliptical cylinder e is estimated by

$$v_e = \pi \cdot x_e \cdot y_e \cdot h_e \,,$$

and the volume of a segment is simply the sum of the volumes of all of its elliptical cylinders. The mass m_e of this elliptical cylinder is its volume times its density (which is assumed to be uniform).

The center of mass of a segment s with n elliptical cylinders is estimated as follows: we assume that all centers of mass of all the elliptical cylinders are on the z-axis of the segment. The 50%-percentile specifies the center of mass of the segment, calculated by adding m_e-values along the segment's z-axis using the value l.

The moments of inertia of the base b (i.e. that's a planar region in 3D space) of such an elliptical cylinder e about its centroidal axes are denoted by $I_{x,b}$, $I_{y,b}$, $I_{z,b}$, and they are estimated by

$$I_{x,b} = \frac{\pi}{4} \cdot x_e \cdot y_e^3 \,, \quad I_{y,b} = \frac{\pi}{4} \cdot x_e^3 \cdot y_e \,, \quad \text{and} \quad I_{z,b} = I_{x,b} + I_{y,b} \,.$$

The moments of inertia of the 3D cylinder e about its centroidal axes are

$$I_{x,e} = I_{x,b} \cdot \rho_e \cdot h_e \,, \quad I_{y,e} = I_{y,b} \cdot \rho_e \cdot h_e \,, \quad \text{and} \quad I_{z,e} = I_{z,b} \cdot \rho_e \cdot h_e \,,$$

where ρ_e is the cylinder's (uniform) density. For the entire segment, the moments of inertia about its centroidal axes are defined with respect to its local xyz-coordinate system, and they are found by applying the parallel axes theorem and summing for $e = 1, ..., n$:

$$I_{x,s} = \sum_{e=1}^{n} (I_{x,e} + m_e \cdot d_e^2) \,,$$

where d_e is the distance between the center of mass (centroid) of the eth elliptical cylinder and the segmental centroid. Due to our assumption that the cylinder centroids are located on the z-axis it follows that the summed cylinder centroid inertia tensor is on the principal segmental axes.

2.2 Shape from Photometric Stereo and Contours

Shape from contours is a method that obtains a 3D model from the occluding contours of an object [10]. The shape from contours approach is a robust method that gives reliable 3D shape estimation. Nevertheless, due to the nature of the approach, surface cavities that are obstructed from the viewing directions by other regions on the surface are unable to be recovered, and normally many contours (say 80 ... 150) are used to complete a 3D shape scan. We will limit our approach to a small number of contours (viewing directions). We use Tsai's calibration method [15,16] for obtaining intrinsic and extrinsic camera parameters.

Furthermore we use a new combination of photometric stereo and shape from contours. The photometric stereo approach for surface recovery calculates local surface orientations according to surface irradiance values [6,8,17]. Local surface orientations are globally integrated to recover the surface depth values. The photometric stereo approach allows to recover 2.5D surfaces in real-time. However, the recovered surface depth values are relatively scaled.

Three light sources successively illuminate the object from directions s_1, s_2 and s_3, and images E_{i1}, E_{i2} and E_{i3} are respectively acquired, where i is the index of current position. After three images have been acquired, the turntable is rotated by ϕ degrees to rotate the object into the next viewing direction, with the index of $i + 1$. The process of image acquisition is repeated for all of the required viewing directions, typically 3 or 4 only. Photometric stereo method is used to recover 2.5D surfaces from the three input images in any viewing direction.

Each layer e in the 3D image data has a height of 1 pixel, which is typically about 2.5 mm on the surface of the mannequin in our set-up. In any layer, all contours from all viewing directions define a convex polygon, as shown by the light gray region in Fig. 2(a). The surface pixels recovered by photometric stereo for the eth layer are fitted into the polygon according to the assumption that all surface pixels must lie within the polygon defined by the contours. The center of mass of the polygon is calculated and each viewing direction is specified as a vector starting from the center of mass. A surface pixel obtained at viewing direction i is accepted if it lies within $\phi/2$ from either side of the viewing direction. Figure 2(b) shows the accepted surface pixels in black. The rejected pixels, which lie outside of the $\phi/2$ threshold, are shown in gray. The accepted surface pixels form a 3D model of the object, as shown by Fig. 2(c).

A region R_e is defined by the accepted surface pixels in one layer as indicated by the dark gray region in Fig. 2(a). The number of pixels that lie within R_e specifies the area of the region, a_e.

The volume of the object at the eth cross section v_e, is estimated by counting the number of pixels that belongs to a_e and multiplied by the pixel to metric ratio k (approximately 2.5 mm in our set-up)

$$v_e = k^3 \cdot a_e .$$

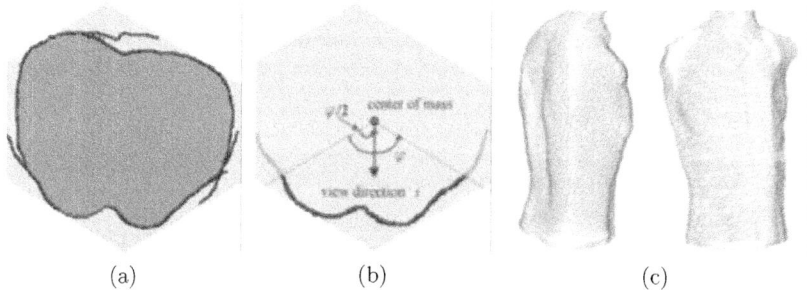

(a)   (b)   (c)

Fig. 2. (a), (b) A layer of the recovered mannequin torso and (c) reconstructed 2.5D surfaces of mannequin torso

The mass of the cross section is its volume multiplied by its density ρ_e, as for the elliptical zones approach. The center of mass (c_x, c_y) for the eth layer is

$$c_x = \frac{1}{a_e} \sum_{(x,y) \in R_e} x \,, \quad c_y = \frac{1}{a_e} \sum_{(x,y) \in R_e} y \,.$$

The segment's center of mass along its z-axis is specified as the 50%-percentile of the cumulative sum of the cross section masses.

The moment of inertia for the base of a cross section about its center of mass is

$$I_{x,b} = k^2 \cdot a_e \cdot \sum_{(x,y) \in R_e} (x - c_x)^2, \quad I_{y,b} = k^2 \cdot a_e \cdot \sum_{(x,y) \in R_e} (y - c_y)^2, \text{ and } I_{z,b} = I_{x,b} + I_{y,b} \,.$$

The moment of inertia for the entire segment about its centroidal axes are as given for the elliptical zones approach.

3   Results

In the situation where mass cannot be measured directly, accurate volume measurement is crucial to the estimation of mass, since mass is obtained by multiplying volume by density. The center of mass and moment of inertia can then be calculated once the mass for any layer of the segment can be estimated.

In this section, the segmental features calculated from different shape recovery techniques are compared with measurements that are obtained by physically measuring the object. Different approaches are discussed with respect to the applicability, performance and possible improvement.

3.1   Comparisons

The measurements are compared with measurements obtained by water displacement, balancing, and pendulum approaches. It is assumed that the density is uniform over the entire object. The values for center of mass are given as distances from the apex of the head. The values of the moment of inertia are calculated with respect to a horizontal axis that passes through the center of mass.

Table 1. Resultant volume measurements and relative percentage error, where the volume obtained by water displacement is 32.82 L

Method	Result (L)	Error (%)
Elliptical zone	30.01	8.56%
Shape from contours (9 views)	32.90	0.24%
Shape from PS and contours (3 view)	27.30	16.82%

Table 2. Resultant center of mass measurements and relative percentage error, where the center of mass obtained by balance is 0.42 m from the apex

Method	Result (m)	Error (%)
Elliptical zone	0.35	16.67%
Shape from contours (9 views)	0.45	7.14%
Shape from PS and contours (3 views)	0.45	7.14%

Table 3. Calculated moment of inertia, where the moment of inertia obtained by pendulum is 0.49 kgm^2. This points out that the density of the mannequin is actually not uniform, which leads to calculated results that do not correspond well with the measurement

Method	Result (kgm^2)
Elliptical zone	1.33
Shape from contours (9 views)	1.31
Shape from PS and contours (3 views)	0.87

3.2   Discussion

From the results it can be seen that shape from contours with 9 viewing directions has provided the closest estimated values to the reference volume, as well as for the reference center of mass. The reason is that shape from contours is robust and reliable for obtaining the 3D shape of the object. Nevertheless, the accuracy of this method is limited, since some cavity regions are irrecoverable from shape from contours alone.

The shape from photometric stereo and contours method has obtained an estimated value for the center of mass that is closest to the reference value. A cause of error in the experiment may be that the depth values obtained by photometric stereo have been incorrectly scaled, leading to larger cavities in the recovered 3D model and thus reducing the total volume of the object.

The traditional elliptical zones model approach has not obtained values that are closest to the reference values in any of the feature measurements. One possible factor which limits the accuracy obtained by the elliptical zones model approach is the assumption that segment cross-sections are ellipsoidal.

Overall, the accuracy of the shape from photometric stereo and contours method is not limited by cavities on the the segments, nor assumptions of the segment shapes. The accuracy is dependent on the reflectance properties of the surface, the number of viewing directions and the resolution of the cameras. The 3D data can further be refined with control points to improve the accuracy of the proposed method.

In the experiment, the density is assumed to be uniform over the entire object for simplicity. In reality, the density is not uniform, and the assumption influences the accuracy for the calculated center of mass, which in turn, effects the calculated moment of inertia. Particularly in the mannequin, where the head is much denser than the torso. Thus the value obtained by physically measuring the moment of inertia does not correspond well to the calculated results under the uniform density assumption.

The ratio for converting from pixel to metric is assumed to be constant throughout the image data and in all x, y and z directions. Nevertheless, due to distortions in image acquisition, the ratio may be varied in different directions.

Water displacement, balancing and pendulum methods have been used to generate the ground truth for volume, center of mass and the moment of inertia. However, in practice, it is not feasible to immerse a person (or part of) in water to obtain the volume of body parts. It is also not possible to locate the center of mass by balancing the object, or obtain the moment of inertia by swinging the object. Furthermore, neither of the physical approaches, nor the elliptical zones method, is applicable for dynamic shape recovery. The shape from photometric stereo and contours method has very low time requirement for image acquisition and shape recovery. It is possible to achieve surface recovery with photometric stereo in real time. The reduction in time requirement reduces errors caused by the object's movement.

4   Conclusion

A new computer vision method for fast 3D shape recovery is proposed and compared with the traditional elliptical zones method in this work. The proposed method combines photometric stereo and shape from contours for 3D shape recovery, and is aimed at recovering the static parameters of body segments. Descriptions are given for calculating body segment parameters, such as mass, center of mass, and moment of inertia, using either method. A full-sized mannequin is used to provide consistent measurements throughout the experiment. The elliptical zones method, shape from contours method, as well as the proposed shape from photometric stereo and contours method are compared with respect to the accuracy in estimating static segmental parameters. From preliminary results it has been seen that the proposed method with only 3 viewing directions has provided results that are comparable with the traditional elliptical zones approach. Possible approaches for improving the accuracy of the shape from photometric stereo and contours method are proposed.

References

1. N. Austin: Estimation of segmental inertial parameter variation in children: Implication for gait analysis. Master thesis, Sport and Exercise Sciences, The University of Auckland (2000).
2. R. F. Chandler, C. E. Clauster, J. T. McConville, H. M. Reynolds, J. W. Young: Investigation of inertial properties of the human body. AMRL Technical Report, TR-74-137 (1975). 43
3. C. E. Clauster, J. T. McConville, J. W. Young: Weight, volume and center of mass of segments of the human body. AMRL Technical Report, TR-69-70 (1969). 43
4. Cyberware, Whole Body Color 3D Scanner: http://www.cyberware.com/products/index.html (1995). 45
5. W. D. Dempster: Space requirements of the seated operator. WADC Technical Report 55-159 (1955).
6. R. Frankot, R. Chellappa: A method for enforcing integrability in shape from shading algorithms. *IEEE Trans. on Pattern Analysis and Machine Intelligence* **PAMI-10** (1988) 439–451. 47
7. R. K. Jensen: Estimation of the biomechanical properties of three body types using a photogrammetric method. *J. of Biomechanics* **11** (1978) 349-358. 45, 46
8. R. Klette, K. Schlüns, *Height data from gradient fields*, Proc. SPIE, 2908, pp.204-215, 1996. 47
9. R. Klette, K. Schlüns, A. Koschan: *Computer Vision: Three-dimensional Data from Images*. Springer, Singapore (1998).
10. J. J. Koenderink: *Solid Shape*. Cambridge, MA, MIT Press (1990). 47
11. R. N. Marshall, R. K. Jensen, G. A. Wood: A general Newtonian simulation of an n-segment open chain model. *J. of Biomechanics* **18** (1985) 359–368. 43
12. B. M. Nigg, W. Herzog: *Biomechanics of the Musculo-Skeletal System*. Wiley, New York (1994). 44
13. Tc2, Body Measurement System: http://www.tc2.com/RD/RDBody.htm (1998). 45
14. S. Tokai, T. Wada, T. Matsuyama: Real time 3D shape reconstruction using PC cluster system. Proc. 3rd Internat. Workshop *Cooperative Distributed Vision*, Kyoto (1999) 171–187. 45
15. R. Y. Tsai: An efficient and accurate camera calibration technique for 3D machine vision. Proc. Internat. Conf. *Computer Vision and Pattern Recognition* (1986) 364–374. 47
16. Tsai Camera Calibration Software: http://www.ius.cs.cmu.edu/afs/cs.cmu.edu/user/rgw/www/TsaiCode.html (1995). 47
17. R. J. Woodham: Photometric method for determining surface orientation from multiple images. *Optical Engineering* **19** (1980) 139–144. 47

Object Identification and Pose Estimation for Automatic Manipulation

Benjamin Hohnhaeuser and Guenter Hommel

Institut f. Technische Informatik, Technische Universität Berlin
Franklinstr. 28/29, D-10587 Berlin, Germany
{bh,hommel}@cs.tu-berlin.de

Abstract. In this paper we present a framework to recognize objects and to determine their pose from a set of objects in a scene for automatic manipulation (bin picking) using *pixel-synchronous* range and intensity images. The approach uses three-dimensional object models. The object identification and pose estimation process is structured into three stages. The first stage is the *feature collection stage*, where the feature detection is performed in an *area of interest* followed by the *hypothesis generation*, which tries to form hypotheses from consistent features. The last stage, the *hypothesis verification*, tries to evaluate the hypotheses by comparing the measured range data to the predicted range data from hypothesis and the model.

1 Introduction

In many industrial assembly applications (i.e. in the automotive industry) the parts are palletized and no recognition is needed. If the parts are delivered in an unsorted way, which is mostly less expensive than the palletizing, a recognition and/or object identification stage must be inserted into the assembly process. Vision based object identification and pose[1] estimation is increasingly more difficult as the complexity of the target object or the complexity of the scene increases. Depending on the object type different features (3D) could be more or less suitable for the identification and determination of its pose. I.e. an arc provides more information about the type and pose of an object than a line and a line more than just a point. Therefore arcs or two adjacent edges with a certain angle are more distinctive features than a single line (or a single edge), because it provides more spatial information and decreases the degrees of freedom.

On the other hand a single arc of a certain radius and a certain length or adjacent edges with a certain angle can provide a very strong hypothesis for the object type and its pose. The more distinctive a feature is the faster the recognition process will lead to a result.

In our application a special sensor system is used (laser-range-camera [11] by ASTRIUM - *European Space Systems Company*) which provides *pixel-synchronous* dense range images and intensity images as well (see Figs. 3 and 4).

---

[1] Pose is the object's position and orientation

R. Klette, S. Peleg, G. Sommer (Eds.): Robot Vision 2001, LNCS 1998, pp. 52–59, 2001.

The great advantage is that the feature extraction process can be performed as well in the range image as in the intensity image and get then the range information for each pixel of each extracted feature later from the range image. Hence it is possible to collect more distinctive features with spatial information in the scene than just from the range image.

1.1   Related Work

Bolles and Horaud [1] developed an object identification and pose estimation system (3DPO) for industrial parts with *smooth/simple* surfaces using the 3D range finder. The hypothesis generation and verification is similar to the one that was used in this approach. Grimson [3] employs 2-D local features to recognize and localize cluttered objects in a scene. His methods are limited to the case where the objects are located on a flat worktable and therefore the appearance distortion from 3-D rotations is not admissible. Rahardja and Kosaka [10] are using stereo vision as sensor system, but they are also modelling complex industrial parts (i.e. alternator covers) with simple features/cues. All indexing or hashing strategies like the *MULTI-HASH* system by Kak and Edwards [6] or the *Geometric Hashing* by Lamdan and Wolfson [8] or even the *Structural Indexing* can not be used in our case, because the memory requirements for a set of different complex objects are extremely high. Faugeras and Hebert [2] describe shape surfaces by curves and patches, which are represented by linear primitives, such as points, lines and planes, but deal only with single objects. Another approach was presented by Zha et al. [13] [4] which uses triangular meshes. The problem again is, that it can be only used for single objects. The *TRIPOD OPERATORS* by Pipitone [9] or the *spin images* by Johnson [5] are not promising applicable for cluttered scenes with many similar and overlapping objects. Very promising approaches to 3D object recognition are presented by T. Stahs in [12], who invented an new framework for the recognition and pose estimation for industrial parts, and by B. Krebs in [7], who analyzed the statistical behavior of features and objects using Bayesian networks.

2   Approach

The object identification and pose estimation process is divided into three stages. After the data acquisition (range and intensity image) is performed in the first stage the "highest point" in the scene (the uppermost point of the most upper object) is determined from the range data. This point defines the center of an *area of interest* in which the feature detection is performed. The types of objects expected in the scene define the kind of features which are searched for, i.e. arcs will be searched if there are cylindric objects expected to be found. If there is no information about the scene and its content available, all supported features will be extracted. The whole feature extraction process is performed on the range-image as well as on the intensity-image. After a sufficient number of features has been collected the hypothesis generation stage starts to form hypotheses from consistent features. Each hypothesis starts with one feature. This feature is used to predict some possible compatible feature. If a second feature is found the two features are used to predict a third one and so on.

In the last stage, the hypothesis verification, all hypotheses which are "strong enough" are validated. In the evaluation process the measured range data are compared to the predicted range data from hypothesis and the model.

If a hypothesis is verified the object type and its pose is ascertained. In a post recognition stage a trajectory for the manipulation can be calculated or an approach-vector for grasping the object can be provided.

Two main ideas were followed. The first is to model complex objects by simple features, thus invariance to the object complexity is established. The second idea is to use features from the range image as well as from the intensity image. This benefit is provided by the pixel-synchronous range and intensity images. I.e. if there is an arc in the scene it will be rather difficult to find the edge of this arc, which is a roof edge (see Fig. 2) in a noisy range image and to determine its position precisely. On the other hand it will be much easier to find the corresponding jump edge (see Fig. 1 in the intensity image, supposed the illumination is not too unfavorable.

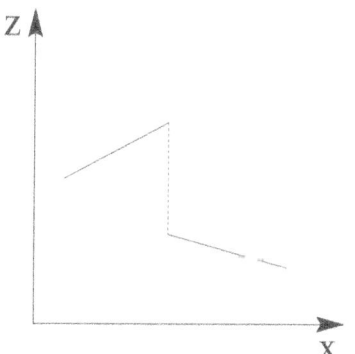
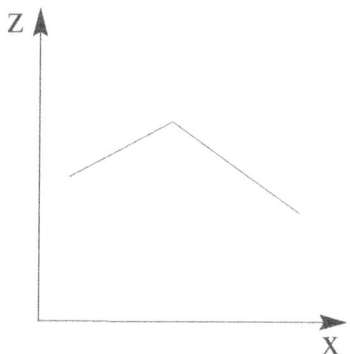

Fig. 1. A jump edge   **Fig. 2.** A roof edge

3   Object Modelling

It is assumed that all objects O in the scene are members of the set of possible objects $\mathbf{O}$ known a priori. All objects in the scene which are not members of the set of possible objects $\mathbf{O}$ are rejected. For each known object O_i a model M_i is provided, which contains all distinctive features $\mathbf{f_{model}}$ and their spatial relation. Each distinctive feature $f_{model,j}$ is of certain type t with its specific parameters $\mathbf{p}$ (i.e. radius of an arc). For each model feature $f_{model,k}$ it is also denoted whether it is intended to be found in the intensity or in the range image or in even both.

So each feature f_{model} is represented by the four-tuple consisting of its type t and its parameters $\mathbf{p}$, a homogenous 4×4 coordinate transformation matrix $\mathbf{T}$ which represents position and orientation regarding to the object coordinate system[2] as well as a description of its occurrence in both images o.

$$f_{model} = (t, \mathbf{p}, \mathbf{T}, o)$$

Each model also provides some surface information as supporting features $\mathbf{s}_{model}$ for the verification of a hypothesis. Supporting features are small, round surface patches with a diameter of approximately five millimeters. These supporting features are used to predict expected range measurements, thus each model is represented by a two-tuple consisting of its distinctive features $\mathbf{f}_{model}$ and its supporting features $\mathbf{s}_{model}$.

$$M = (\mathbf{f}_{model}, \mathbf{s}_{model})$$

Fig. 3. Range image of a scene with five simple objects

Fig. 4. Intensity image of the same scene

4   Feature Extraction

Regarding the rather noisy range images (see Fig. 3) that the laser-range-camera [11] by ASTRIUM provides, some image rectification steps must be performed before the real work can be done. The feature detection starts determining the center of an *area of interest*, as described above. This area is a hemisphere in the range data and a circle in the intensity image. The radius in both cases is defined by the greatest elongation of all expected objects $\mathbf{O}$. It can be calculated using the Cartesian coordinates of the "highest point" in the range image and the parameters of the camera.

---

[2] The spatial relation of distinctive features among each other can be calculated by multiplying and inverting the matrices.

Depending on the expected objects all legitimately possible features $\mathbf{f}_{\text{scene}}$ in the *area of interest* are extracted. All features f_{scene} are represented by a four-tuple, like the model features, which consists of its type t and its parameters $\mathbf{p}$, a transformation $\mathbf{T}$ which represents position and orientation regarding to the sensor coordinate system as well as a description of its occurrence in both images o.

$$f_{scene} = (t, \mathbf{p}, \mathbf{T}, o)$$

The feature extraction process is performed in the range image as well as in the intensity image (see Figs. 5 and 6).

Fig. 5. Range image of the scene with detected lines and arcs

Fig. 6. Intensity image of the scene with detected lines and arcs

5   Hypothesis Generation and Feature Prediction

If either a sufficient number of features in the *area of interest* is detected, i.e. about 20 features in a tidy or about a 100 in a cluttered scene, or even all features are extracted, the hypothesis generation starts. A hypothesis H consists of a three-tuple which is composed of a model M_i, a transformation $\mathbf{T}$ that determines the position and orientation of the objects regarding to a world coordinate system and a set of consistent features f_{scene} belonging to the object respectively to its model M_i.

$$H = (M_i, \mathbf{T}, \mathbf{f}_{\text{scene}})$$

First every hypothesis H_i consists only of one distinctive feature $f_{scene,i}$. No model and no transformation is assigned. In the next step for each model M_j including features from the same type t with the same parameters as $f_{scene,i}$, a new set of hypotheses is generated consisting of one feature, the model M_j and the transformation $\mathbf{T}$, that determines the position and orientation of the object regarding to the world coordinate system.One hypothesis is generated for each matching feature of the model. Now for all hypotheses a prediction of other features is possible, since the spatial relation of all model features regarding to the model coordinate system is known. If one of the predicted features is found in the scene the hypothesis grows. If the pose, defined by the two features, is non-ambiguous the hypothesis can enter the verification stage. In case the hypothesis provides an ambiguous interpretation of the pose a third feature must be predicted and found. Then it must be tested again if the pose, defined by the three features, is non-ambiguous and so on. If no more consistent features for an ambiguous hypothesis can be found the hypothesis is rejected.

6   Hypothesis Verification

If a hypothesis H_i contains a sufficient number of consistent distinctive features $f_{scene,j}$ (typically three to six, depending on the type of object and the type of features), the hypothesis verification starts. For the evaluation of a hypothesis H_i the supporting features (surface patches) are used. The hypothesis verification works as follows. Since the hypothesis provides the pose of the object it can be determined which parts of the surface are potentially "visible" for the sensor and which are not. Occlusion of other objects is so far not taken into account. All surface patches, that are visible and whose normal vector points approximately into the direction of the sensor are selected. Any of these can be used to predict a range measurement.

A hypothesis is verified if no measured value is bigger than the predicted one (the object is farther away than assumed) and at least a certain number of predictions are correct. If the measured value is smaller than the predicted measurement, it is an indicator for occlusion.

7   Experimental Results

The experimental results show that combining features of the gray image and the intensity image leads to a benefit in accuracy of position and orientation. The knowledge of exact position of the features in intensity image empowers the system to find the corresponding range information by using the pixel-synchronism of the range and intensity images. An improvement of the recognition time of simple objects (barbells) could not be observed. The reason for this is, that there are not significantly different or more features in the intensity image than in the range image.

In the case of occlusion between several object a decrease of the accuracy of the estimated pose could be observed for the occluded objects. Occluded objects can not be identified and their pose can not be estimated if the number of visible and extracted distinctive feature of this object is too small, i.e. less than the above mentioned number of features that are necessary for a hypothesis to enter the verification stage. In this context the problem of self-occlusion is treated as the occlusion by other objects. The experiments were done with simple, single-colored objects (barbells) in tidy scenes like shown in Fig. 4 and showed, that the accuracy of the pose estimation process can be increased between 12% and 15% for the translatorial deviation depending on the object type, using range information as well as intensity information. An increase of the angular precision could not be observed. The overall accuracy is between 6 millimeters and 9 millimeters and the angular precision is about 5 degrees.

8   Summary and Conclusions

As the experimental results are showing the accuracy of the recognition using range information as well as the intensity image increases. The approach to collect object features in the intensity image and then get the range information form the pixel-synchronous range image is suitable to increase the recognition accuracy of the position and orientation for objects in noisy range images. A further increase of the recognition accuracy and a decrease of recognition time can be expected if objects with a special texture or artificial markers are used. The obtained accuracy is sufficient for bin picking operations for industrial assembly.

Acknowledgements

This work has been performed in cooperation with DaimlerChrysler AG, Research and Technology, Intelligent Systems Group, Berlin. Special thanks to Dr. A. Stopp for his helpful comments and annotations.

References

1. R. C. Bolles, P. Horaud: 3dpo: A threedimensional part orientation system. *J. of Robotics Research*, **5** (1986) 3–26. 53
2. O. D. Faugeras, M. Hebert: The representation, recognition and location of 3-d objects. *J. of Robotics Research*, **5** (1986) 27–52. 53
3. W. E. L. Grimson: *Object Recognition by Computer*. MIT Press, (1990). 53
4. K. Hara, H. Zha, T. Hasegawa: Regularization-based 3-d object modeling from multiple range images. In: IEEE-Proceed. of the Internat. Conf. on *Pattern Recognition*, ICPR'98, Queensland, Australia, (August 1998). 53
5. A. E. Johnson: Spin-images: a representation for 3-D surface matching. PhD thesis, Carnegie Mellon University, Pittsburgh, Pennsylvania, (August 1997). 53
6. A. C. Kak, J. L. Edwards: Experimental state of the art in 3d object recognition and localization using range data. In: Proceed. of Workshop on *Vision for Robots in IROS'95 Conference*, Pittsburgh, Pennsylvania, (1995). 53
7. B. Krebs: Probabilistische Erkennung von 3d Freiformobjekten mit Bayesschen Netzen. PhD thesis, TU Braunschweig, (October 1999). 53
8. Y. Lamdan, H. J. Wolfson: Geometric hashing: a general and efficient model-based recognition scheme. In: IEEE-Proceed. of the Internat. Conf. on *Computer Vision*, Tarpon Springs, Florida, (1988) 238–249. 53

9. F. Pipitone, W. Adams: Tripod operators for recognizing objects in range images; rapid recognition of library objects. In: IEEE-Proceed. of the Internat. Conf. on *Robotics and Automation*, ICRA92, Nice, France, (1992) 1596–1601. 53

10. K. Rahardja, A. Kosaka: Vision-based bin-picking: Recognition and localization of multiple complex objects using simple visual cues. In: Proceed. of the IEEE/RSJ Internat. Conf. on *Intelligent Robotics and Systems*, IROS'96, Osaka, Japan, (1996). 53

11. W. Schroeder, E. Forgber, G. Roeh: Laser range camera application. Technical report, 1999. 52, 55

12. T. Stahs: Objekterkennung mit einem aktiven 3D-Robotersensorsystem. PhD thesis, TU Braunschweig, (June 1994). 53

13. H. Zha, S. Tahira, T. Hasegawa: Multi-resolution surface descripion of 3-d objects by shape-adaptive triangular meshes. In: IEEE-Proc. of the Internat. Conf. on *Pattern Recognition*, ICPR'98, Queensland, Australia, (August 1998). 53

Toward Self-calibration of a Stereo Rig from Noisy Stereoscopic Images

Slimane Larabi

Computer Science Institute of U.S.T.H.B University
BP 32, El Alia, Algiers, Algeria
larabi@ist.cerist.dz

Abstract. This paper deals with the analysis of uncertainty of epipole localizations in case of noisy stereo images. Initial uncertainty in point locations can be propagated through to an uncertainty in epipole localization, resulting in a region in the image called *epipolar zone*.

1 Introduction

Epipole computation is a step of major importance in self-calibration of stereo-scopic systems. The procedure for estimating the epipolar geometry without previous camera calibration is actually well known [3,7,9,12]. It depends upon estimations of some image point correspondences (typically using correlation methods) to apply a fundamental matrix calculation algorithm. The results obtained show that image noise influences the uncertainty in the epipole localization, and more matches need to be available to improve the precision in this computation.

Statistical and analytical methods have been presented [2] to analyze the uncertainty of fundamental matrices. The results obtained show that the esti-mation of the epipoles is good enough even when the noise reaches 3 pixels with the analytical method. In case of the statistical method, a good estimation for the epipoles is obtained only if the noise's standard deviation satisfies $\delta \leq 2$.

In the case where the images are very noisy (noise greater than 3 pixels), we can not apply this approach. We show in this paper that it is possible to analyze the uncertainty in the epipole localization for any noise affecting the 2D primitives. From a set of 2D line segments that have been matched on a pair of images using a stereoscopic algorithm, we propose a geometric analysis of the uncertainty measure that assumes noisy 2D edge points and propagates the effect of this edge detection noise to various stages of the epipole localization method of Mohr and Arbogast [11]. This allows to locate a zone containing the theoretical epipole that will be denoted as *epipolar zone*.

2 Hypothesis

The stereoscopic vision system used is modeled on two image planes P_1 and P_2, on which scenes are projected through two centers L_1 and L_2, see Fig. 1. We as-sume that the stereoscopic images are segmented and line segments approximate

R. Klette, S. Peleg, G. Sommer (Eds.): Robot Vision 2001, LNCS 1998, pp. 60–68, 2001.
© Springer-Verlag Berlin Heidelberg 2001

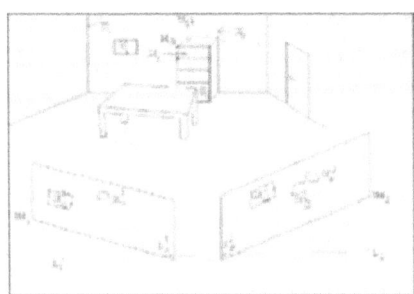

Fig. 1. Epipole localization process

their contour points. We assume also that the observed scene contains at least one planar surface with four or more vertices located in each image plane as a chain of closed 2D line segments [5]. We assume that P_1, P_2 are matched without calibration using correlation-based algorithms [13]. Our aim is to calculate the epipole position using only the relative disposition of different features in the two image planes.

3   Epipole Localization from Correspondences

Using projective geometry [1,10] Mohr and Arbogast in [11] proposed a method for epipole localization. We briefly cite its basic principles as follows:

We denote S_i^1 (resp. S_j^2) as the ith (resp. j^{th}) 2D surface region in P_1 (resp. P_2). Let $(S_{i_1}^1, S_{i_2}^2), (S_{j_1}^1, S_{j_2}^2)$ be the images on P_1, P_2 of two 3D surfaces patches S_i, S_j belonging to different planes Π_i and Π_j (see Fig. 1). Let S_p^2 be the 2D surface homologous to $S_{j_1}^1$ so that its preimage in the three-dimensional space belongs to the plane containing S_i.

We can deduce from this construction (see Fig. 1) that the lines $M_{j_k} M_{p_k}$ ($k = 1, ..., n_v$, where n_v is the number of vertices of S_j) pass through L_1 and their images on P_2 converge toward the same point E_p^2: the image of L_1 on P_2 is called the *epipole*.

To calculate the vertices of the hypothetical surface S_p^2 knowing the correspondence between $S_{i_1}^1, S_{j_1}^1$ and $S_{i_2}^2, S_{j_2}^2$, we use the cross-ratio invariance by

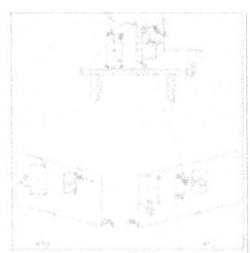

Fig. 2. Localization of S_p^2

projection [1,10]. We assume that (see Fig. 2) $M^1_{i_1,k}, k = 1, ..., 4$, and $M^2_{i_2,k}, k = 1, ..., 4$, are some vertices of $S^1_{i_1}$ and $S^2_{i_2}$, respectively, $S^1_{j_1}$ and $S^2_{j_2}$ are defined by n_v vertices $M^1_{j_1,k}$ and $M^2_{j_2,k}, k = 1, ..., n_v$, respectively, and for each vertex $M^1_{j_1,k}$ of $S^1_{j_1}$ we calculate its projective coordinates r_{1k}, r_{2k} relatively to the projective basis $(M^1_{i_1,1}, M^1_{i_1,2}, M^1_{i_1,3}, M^1_{i_1,4})$.

The localization on P_2 of the corresponding vertex $M^2_{p_k}$ so that its projective coordinates relatively to the projective basis $(M^2_{i_2,1}, M^2_{i_2,2}, M^2_{i_2,3}, M^2_{i_2,4})$ are equal to r_{1k}, r_{2k} allows to assert that the preimage of $(M^1_{j_1,k}, M^2_{p_k})$ belongs to the plane containing the preimage of $(S^1_{i_1}, S^2_{i_2})$. According to this result, the procedure operating over $(S^1_{i_1}, S^2_{i_2}, S^1_{j_1}, S^2_{j_2})$ starts to calculate for each $M^1_{j_1,k}$ of $S^1_{j_1}$ its projective coordinates (r_{1k}, r_{2k}) relatively to the projective basis $(M^1_{i_1,1}, M^1_{i_1,2}, M^1_{i_1,3}, M^1_{i_1,4})$. It computes after the position of $M^2_{p_k}$ on P_2 which has the same projective coordinates (r_{1k}, r_{2k}) relatively to the projective basis $(M^2_{i_2,1}, M^2_{i_2,2}, M^2_{i_2,3}, M^2_{i_2,4})$. Once the S^2_p vertices are calculated, the E^2_p point is located as the intersection of the n_v lines $M^2_{j_2,k}M^2_{p_k}$, see Fig. 1.

So, as the E^2_p position is unique for each stereoscopic system, the repetition of this procedure with all 2D surfaces $(S^1_{i_1}, S^2_{i_2})$ and $(S^1_{j_1}, S^2_{j_2})$ for each (i_1, i_2) and (j_1, j_2) produces a group of bundles of lines passing through the same epipolar point E^2_p.

We note that if S_i and S_j are in the same plane, the S^2_p located coincides with $S^2_{j_2}$ and therefore E^2_p can not be calculated.

With the following algorithm, the epipolar point is calculated $N_{bS} \cdot (N_{bS} - 1)$ times where N_{bS} is the number of 2D surfaces in correspondence:

```
 1 :   Begin
 2 :   for  each (S¹_{i₁}, S²_{i₂}) of (P₁ × P₂)
 3 :      do for  each  (S¹_{j₁}, S²_{j₂}) of (P₁ − {S¹_{i₁}} × P₂ − {S²_{i₂}})
 4 :        do for  each   M¹_{j₁,k} of S¹_{j₁} (k = 1, ..., n_v)
 5 :        do
 6 :           Calculation of (r¹_k, r²_k): projective coordinates of M¹_{j₁,k}
 7 :             relatively to Si₁¹.
 8 :           Location on P₂ of M²_{p_k} having (r¹_k, r²_k) as projective
 9 :             coordinates relatively to S²_{i₂}.
10 :        enddo
11 :        Calculation of Set : the intersection of n_v lines joining
12 :           M²_{p_k} and S²_{j₂} vertices
13 :        if  Card(Set)=1  then
14 :          Set={E²_p}
15 :        else
16 :          Set={E²_1, E²_2, ..., E²_{n_e}}, n_e ≤ 6
17 :        endif
18 :      enddo
19 :   enddo
20 :   End
```

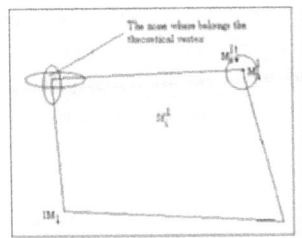

Fig. 3. Noise Modeling

4   Adaptation of the Algorithm to Noise

It is well known that it is necessary to adapt each method based on exact data to real data where there is positional uncertainty. Concerning the previous algorithm, its sensitivity to noise appears at the level of:

- Calculation of r_k^1, r_k^2 cross-ratios for each image point $M_{j_1,k}^1$ of $S_{j_1}^1$ relatively to the projective basis $S_{i_1}^1$. Each value r_k^1 and r_k^2 is obtained from four noisy points and the calculated points I^1, J_k^1, K_k^1, see Fig. 2.
- Localization of the image point Mp_k^2 that corresponds to $M_{j_1,k}^1(S_p^2$ vertex) using r_k^1, r_k^2 and $S_{i_2}^2$. As r_k^1, r_k^2 are calculated with uncertainty, and the $S_{i_2}^2$ vertices are noisy, the estimated image point vertex of S_p^2 is also noisy.
- Calculation of the intersection of lines joining S_p^2 and $S_{j_2}^2$ vertices. As the vertices of the calculated surface S_p^2 and $S_{j_2}^2$ are noisy, so are the lines joining them. The intersection point of two lines altered by a slight noise could be very far away from the theoretical intersection point.

Based on a Gaussian model of the noise, Deriche et al. show in [4] that each extremity of a line segment, which fits a set of noisy edge points, appertains to an ellipse with a probability following a χ_2 distribution.

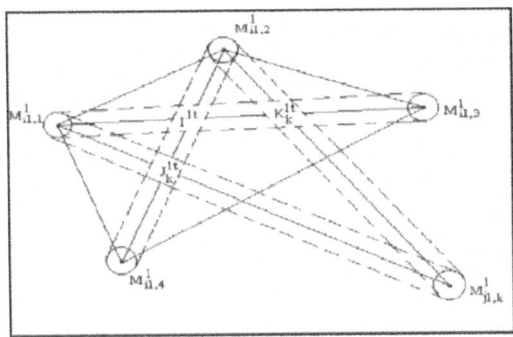

Fig. 4. Calculation of the theoretical cross-ratios

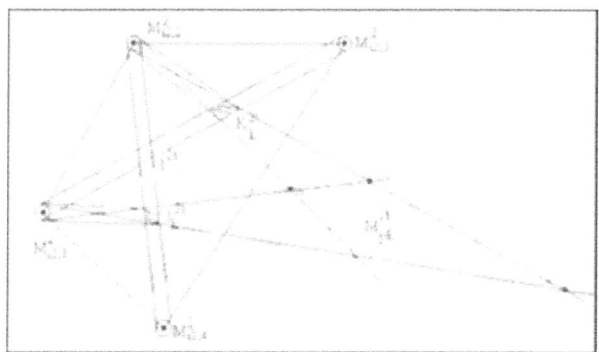

Fig. 5. Localization of the theoretical point $M_{p_k}^{2t}$

If we consider two connected line segments, a theoretical vertex belongs with a certain probability to the intersection of two ellipses as shown in Fig. 3. The form of its intersection zone can be assimilated to a disc. This hypothesis corresponds to the model of noise used by Grimson et al. in [8], where each theoretical image point M_i^{1t} (or M_i^{2t}) is supposed to belong to the δ -disc denoted by $D(M_i^1, \delta)$ (or $D(M_i^2, \delta)$) with a center located at image point M_i^1 (or M_i^2), see Fig. 3.

To ensure that the previous algorithm functions correctly in spite of noise affecting the vertices, it is necessary to include all of the theoretical image points in the different calculations. Thus the adaptation strategy consists to apply each stage of the algorithm with the disc that encompasses the noisy image in such a way that the obtained result is valid for the theoretical image points. The steps of the algorithm require modifications that must be carried out:

Calculation of theoretical cross-ratios r_k^1 and r_k^2: Theoretical cross-ratios r_k^{1t}, r_k^{2t} will be calculated using theoretical image points, see Fig. 4.

– Firstly it is necessary to locate the zone containing the theoretical line joining two theoretical vertices: it is the bundle of lines joining the two discs

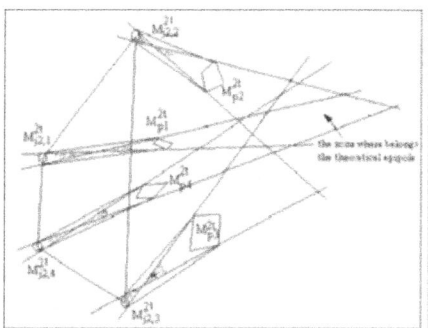

Fig. 6. The intersection of theoretical lines

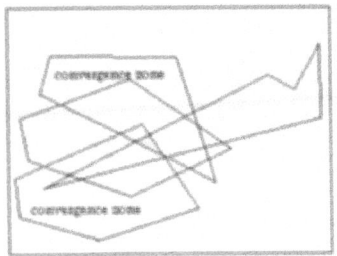

Fig. 7. Intersection of set of convergence zones

encompassing the two noisy vertices and defined from the parallel pair of tangent to the two discs.
- The next step is the localization of the zone containing both theoretical points I^{1t}, J_k^{1t} and K_k^{1t} obtained as intersection of corresponding bundles. The geometric shape of the three zones is a parallelogram.

The cross-ratios r_k^{1t} and r_k^{2t} can be limited between the maximum and minimum values of r_k^1 and r_k^2 calculated using intersection points between the different bundles: $r_{kmin}^1 \leq r_k^{1t} \leq r_{kmax}^1$ and $r_{kmin}^2 \leq r_k^{2t} \leq r_{kmax}^2$

Localization of the theoretical point $M_{p_k}^{2t}$: The theoretical point $M_{p_k}^{2t}$ will be located on the image plane P_2 from its projective coordinates r_k^{1t} and r_k^{2t} relatively to the basis projective $(M_{i_2,1}^{2t}, M_{i_2,2}^{2t}, M_{i_2,3}^{2t}, M_{i_2,4}^{2t})$, see Fig. 5.

The first step is the localization of the zones containing both theoretical image points I^{2t}, J_k^{2t} and K_k^{2t}. Each zone is of quadrilateral form. The next step is the localization of the zone containing the theoretical line, joining two noisy points encompassed by a disc and a quadrilateral. It is a bundle of lines delimited by the two pairs of tangents to the disc and passing through the vertices of the quadrilateral. Each theoretical image point $M_{p_k}^{2t}$ is then obtained as the intersection of the theoretical lines $M_{i_2,1}^{2t} J_k^{2t}$ and $M_{i_2,2}^{2t} K_k^{2t}$, thus it belongs to the intersection zone of the two bundles of lines $B(M_{i_2,1}^{2t}, J_k^{2t})$ and $B(M_{i_2,2}^{2t}, K_k^{2t})$. Its form is a quadrilateral.

Table 1. $(dx - dy)$ of each epipolar zone (unity=pixel)

S_{fNb}	$\delta_n = 0.01$	$\delta_c = 0.1$	$\delta_c = 0.5$	$\delta_c = 1.0$	$\delta_c = 2.0$
2	152.3-3.93	2667-42	11761-414	17700-1644	17932-2997
5	91.5-1.33	957-23	4222-98.6	8921-618.6	5829-430
10	46.6-1.73	178-16	1234-41.13	4348-294.6	1902-55.4
15	3.55-.86	46-4.66	124.5-17.33	538.5-40.73	899-41
20	3.55-0.55	46-4.66	124.5-17.33	538.5-40.73	898-41

Table 2. $(dx - dy)$ of each epipolar zone (unity=pixel)

δ_n (pixels)	$\Phi = 30.21°$	$\Phi = 19.15°$	$\Phi = 15.37°$	$\Phi = 10.28°$
0.01	29-4.13	113.5-4.66	129-6.5	137-8
0.1	178-16.6	1122-47	1121-62	1325-48
0.2	538-25.3	1788-73	2045-114	2845-87
0.5	792-116	1865-85	2140-166	3551-267
1.0	974-110	2120-266	2961-220	5079-272
2.0	974-134	2412-522	3234-981	5665-322

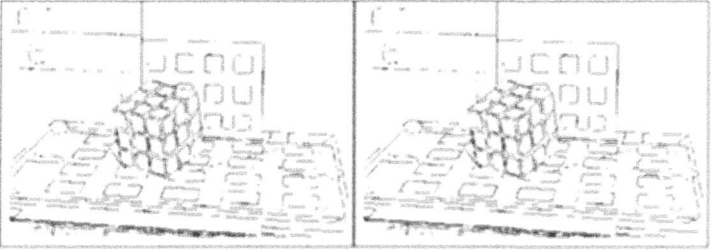

Fig. 8. Stereoscopic images of scene 1

Theoretical lines joining the vertices of S_p^2 and $S_{j_2}^2$: As each theoretical image point of S_p^2 belongs to a zone of a quadrilateral form, and each theoretical vertex of the surface $S_{j_2}^2$ belongs to a zone of a disc form, each theoretical line $M_{j_2,k}^{2t} M_{p_k}^{2t}$ belong to the bundle delimited by the two pairs of tangent to the disc and passing by the quadrilateral, as is illustrated by the Fig. 6. From this geometric construction, the theoretical epipole belongs to the zone located as the intersection of the set of bundles $\mathrm{B}(M_{j_2,k}^{2t}, M_{p_k}^{2t})$, $k = 1,...,n_v$ where n_v is the number of $S_{j_2}^2$ vertices.

To improve the precision in the localization of the theoretical epipole, we must repeat this procedure with all quadruplets $(S_{i_1}^1, S_{i_2}^2, S_{j_1}^1, S_{j_2}^2)$. This allows to produce a set of convergence zones,where everyone contains the theoretical epipole. Their intersection contains the theoretical epipole and will be noted epipolar zone. Its area decreases when the number of 2D surfaces used increases.

5   Experimental Results

We first report on noisy artificial images.

Determination of δ_c: In practice, the noise δ_n affecting the 2D surfaces is unknown, the use of δ_c noise supposed to affect the image points has an influence on convergence zone coherence. For this, if $\delta_c \geq \delta_n$, the convergence zone contains the epipolar point as we have shown in the proposed geometric approach. For all tests done over noisy artificial images the area of epipolar zone obtained with

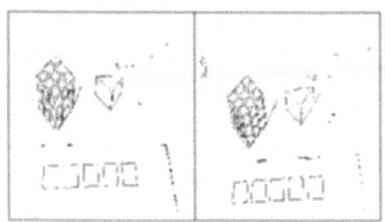

Fig. 9. Stereoscopic images of scene 2

$(\delta_c \geq \delta_n)$ is always greater than the area of epipolar zone obtained with $(\delta_c \leq \delta_n)$. The minimal area of epipolar zone is obtained with $\delta_c = \delta_n$. To determine the optimal computation noise δ_c, we begin the calculation with $\delta_c = 4$ pixels and we repeat this process decreasing δ_c until to obtain the minimal area of epipolar zone.

Influence of noise and surface numbers on epipolar zone area: Table 1 illustrates, using $(dx - dy)$ where dx (resp. dy) is the maximal distance (pixels) between the x (resp. y) coordinates of all vertices of this zone, the variation of epipolar zone area in terms of the 3D surfaces number S_{fNb} and the noise affecting its vertices (we use $\delta_c = \delta_n$). It is clear that to reduce the epipolar zone, we must use a maximum of 2D surfaces and the epipolar zone area is disconcerted to one $pixel^2$ for $S_{fNb} = 20$ and $\delta_c = 0.01$ pixels.

Influence of the convergence angle (Φ) on the epipolar zone area: Table 2 illustrates using $(dx - dy)$ the area variation of epipolar zone in terms of convergence angle Φ for $\delta_c = \delta_n$ and the focal length, distance between the two cameras are equal to $9mm$ and $135mm$. These results show that the decrease of convergence angle Φ implies the increase of the epipolar zone area because the epipolar point is moving away from the image center and the dx of the epipolar zone becomes larger.

Finally we briefly report on applications using real images, see Figs. 8 and 9. As the algorithm treats the vertices of the 2D surfaces, it is necessary to recover the vertices as intersection of neighboring lines. Each connected contour graph is considered as potential correspondent to a flat surface. The number of 2D surfaces in the left and right images are 42, 37 for scene 1 and 26, 26 for scene 2. Applying this method to the two pairs of stereoscopic images pair with various dc beginning with $\delta_c = 1.5$ to 0.1 pixels, the area of epipolar zones located contains the theoretical epipole. The coordinates of the epipolar point are ($x = 8900$ pixels, $y = 250$ pixels). To compute the epipolar zone we have used different numbers of 2D surfaces.

References

1. H. S. M. Coxeter: *Projective Geometry.* Springer, New York (1987).   61

2. G. Csurka, C. Zeller, Z. Zhang, O. Faugeras: Characterizing the uncertainty of the fundamental matrix. *Computer Vision and Image Understanding*, **68** (1997) 18–35. 60

3. R. Deriche, Z. Zhang, Q. T. Luong, O. D. Faugeras: Robust recovery of the epipolar geometry for an uncalibrated stereo rig. In: *Proceed. ECCV'94*, Stockholm (1994). 60

4. R. Deriche, R. Vaillant, O. D. Faugeras: From noisy edge points to 3D reconstruction of scenes: a robust approach and its uncertainty analysis. In: *Proceed. 7th Scandinavian Conf. on Image Analysis*, Alborg, Denmark, (August 1991) 225–232. 63

5. R. Deriche, J. P. Cocquerez: Extraction de composants connexes basé sur une détection optimale des contours. In: *CESTA*, Paris (1987).

6. O. D. Faugeras: What can be seen in three dimensions with a uncalibrated stereo rig. In: *Proceed. ECCV'92*, Santa Margherita Ligure, Italy, (1992).

7. O. D. Faugeras, Q. T. Luong, S. J. Maybank: Camera self-calibration: theory and experiments. In: *Proceed. ECCV'92*, Santa Margherita Ligure, Italy, (1992). 60

8. W. E. L. Grimson, D. P. Huttenlocher, D. W. Jacobs: A study of affine matching with bounded sensor error. In: *Proceed. ECCV'92*, Santa Margherita Ligure, Italy, (1992). 64

9. S. J. Maybank, O. D. Faugeras: A theory of self calibration of a moving camera. *Internat. J. of Computer Vision*, **8** (1992) 123–151. 60

10. R. Mohr, B. Triggs: Projective geometry for image analysis. Tutorial given at IS-PRS, Vienna, (July 1996). 61

11. R. Mohr, E. Arbogast: It can be done without camera calibration. *Pattern Recognition Letters*, **12** (1991) 39–43. 60, 61

12. P. R. S. Mendonca, R. Cipolla: Estimation of epipolar geometry from apparent contours: affine and circular motion cases. In: *Proceed. CVPR'99*, (1999). 60

13. P. Remagnino, P. Brand, R. Mohr: Correlation techniques in adaptive template matching with uncalibrated cameras. In: *SPIE Proceed. Vision Geometry III*, Boston Mass., (November 1994) 252–263. 61

A Color Segmentation Algorithm for Real-Time Object Localization on Small Embedded Systems

Philippe Leclercq and Thomas Bräunl

Center for Intelligent Information Processing Systems
Department of Electrical and Electronic Engineering
The University of Western Australia
leclercq@tartarus.uwa.edu.au
braunl@ee.uwa.edu.au

Abstract. We present an approach to color segmentation and object localization for small-embedded systems. The algorithm defines sub-spaces in the main color space (RGB), each sub-space being a color class. A look-up table is used to speed up pixel classification in color classes. A binary image is computed for every color class. After noise reduction, we obtain a color class pixel list; after each object's image coordinates are found and transformed to world-coordinates depending on the camera position.

1 Introduction

We introduce an algorithm for real-time color segmentation on small-embedded systems, i.e. an electronic platform (EyeBot MK 3, [1]) based on a Motorola 68332. A *normalized RGB space* is used, as the conversion from the standard RGB space is much easier than the conversion into HSI or HSV spaces. The *normalized RGB space* is also robust to different lighting conditions for color detection. In order to achieve color segmentation, we divide the main color space into sub-spaces, each sub-space being a color class. A *color class* look-up table is filled to associate every (R, G, B) triplet with a class. Using this table, we obtain one *binary image* for each class. We reduce its noise using an "erosion" (morphological image operator) routine and localize object boundaries. Object image coordinates are then transformed back to world coordinates, which requires a camera calibration for angle offsets and relative distance. We implemented the color segmentation algorithm presented here for mobile robot systems using on-board vision at the "RoboCup 2000" robot soccer competition. Therefore, our examples are based on distinguishing the three colors used at this competition (blue and yellow for the goals and orange for the ball). We also present a general color class segmentation approach, in order to be able to use the presented algorithm in any general cases. A similar, but more computation-intense approach to color segmentation on larger computer systems has been proposed as "CMVision" [2]

R. Klette, S. Peleg, G. Sommer (Eds.): Robot Vision 2001, LNCS 1998, pp. 69–76, 2001.
© Springer-Verlag Berlin Heidelberg 2001

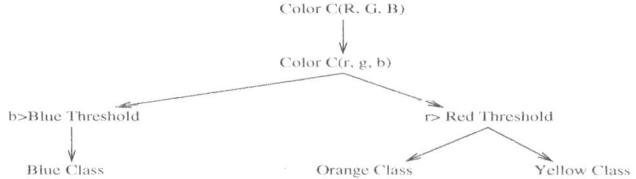

Fig. 1. Color Segmentation Algorithm

2   Color Space and Color Classes

The use of a color space like HSV (Hue, Saturation, Value) would be most appropriate to solve the task of color segmentation. However, we opted for a *normalized RGB* [3] color space instead for efficiency reasons, since the camera delivers only RGB data and because it is robust to lighting condition changes [4]. The *normalized RGB* color space (*rgb*) is defined by equation 1:

$$r = \frac{R}{R+G+B} \quad , \quad g = \frac{G}{R+G+B} \quad , \quad b = \frac{B}{R+G+B} \tag{1}$$

Assuming the luminosity $L = R + G + B$, we now have $L' = r + g + b = 1$ for all (r, g, b). A color C would have different components in the RGB space for different lighting conditions, where the luminosity is always 1 for the *rgb* space. As we want to process as quickly as possible, we need to be able to build a *"color-class" look-up table*, i.e. knowing for each (R, G, B) triplet, which color-class it belongs to [2]. This table will associate a class (i.e. a color) to every (R, G, B) triplet in the space. If the R, G and B values are coded on 8 bits, then, this table would be $(2^8)^3 = 16Mb$, which is too large for our platform. Practically we use only the five most significant bits, i.e. $(2^5)^3 = 32kb$, which we have found to be sufficient by experiment. We give an example on how to differentiate three colors: blue, orange and yellow. As we do not have much processing power, we need to use ideas as easy as possible, i.e. the blue class will be for colors containing mostly blue. Yellow and orange will be for colors having mostly red. Figure 1 shows the color segmentation algorithm. The orange and yellow classes need two criteria: red and green (we make the difference between yellow and orange by the amount of green). This becomes clear by observing a straight line between red and green on the C.I.E. chromaticity diagram. Threshold denotes the initially calibrated constant threshold values. We are only comparing the normalized color component to a certain level to find out if the color belongs to one class or not. In order to improve the speed of the algorithm we can use R, G and B values (camera input) in equation 2, i.e. we avoid a complete image conversion from RGB to *rgb* space. The relation $b > \text{Threshold}_{\text{blue}}$ can be expressed as $B > \text{Threshold}_{\text{blue}} \cdot (R + G + B)$, i.e. $B > \text{Threshold}_{\text{BlueFloat}} \cdot L$, where $L = R + G + B$ is the luminosity. And finally,

$$100 \cdot B > \text{Threshold}_{\text{BlueInteger}} \cdot L \qquad \text{First Class Condition,} \qquad (2)$$

If the result of this relation is true, then the color $C(R, G, B)$ belongs to the first color class (blue). By multiplying B by 100, $\text{Threshold}_{\text{Blue}}$ is now an integer between 0 and 100 instead of a floating-point between 0 and 1. Now, this relation only uses integers with multiplications and additions. Equations 3 and 4 show the same kind of relation for the second and third colors:

$$(100 \cdot R > \text{Threshold}_{\text{Red}} \cdot L) \quad \text{AND} \quad (100 \cdot G > \text{Threshold}_{\text{Green}} \cdot G), \qquad (3)$$

$$(100 \cdot R > \text{Threshold}_{\text{Red}} \cdot L) \quad \text{AND} \quad (100 \cdot G < \text{Threshold}_{\text{Green}} \cdot G). \qquad (4)$$

3   Color Calibration

We now have a way to distinguish the three colors we are looking for. For this method to work, we need to define the thresholds for red, green and blue. The blue class only needs the $\text{Threshold}_{\text{Blue}}$ value, so we put the camera at a known distance of the blue object, take an image, process it incrementing the $\text{Threshold}_{\text{Blue}}$ value, unless we find a rectangle of the size this object should be at that distance (d_{pix}). For the $\text{Threshold}_{\text{Red}}$, we can use exactly the same algorithm, i.e. looking at the yellow object while incrementing the threshold value until we find it; this will give us the red value for both yellow and orange classes. We need then to find the $\text{Threshold}_{\text{Green}}$ to make the difference between these two classes. We take an image of the yellow object, applying the red condition with a decrementing $\text{Threshold}_{\text{Green}}$ as long as we find the object. That way, we find the lowest $\text{Threshold}_{\text{Green}}$ for the yellow color, i.e. the highest for the orange color.

Figure 2: $d_{\text{pix}} = \alpha_{\text{pix}} \cdot \frac{\theta_{\text{real}}}{\alpha_{\text{real}}}$, and finally:

$$d_{\text{pix}} = \alpha_{\text{pix}} \cdot \frac{\arctan(d_{\text{real}}/D_{\text{real}})}{\alpha_{\text{real}}}. \qquad \text{Object Width in Pixels} \qquad (5)$$

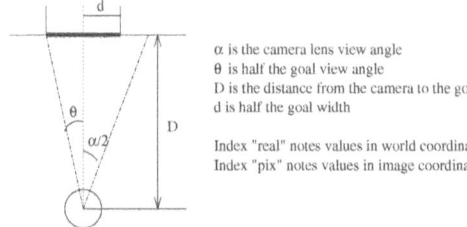

α is the camera lens view angle
θ is half the goal view angle
D is the distance from the camera to the goal
d is half the goal width

Index "real" notes values in world coordinates
Index "pix" notes values in image coordinates

Fig. 2. Object Size Determination

During the calibration the camera is at a known distance (D) from the object (See Fig. 2). We also know the width, d, of the object. α is known as it depends on the camera. Then we know the width in pixels d_{pix} (See equation 5) of the object on the image. Therefore, we know when to stop incrementing or decrementing the thresholds while calibrating. Once these three levels have been set up, we fill in the color class look-up table. For each color $C(R, G, B)$ we process all three relations. If none is true, then the color belongs to no color class, if only one relation is true, then the color belongs to that class. If several relations are true, the color belongs to no class to avoid conflicts. As we are not describing each color unambiguously, it may happen that several relations are true for the same R, G and B values. Practically, this problem occurs for low levels of R, G and B (black, dark grey) or for high levels of R, G and B (white, light grey); i.e. for levels where the colors are uncertain and not clearly defined. In practice, our approach works for separating three classes. For more color classes, we would need a more precise description of every color. As an example, equation 6 gives a more precise rectangle around the selected color and one can think of any other topological way of describing a color in the RGB color space.

$$(100 \cdot R > \text{ThresMin}_{\text{Red}} \cdot L) \;\; \text{AND} \;\; (100 \cdot R < \text{ThresMax}_{\text{Red}} \cdot L),$$
$$(100 \cdot G > \text{ThresMin}_{\text{Green}} \cdot L) \;\; \text{AND} \;\; (100 \cdot G < \text{ThresMax}_{\text{Green}} \cdot L), \quad (6)$$
$$(100 \cdot B > \text{ThresMin}_{\text{Blue}} \cdot L) \;\; \text{AND} \;\; (100 \cdot B < \text{ThresMax}_{\text{Blue}} \cdot L).$$

4 General Approach to Color Class Segmentation

We described an easy but specific approach to color segmentation in a particular case. In order to give to this algorithm a general use, we need to be able to define any color class easily. In the HSV (Hue, Saturation, Value) color space, the color information is contained by the hue value. The conversion from RGB to Hue is described by equation 7 (see [3]):

$$H = \arccos\left(\frac{(R - G) + (R - B)}{2 \cdot \sqrt{(R - G)^2 + (R - B)(G - B)}} \right) \quad \text{RGB to Hue} \quad (7)$$

As this conversion is not trivial, we will need to build a conversion table. As previously, we decided to use only the five most significant bits. We can now have a single hue information out of R, G and B values. The idea to characterize the color we want to define is to put the object in front of the camera and detect its hue. For this, we convert the middle area of the image (4×4 pixels for example) from RGB to hue; then we compute the median hue. We also need to specify a *hue range* to build the color class. To fill the RGB color class lookup table, we process every (R, G, B) triplet, convert it into hue, and check if this hue is within the specified range. This way, every single cell of the RGB space is filled. We can add more classes by looking at other objects, finding their median hue and fill the RGB space. Again, in case of conflict we describe the color as belonging to no class. Remark: in the HSV space, hue is only describing the color. The saturation describes the amount of white the color is mixed with (i.e. full color → no color i.e. white). That means, that in this space, white is only described by saturation (and intensity...) value and may have any hue value. So, we either have to treat white as separate color class (that we can directly fill in the RGB space: $R = G = B$), or have to consider a saturation threshold while building the color class.

5   Noise Reduction of Segmented Images

We now have a color class look-up table to compute binary images (one for each color class) out of the initial color image. Each binary image is filled with 1 if a pixel belongs to a color class and 0 if not. To find the object limits we have to reduce the noise on these binary pictures. The noise is mainly random dots or lines, so we use a single "erosion" method (morphological operator), i.e. we keep only the pixels having top, bottom, right and left neighbors of the same color class. Each time we find a valid pixel, we add its coordinates into an *object pixel list*. We process the image in a known order, from top to bottom and from left to right. By knowing this, we accelerate finding the object limits: the first pixel in the object pixel list will always be the topmost pixel and the last pixel will always be the bottommost pixel. Finally, we just need to perform maximum and minimum tests to find the rightmost and leftmost values. We are assuming that there will be only one object of a certain color in the image. Therefore we can use "color class" or "object" to describe the same thing. We now have, for every color class, the coordinates of the four points of the rectangle surrounding them. We need to adjust these values, as the erosion noise reduction method destroys one pixel around the object.

6   Camera Calibration

In order to localize objects found in the image, we need to have a reference position for the camera (see Fig. 3). Since we use a panning camera, we need to determine for which position the camera is looking straight towards the front. To know the horizontal offset angle we place an object in front of the camera and rotate until we find it in the middle of the picture. We also need to know the vertical initial angle, i.e. the angle of the normal middle axis (β_{norm}) compared to the camera middle axis (β_{cam}). We place an object (here: the ball) in front of the camera (see Fig. 3) and we need to find the camera middle axis position offset, $\beta_{\text{offset}} = \beta_{\text{norm}} - \beta_{\text{cam}}$, where:

$$\beta_{\text{norm}} = \arcsin \left(\frac{\text{Cam}_{\text{height}} - \text{Obj}_{\text{radius}}}{\text{Distance}} \right) \quad \beta_{\text{cam}} = \frac{b_{\text{col}} - b_{\text{mid}}}{Pic_{\text{Height}}/2} \cdot \frac{Cam_{\text{ViewAngle}}}{2}$$

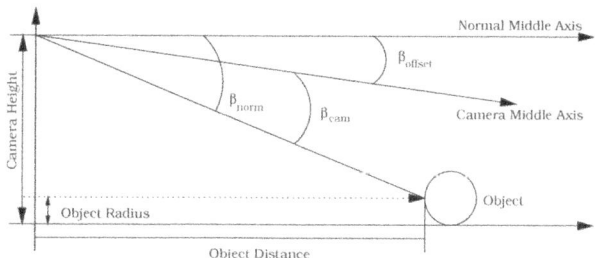

Fig. 3. Horizon Offset Calibration

7 Determination of Object Position

Once we know the object limits in the image coordinates (see Fig. 4), we can easily estimate its position in world coordinates by projection (see Fig. 5). The distance between the robot and the object is $|C'B|$. We know the height of the center of the camera, C, and the center of the object, B. We determine the angles and distances by projection. We will determine α and β' by the coordinates of the ball on the picture (see Fig. 4). We are only taking into account the middle pixel of the detected object $P(row, col)$. If we put the middle of the axes in the middle of the screen, the new value is $P(\text{Object}_{\text{row}} - \text{mid}_{\text{row}}, \text{Object}_{\text{col}} - \text{mid}_{\text{col}})$. Knowing the view angle of the lens, we determine from Fig. 4:

$$\alpha = \frac{b_{\text{row}} - b_{\text{mid}}}{\text{Pic}_{\text{width}}/2} \cdot (\text{ViewAngle}/2), \quad \beta' = \frac{b_{\text{col}} - b_{\text{mid}}}{\text{Pic}_{\text{height}}/2} \cdot (\text{ViewAngle}/2).$$

Knowing α and β' we can determine $C'B'$. We project on the x axis. We find β' in the $C''B'B$ triangle. From Fig. 5, we have $BB' = C''B \cdot \cos\alpha$. We obtain finally:

$$C'B = \frac{C_{z''} - b_r}{\sin\beta' \cdot \cos\alpha}. \qquad \text{Relative Distance,} \qquad (8)$$

where b_r is the ball radius. We can now get the distance of the object for every position of its center on the image. We can build several 2D look-up tables

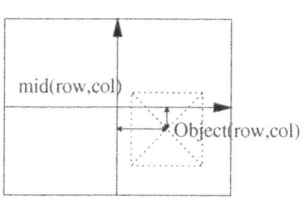

Fig. 4. Object Position

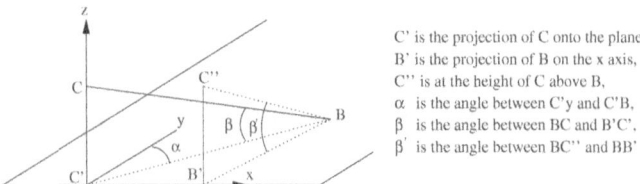

Fig. 5. Camera to Object Distance

containing the object distance (if we know its size) as well as the values of α and β': so we can easily have the angle position of any object without knowing its size as α and β' are only functions of the position of the middle pixel in the image. Each table would be *width* $\times$ *length* $\times$ *size* bytes long. In our case, we used 16 bit integers and the image size is 80×60 which is 9600 bytes.

8   Experiments and Results

We conducted a number of object recognition experiments in the context of robot soccer. However, the algorithm is very general and can be applied to a number of other areas by changing the color class description and calibration method. The platform used is an EyeBot MK3 [1], based on a 25 MHz 32 bit controller (Motorola 68332), 1MB RAM, 512KB ROM for system and user programs. On this system, color segmentation and object recognition for three classes (two soccer goals and one ball) takes between 0.02 and 0.03 seconds: 0.02 if no object is present in the picture, 0.03 if a large object is found. The ball can be detected from a distance up to 60 cm – 70 cm in front of the robot, the goals up to 1,50 m. The example, on the top of Fig. 6 shows, on the left, the initial image; the center picture shows the binary image resulting from the yellow segmentation; finally, the picture on the right shows the result after noise reduction. The right picture is in fact only a pixel-list among which we look for the topmost, bottommost, rightmost and leftmost position to define the object coordinates (represented by the grey rectangle). We can observe that the image is having several lines of noise (the upper ones, with for example a computer screen) and that the algorithm is finding the goal quite precisely at the good position. The example on the bottom part of Fig. 6 is containing two different classes : the yellow goal (black pixels) and the orange ball (medium grey pixels). The binary image (in the center) is in fact the superposition of the two binary images from the two classes. The right image shows, in light grey the found boxes around the two objects to detect. Again the noise here is coming from the upper lines and from one of our robots in front of the goal. We can observe that as long as the robot is not completely hiding one side of the goal, we still manage to find the complete length of the goal.

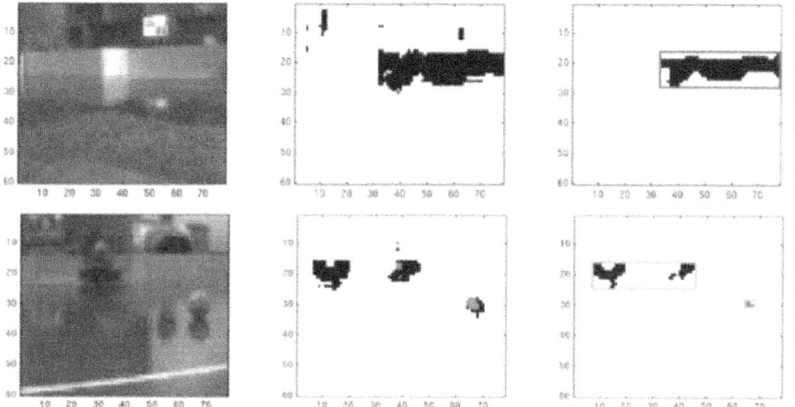

Fig. 6. Sample of the LibVision with one Class (up) and two Classes (down)

9   Conclusion

We have demonstrated a simple and effective real-time color segmentation algorithm that can be implemented on embedded systems with limited computing performance. This approach is much more efficient but not as robust as some others based on the mean shift algorithm [5] needing several seconds on a standard workstation for bigger pictures (512 × 512) or neural networks [6] used by the Azzura Robot Team in the middle size league. However, it brings real time for the small size league robot vision and is reliable enough for a limited number of color classes.

References

1. T. Bräunl, B. Graf: Small robot agents with on-board vision and local intelligence. *Advanced Robotics*, **14** (2000) 51–64. 69, 75
2. J. Bruce, T. Balch, M. Veloso: Fast and inexpensive color image segmentation for interactive robots. In: Proceed. of the IEEE/RSJ Internat. Conf. on *Intelligent Robots and Systems*, (2000). 69, 70
3. R. C. Gonzalez, R. E. Woods: *Digital Image Processing.* Addison Wesley, (1992). 70, 72
4. J. Brusey, L. Padgham: Techniques for obtaining robust, real-time, color-based vision for robotics. In: *IJCAI-99: Proceed. of the 3rd Internat. Workshop on RoboCup*, M. Veloso, E. Pagello, H. Kitano, eds., (July 1999). 70
5. D. Comaniciu, P. Meer: Robust analysis of feature spaces : color image segmentation. In: IEEE-Proceed. of the Internat. Conf. on *Computer Vision and Pattern Recognition*, (June 1997) 750–755. 76
6. C. Amoroso, A. Chella, V. Morreale, P. Storniolo: A segmentation system for soccer robot based on neural networks. In: *IJCAI-99: Proceed. of the 3rd Internat. Workshop on RoboCup*, M. Veloso, E. Pagello, H. Kitano, eds., (July 1999). 76

EYESCAN – A High Resolution Digital Panoramic Camera

Karsten Scheibe, Hartmut Korsitzky, Ralf Reulke,
Martin Scheele, and Michael Solbrig

German Aerospace Center (DLR)
Institute of Space Sensor Technology and Planetary Exploration
Rutherfordstrasse 2, D-12489 Berlin, Germany
karsten.scheibe@dlr.de

Abstract. A digital panoramic camera system is introduced consisting of a CCD line scanner and a high precision turntable. This combination allows the use of such a digital imaging systems for photogrammetric, robot vision, artistic and other applications. Additionally, these images with a reduced resolution or parts of it can be used for internet applications. Typical fields of application are photogrammetry in architecture, digital archiving of cultural objects and virtual reality. The imaging geometry causes panoramic distortions, therefore the images must be transformed into plane coordinate systems in order to work with them. Basic equations are given for these projections.

1 Introduction

EYESCAN, a digital panoramic camera, is a joint development between the German Aerospace Center and KST (Kamera Systemtechnik) Dresden. The camera will be preliminary used as a measurement system to create high resolution 360 degrees panoramic images for photogrammetric purposes. The sensor principle is based on a CCD line, which is mounted on a turntable parallel to the rotation axis. The second image dimension is generated by rotating the turntable. To reach highest resolution and a large field of view a CCD line with about 10.000 detector elements is used. This CCD is a RGB triplet and allows to acquire true color images. A high SNR electronic design allows a short capture time for a 360^0 scan.

EYESCAN is designed for field experiments as well as for laboratory measurements. Combined with a robust and powerful portable PC it becomes easy to capture seamless digital panoramic images.

This paper describes the camera modules, the control and processing software as well as some important applications for the camera.

2 The Sensor System

The sensor system consists of three main parts, the

R. Klette, S. Peleg, G. Sommer (Eds.): Robot Vision 2001, LNCS 1998, pp. 77–83, 2001.
© Springer-Verlag Berlin Heidelberg 2001

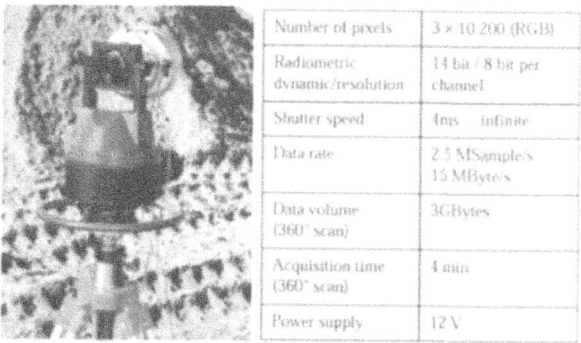

Number of pixels	3 × 10.200 (RGB)
Radiometric dynamic/resolution	14 bit / 8 bit per channel
Shutter speed	4ms ... infinite
Data rate	2.5 MSamples/s 15 MByte/s
Data volume (360° scan)	3GBytes
Acquisition time (360° scan)	4 min
Power supply	12 V

Fig. 1. EYESCAN camera and its basic parameters

- camera head,
- optical part,
- high precision turntable with a DC-gearsystem motor.

The following sections describe these modules more in detail.

- *Camera head module - measurement unit*
  Figure 1 shows the EYESCAN-camera. The essential parameters of the camera head are listed in the table. The camera head is connected to the PC with a bidirectional fiber link for data transmission and camera control. The camera head is mounted on a tilt unit for vertical tilts of $\pm 30^0$ (steps of 15^0). The axis of tilt and rotation are in the needlepoint.
- *Optics*
  The system uses a high performance Rhodenstock lens APO-Sironar digital HR 4/60 with a optical adjustment in 5 steps for distances of 20m, 9m, 6m, 4m, 2.8m. Other measurement lenses can be used by applying adjustment rings. The camera is fully geometrically calibrated.
- *Turntable*
  The camera head is mounted on a high precision turntable with a sinus-commutated DC-gearsystem motor, internal motion control and direct controlling by the PC. Rotation speed and scan angle are pre-selectable and correspond to the shutter speed, image size and focal length of the lens.

3 Camera Control and Data Preprocessing

The data processing and the complete control of the camera is integrated in a graphical user interface. The next sections will describe it.

- *Camera head control*
  This software initializes the camera head and selects all parameters being essential for capturing an image, like clock time and integration time.

- *Turntable control*
  This tool realizes an automatic control for the motor, e.g. set rotation angle
  and speed, read status of the motor, etc
- *Data reception, caching and storage*
  The framegrabber writes the image data in the PC memory. With a caching
  approach, which is realized as a multithredding program, the data can be
  written on harddisk.
- This part consists of data correction (photo response non-uniformity, dark
  signal non-uniformity, offsets) and a (non linear) radiometric normalization
  to cast the data from 16 to 8 bit. All this procedures can be run in real time
  or off-line. Additional software packages allow a visualisation of the data
  of the current scan line, a fast preview for scene selection and a quicklook
  during data recording. Typical commercial image processing programs are
  not able to handle the huge raw data amount for one 360^0 panoramic scan
  of about 3 GByte. Therefore some routines have been developed for
  - spatial shift between the RGB channels to correct the effect of the triplet
    view,
  - 90^0 rotation of the whole image,
  - visualisation of selected image parts,
  - histogram and contrast equalization,
  - correction of the panoramic effect.

  The last processing step is necessary for real photogrammetric approaches
  to transform image data from the cylindrical panorama view into a classical
  plane view. This procedure will be explained in the next section.

4   Post Processing and Application

Based on the recording geometry it is necessary to transform the images for the
different applications. Analog to the transformation of images acquired by air-
planes (correction of roll, pitch and yaw) it is possible to transform the panoramic
images into different cartesian coordinate systems. Examples for such transformation are

- from a cylindrical panorama view into a classical plane view,
- from a conical panorama view into a classical plane view

The basics of these transformations are well-known [1,2]. The following equation
can be derived easily:

$$r = r_0 + t.A(\phi_2) \cdot A(j.\Delta\Theta).A(\phi_1) \cdot r_d \tag{1}$$

$[r]$ - vector of the object point with $r = (x, y, z)^T$
$[r_0]$ - vector of the camera position with $r_0 = (x_0, y_0, z_0)^T$
$[t]$ - scale
$[A]$ - rotation matrix
$[r_d]$ - vector of the pixel position in the focal plane with $r_d = (0, y_d - f)^T$

Fig. 2. Gendarmenmarkt (Berlin), cylindrical image coordinates

$[j]$ - line number of the image
$[\Delta\Theta]$ - rotation angle increment
$[\phi_1]$ - inclination of the optical axis (constant for a scan)
$[\phi_2]$ - - inclination of the rotation axis

Figure 2 shows the Gendarmenmarkt in Berlin (Germany). This image is not modified after capturing, the image coordinates are cylindrical. The angles in the image space are the same as in the object space. This means that straight lines in the object space are curved in the image. This effect can be observed in this image on the pavement. Figure 3 shows the transformation from a cylindrical panorama view into a plane view. This image corresponds now to a matrix camera view. Now it is possible to measure in these images (stereo pairs) and incorporate this data in conventional photogrammetric stereo systems. Capturing panorama image data with a suitable camera and $\phi_1 \neq 0, \phi_2 = 0$ one is able to record floors and ceilings of rooms in a practical way. Based on equation (1) the following equation of transformation are obtained [3]

$$j' = Q \cdot \sin(j \cdot \Delta\Theta) \cdot (i \cdot \delta \cdot \sin\phi_1 + f \cdot \cos\phi_1) \tag{2}$$

$$i' = Q \cdot \cos(j \cdot \Delta\Theta) \cdot (-i \cdot \delta \cdot \sin\phi_1 - f \cdot \cos\phi_1) \tag{3}$$

$$Q = \frac{f}{\delta' \cdot (i \cdot \delta \cdot \cos\phi_1 - f \cdot \sin\phi_1)} \tag{4}$$

$[i]$ - sample in the original image,
$[j]$ - line in the original image,
$[i']$ - sample in the transformed image,
$[j']$ - line in the transformed image,
$[\delta]$ - scaling factor in the original image,
$[\delta']$ - scaling factor in the transformed image

Fig. 3. Gendarmenmarkt (Berlin), cartesian image coordinates

Figure 4 presents a 360^0 scan in the Grottensaal (Neues Palais, Potsdam Sanssouci) with $f = 45^0$. For the acquisition of these images the WAAC camera from German Aerospace Center DLR was used [4]. This camera had a configuration with no antiblooming electronic. This is the reason for the saturation in region of the windows. By using equations (2) to (4) the transformed image in Fig. 5 is obtained. Such a kind of panoramic views allows to generate 4π spherical data sets. This transform is executed as follows

$$\Theta = j\Delta\Theta \qquad (5)$$

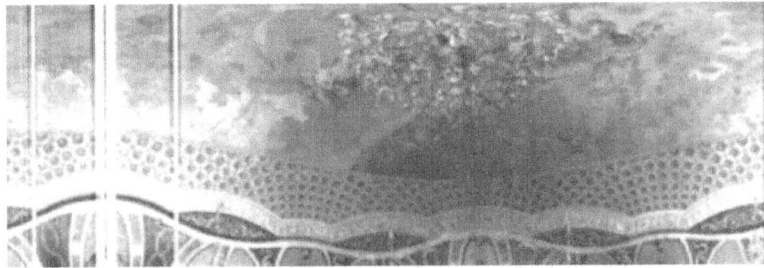

Fig. 4. Ceiling of the Grottensaal (Potsdam), conical image coordinates

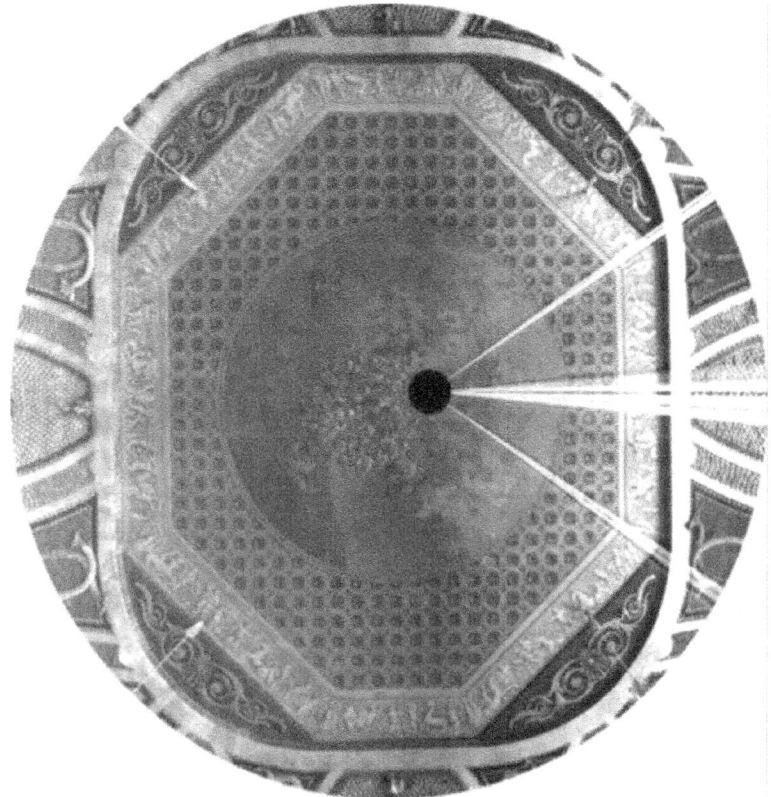

Fig. 5. Ceiling of the Grottensaal (Potsdam), cartesian image coordinates

$$\Phi = \arctan\left(\frac{i \cdot \delta \cdot \cos\phi_1 + f \cdot \sin\phi_1}{f \cdot \cos\phi_1 - i \cdot \delta \cdot \sin\phi_1}\right) \tag{6}$$

$[\Theta]$ - longitude angle,
$[\Delta\Theta]$ - increment of the longitude angle,
$[\Phi]$ - latitude angle.

Figure 6 shows an image being transformed from two separate scans into one spheric 4π image. By using suitable software it is possible to project parts of this image onto a plane view. So the whole hemisphere can be depicted into a field of view of a virtual observer.

5   Conclusions

We introduced a CCD line camera in combination with a high precision turntable. This approach makes possible the use of such a digital imaging systems for

Fig. 6. Ceiling, floor and walls of the Grottensaal, 4π hemisphere

photogrammetric, artistic and other applications. In addition, the same images with reduced resolution can be used for internet applications. The mathematic basics are described in order to project the images into cartesian coordinate systems. Current work is focused on applying this system in the field of photogrammetry.

References

1. Th. Luhmann: *Nahbereichsphotogrammetrie.* Herbert Wichmann Verlag, Heidelberg (2000). 79
2. W. Lisowski, A. Wiedemann: Auswertung von Bilddaten eines Rotationszeilenscanners. *DGPF-Tagungsband 7* (1999) p. 183. 79
3. R. Reulke, M. Scheele: CCD-Line Digital Imager for Photogrammetry in Architecture. *Internat. Archives of Photogrammetry and Remote Sensing* **XXXII** Part 5C1B (1997) p. 195. 80
4. R. Sandau, A. Eckardt: The stereo camera family WAOSS/WAAC for spaceborne/airborne applications. *ISPRS* **XXXI** Part B1, Commission I, (1996) 170–175. 81

A Wavelet-Based Algorithm
for Height from Gradients

Tiangong Wei and Reinhard Klette

CITR, University of Auckland, Tamaki Campus
Building 731, Auckland, New Zealand
{rklette,twei001}@citr.auckland.ac.nz

Abstract. This paper presents a wavelet-based algorithm for height from gradients. The tensor product of the third-order Daubechies' scaling functions is used to span the solution space. The surface height is described as a linear combination of a set of the scaling basis functions. This method efficiently discretizes the cost function associated with the height from gradients problem. After discretization, the height from gradients problem becomes a discrete minimization problem rather than discretized PDE's. To solve the minimization problem, perturbation method is used. The surface height is finally decided after finding the weight coefficients.

1 Introduction

Shading-based 3D shape recovery techniques, e.g. shape from shading (SFS), photometric stereo method (PSM), normally provide gradient values (vector field) for a discrete set of visible points on object surfaces. These gradient values have to be integrated to achieve relative height or depth values. See Fig. 1 for an illustration of such a mapping of a vector field into a depth map. However, no much work was done so far in integration techniques for the gradient vector field.

Essentially there are two main classes of integration techniques for discrete gradient vector field: *local integration techniques* and *global integration techniques* (for a review, see Klette and Schlüns [8]). Suppose that a surface $Z(x,y)$ is defined over a region Ω which is either the real plane R^2 or a bounded subset of this plane, and that the gradient values of this surface at discrete points $(x,y) \in \Omega$

$$p(x,y) = \frac{\partial Z(x,y)}{\partial x} = Z_x, \quad q(x,y) = \frac{\partial Z(x,y)}{\partial y} = Z_y \tag{1}$$

are only available as input data, for instance, in the form of a *needle diagram*(see the left picture in Fig. 1). Local integration methods [2,5,12] are based on the following curve integrals:

$$Z(x,y) = Z(x_0,y_0) + \int_{\gamma} p(x,y)dx + q(x,y)dy. \tag{2}$$

R. Klette, S. Peleg, G. Sommer (Eds.): Robot Vision 2001, LNCS 1998, pp. 84–90, 2001.

where γ is an arbitrarily specified integration path from (x_0, y_0) to $(x, y) \in \Omega$. Starting with initial height values, the methods propagate height values according to a local approximation rule (e.g., based on the 4-neighborhood) using the given gradient data. Such a calculation of relative height values can be repeated by using different scan algorithms. Finally, resulting height values can be determined by averaging operations. Generally, local integration methods are easy to implement and do not explicitly implement any assumption of the integrability condition. However, initial height values have to be provided. The locality of the computations strongly depends on data accuracy, and the propagation of errors may occur due to the propagation of height increments along paths. Therefore, local integration techniques perform badly when the data are noisy.

Integration of discrete gradient vector fields is thought to be an optimization problem in global integration techniques [4,6,7]. That is, the problem of finding Z from p and q can be solved by minimizing the following functional (cost function):

$$E = \iint_{\Omega} [(Z_x - p)^2 + (Z_y - q)^2] dx dy. \tag{3}$$

Comparing with the local methods, the *Frankot-Chellappa algorithm* [4] is more robust against noise and leads to considerably better results for the task of calculating height from gradients (see Klette et al. [9]). Figure 1 shows a result of the global method. Nevertheless, the height values obtained in the algorithm

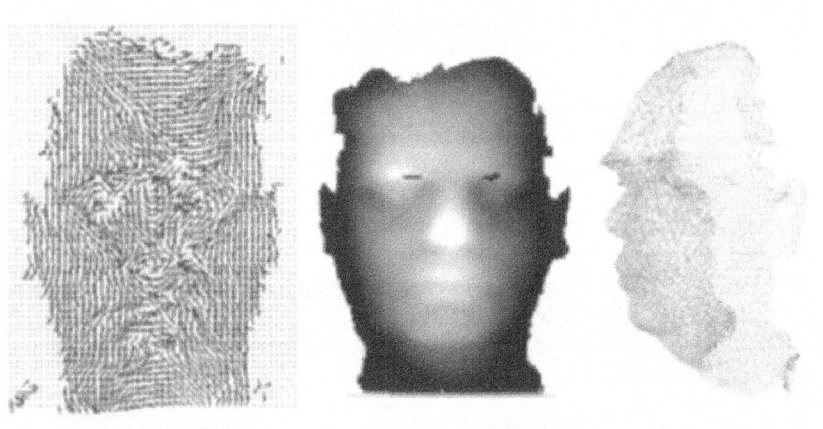

Fig. 1. The left picture shows a needle map representation of surface normals of a human face calculated based on photometric stereo. The middle image shows a depth map obtained from the normals using the global integration method by Frankot-Chellappa. The right image visualizes the recovered 3D shape of the human face

may suffer from high frequency oscillations and this method needs slightly more computing time.

Wavelets theory has proved to be a powerful tool in various applications such as numerical analysis, pattern recognition, signal and image processing. Using wavelet decomposition it is possible to detect singularities, irregular structure and transient phenomena exhibit by a function. This paper presents a new integration technique for discrete gradient fields. The tensor product of the third-order Daubechies' scaling functions is used to span the solution space. The surface height is described as a linear combination of a set of the scaling basis functions. This method efficiently discretize the cost function associated with the height from gradients problem. After discretization, the height from gradients problem becomes a discrete minimization problem rather than discretized PDE's. To solve the minimization problem, perturbation method is used. The surface height is finally decided after finding the weight coefficients.

The rest of this paper is organized as follows. In the next section, the basic concepts of the wavelet transform and the relevant properties of Daubechies wavelet will be briefly addressed. Then,in Section 3, the proposed wavelet-based algorithm for height from gradients will be described. In this short note the pertinent results are presented only. Finally, a conclusion is given in Section 4.

2   Daubechies Wavelet Basis and Connection Coefficients

In this section, we will briefly describe the basic idea of *wavelet transform*. Wavelets are mathematical functions that cut up data into different frequency components, and then study each component with a resolution matched to its scale. They have advantages over traditional Fourier methods in analyzing physical situations where the signal contains discontinuities and sharp spikes. They provide the methods for representing a set complex phenomena in a simpler, more compact, and thus more efficient manner.

Let $\phi(x)$ and $\psi(x)$ are the Daubechies *scaling function* and *wavelet*, respectively. They both are implicitly defined by the following two-scale equation [3]

$$\phi(x) = \sum_{k \in Z} a_k \phi(2x - k), \quad \psi(x) = \sum_{k \in Z} (-1)^k a_{1-k} \phi(2x - k),$$

where a_k are called the Daubechies wavelet filter coefficients. Denote by $L^2(R)$ the space of square integrable functions on the real line. Let V_j be the closure of the function subspace spanned by $\phi_{j,k}(x) = 2^{j/2} \phi(2^j x - k)$, $j, k \in Z$, and suppose that W_j, the orthogonal complementary of V_j in V_{j+1}, be the closure of the function subspace generated by $\psi_{j,k}(x) = 2^{j/2} \psi(2^j x - k)$, $j, k \in Z$. Then the function subspaces V_j and W_j have the following properties: $V_j \subseteq V_{j+1}$, for all $j \in Z$;

$$\bigcap_{j \in Z} V_j = \{0\}; \quad \bigcup_{j \in Z} V_j = L^2(R); \quad V_{j+1} = V_0 \oplus W_0 \oplus W_1 \oplus \cdots \oplus W_j,$$

where $\oplus$ denotes the orthogonal direct sum. On each fixed scale j, the scaling functions $\{\phi_{j,k}(x), k \in Z\}$ form an orthonormal basis of V_j and the wavelets $\{\psi_{j,k}(x), k \in Z\}$ form an orthonormal basis of W_j. The set of subspaces V_j is called a *multiresolution analysis* of $L^2(R)$.

Let J be a positive integer. A function $f(x) \in V_J$ can be represented by the wavelet series

$$f(x) = \sum_{k \in Z} c_{J,k} \phi_{J,k}(x),$$

where the expansion coefficients $c_{J,k}$ are specified by $c_{J,k} = \int f(x) \phi_{J,k}(x) dx$. Since $V_J = V_0 \oplus W_0 \oplus W_1 \oplus \cdots \oplus W_{J-1}$, $f(x)$ can be alternatively represented by

$$f(x) = \sum_{k \in Z} c_{0,k} \phi_{0,k}(x) + \sum_{j=0}^{J-1} \sum_{k \in Z} d_{j,k} \psi_{j,k}(x).$$

The wavelet series expansion coefficients $c_{0,k}$ and $d_{j,k}$ can be computed via the decomposition algorithm [3].

The *connection coefficients* [1,10] play an important role in representing the relation between the scaling function and differential operators. Throughout this paper, we assume that the scaling function $\phi(x)$ has N *vanishing moments*. For $k \in Z$, we define that

$$\Gamma_k^0 = \int \phi(x) \phi(x-k) dx,$$

$$\Gamma_k^1 = \int \phi^{(x)}(x) \phi(x-k) dx,$$

$$\Gamma_k^2 = \int \phi^{(x)}(x) \phi^{(x)}(x-k) dx.$$

Then we have the following properties: $\Gamma_0^1 = 0$; for the scaling function $\phi(x)$ which has N vanishing moments, $\Gamma_k^1 = \Gamma_k^2 = 0, k \notin [-2N+2, 2N-2]$; and

$$\Gamma_k^0 = \begin{cases} 1, & k = 0, \\ 0, & \text{otherwise.} \end{cases}$$

The connection coefficients for Daubechies' wavelet with $N = 3$ vanishing moments are shown in the following table [11]:

3   Wavelet-Based Height from Gradients

In this section, we will derive a new wavelet-based algorithm for solving the height from gradients. First of all, we assume that the size of the domain of the surface $Z(x, y)$ is $M \times M$, and the surface $Z(x, y)$ is represented by a linear combination of a set of the third-order Daubechies scaling basis functions in the following format:

$$Z(x,y) = \sum_{m=0}^{M-1} \sum_{m=0}^{M-1} z_{m,n} \phi_{m,n}(x, y), \tag{4}$$

Table 1. Connection Coefficients with $N = 3$

Γ_k^1	Γ_k^2
$\Gamma_{-4}^1 = 0.00034246575342$	$\Gamma_{-4}^2 = -0.00535714285714$
$\Gamma_{-3}^1 = 0.01461187214612$	$\Gamma_{-3}^2 = -0.11428571428571$
$\Gamma_{-2}^1 = -0.14520547945206$	$\Gamma_{-2}^2 = 0.87619047619052$
$\Gamma_{-1}^1 = 0.74520547945206$	$\Gamma_{-1}^2 = -3.39047619047638$
$\Gamma_0^1 = 0.0$	$\Gamma_0^2 = 5.26785714285743$
$\Gamma_1^1 = -0.74520547945206$	$\Gamma_1^2 = -3.39047619047638$
$\Gamma_2^1 = 0.14520547945206$	$\Gamma_2^2 = 0.87619047619052$
$\Gamma_3^1 = -0.01461187214612$	$\Gamma_3^2 = -0.11428571428571$
$\Gamma_4^1 = -0.00034246575342$	$\Gamma_4^2 = -0.00535714285714$

where $z_{m,n}$ are the weight coefficients, $\phi_{m,n}(x, y)$ are the tensor product of the third-order Daubechies scaling functions, i.e., $\phi_{m,n}(x, y) = \phi(x - m)\phi(y - n)$. For the known gradient values $p(x, y)$ and $q(x, y)$, we assume that

$$p(x, y) = \sum_{m=0}^{M-1} \sum_{m=0}^{M-1} p_{m,n}\phi_{m,n}(x, y), \tag{5}$$

$$q(x, y) = \sum_{m=0}^{M-1} \sum_{m=0}^{M-1} q_{m,n}\phi_{m,n}(x, y), \tag{6}$$

where the weight coefficients $p_{m,n}$ and $q_{m,n}$ can be determined by

$$p_{m,n} = \int\int p(x,y)\phi_{m,n}(x,y)dxdy, \quad q_{m,n} = \int\int q(x,y)\phi_{m,n}(x,y)dxdy.$$

Substituting (4), (5) and (6) into (3), we have

$$E = \int\int \left[\left(\sum_{m,n=0}^{M-1} z_{m,n}\phi_{m,n}^{(x)}(x, y) - \sum_{m,n=0}^{M-1} p_{m,n}\phi_{m,n}(x, y) \right)^2 \right.$$

$$\left. + \left(\sum_{m,n=0}^{M-1} z_{m,n}\phi_{m,n}^{(y)}(x, y) - \sum_{m,n=0}^{M-1} q_{m,n}\phi_{m,n}(x, y) \right)^2 \right] dxdy$$

$$= E_1 + E_2, \tag{7}$$

where $\phi_{m,n}^{(x)}(x, y) = \partial\phi_{m,n}(x, y)/\partial x$ and $\phi_{m,n}^{(y)}(x, y) = \partial\phi_{m,n}(x, y)/\partial y$.

In order to derive the iterative scheme for Z, let $\Delta z_{i,j}$ represent the updating amounts of $z_{i,j}$ in the iterative equation, $z'_{i,j}$ be the value after update. Then $z'_{i,j} = z_{i,j} + \Delta z_{i,j}$. Substituting $z'_{i,j}$ into E_1, E_1 will be changed by an amount ΔE_1, that is,

$$E_1' = E_1 + \Delta E_1$$

$$= \int \int \left[\left(\sum_{m,n=0}^{M-1} z_{m,n} \phi_{m,n}^{(x)}(x,y) - \sum_{m,n=0}^{M-1} p_{m,n} \phi_{m,n}(x,y) \right) \right.$$

$$\left. + \Delta z_{i,j} \phi_{i,j}^{(x)}(x,y) \right]^2 dxdy$$

$$= E_1 + 2\Delta z_{i,j} \sum_{m,n=0}^{M-1} z_{m,n} \Gamma_{i-m}^2 \Gamma_{j-n}^0$$

$$- 2\Delta z_{i,j} \sum_{m,n=0}^{M-1} p_{m,n} \Gamma_{i-m}^1 \Gamma_{j-n}^0 + \Delta z_{i,j}^2 \Gamma_0^2. \tag{8}$$

Using the same derivation, we have

$$E_2' = E_2 + \Delta E_2$$

$$= E_2 + 2\Delta z_{i,j} \sum_{m,n=0}^{M-1} z_{m,n} \Gamma_{i-m}^0 \Gamma_{j-n}^2$$

$$- 2\Delta z_{i,j} \sum_{m,n=0}^{M-1} q_{m,n} \Gamma_{i-m}^0 \Gamma_{j-n}^1 + \Delta z_{i,j}^2 \Gamma_0^2. \tag{9}$$

Substituting (8) and (9) into (7), it is shown that

$$\Delta E = \Delta E_1 + \Delta E_2$$

$$= 2\Delta z_{i,j} \sum_{m,n=0}^{M-1} z_{m,n} \left(\Gamma_{i-m}^2 \Gamma_{j-n}^0 + \Gamma_{i-m}^0 \Gamma_{j-n}^2 \right)$$

$$- 2\Delta z_{i,j} \sum_{m,n=0}^{M-1} p_{m,n} \Gamma_{i-m}^1 \Gamma_{j-n}^0$$

$$- 2\Delta z_{i,j} \sum_{m,n=0}^{M-1} q_{m,n} \Gamma_{i-m}^0 \Gamma_{j-n}^1 + 2\Delta z_{i,j}^2 \Gamma_0^2.$$

In order to make the cost function decrease as fast as possible, ΔE must be maximized. From $\partial \Delta E / \partial \Delta z_{i,j} = 0$, we have

$$\Delta z_{i,j} = \frac{1}{2\Gamma_0^2} \sum_{k=-2N+2}^{2N-2} \left[(p_{i-k,j} + q_{i,j-k}) \Gamma_k^1 - (z_{i-k,j} + z_{i,j-k}) \Gamma_k^2 \right]$$

From the above results, the iterative equation can be represented as follows:

$$z_{i,j}^{[t+1]} = z_{i,j}^{[t]} + \Delta z_{i,j}, \qquad t = 0, 1, \dots \tag{10}$$

The initial values are zero.

4   Conclusions

In this paper, we presented a new iterative algorithm for solving the height from gradients problem. Wavelet transform is a generalization of Fourier transform, and a power tool for efficiently representing images. Therefore, the proposed method takes the advantages of wavelet transform. By applying the wavelet transform, the objective function associated with the original height from gradients problem is converted into the wavelet-based format. In the new iterative algorithm, the step size can be easily determined by maximizing the decrease of the objective function. we only presented the pertinent results in this short note. In the future the new algorithm should be studied in combining and comparing it with existing height from gradients techniques.

References

1. G. Beylkin: On the representation of operators in bases of compactly supported wavelets. *SIAM J. Numer. Anal.*, **29** (1992) 1716–1740. 87
2. N. E. Coleman, Jr. and R. Jain: Obtaining 3-dimensional shape of textured and specular surfaces using four-source photometry. *CGIP*, **18** (1982) 439–451. 84
3. I. Daubechies: Orthonormal bases of compactly supported wavelets. *Commun. Pure Appl. Math.*, **41** (1988) 909–996. 86, 87
4. R. T. Frankot and R. Chellappa: A method for enforcing integrability in shape from shading algorithms. *IEEE Trans. on Pattern Analysis and Machine Intelligence*, **10** (1988) 439–451. 85
5. G. Healey and R. Jain: Depth recovery from surface normals. *ICPR'84*, Montreal, Canada, Jul. 30 – Aug. 2 **2** (1984) 894-896. 84
6. B. K. P. Horn and M. J. Brooks: The variational approach to shape from shading. *Computer Vision, Graphics, and Image Processing*, **33** (1986) 174–208. 85
7. B. K. P. Horn. Height and gradient from shading. *International Journal of Computer Vision*, **5** (1990) 37–75. 85
8. R. Klette and K. Schlüns: Height data from gradient fields. SPIE-Proceed on *Machine Vision Applications, Architectures, and Systems Integration*, Boston, Massachusetts, USA. **2908** (1996) 204–215. 84
9. R. Klette, K. Schlüns and A. Koschan: *Computer Vision - Three-dimensional Data from Images*. Springer, Singapore, (1998). 85
10. Y. Meyer: *Wavelets and Operators*. Cambridge Univ. Press, Cambridge, UK (1992). 87
11. H. L. Resnikoff and R. O. Wells: *Wavelet Analysis: The Scalable Structure of Information*. Springer-Verlag New York, USA (1998). 87
12. Z. Wu and L. Li: A line-integration based method for depth recovery from surface normals, *Computer Vision, Graphics, and Image Processing*, **43** (1988) 53–66. 84

Enhanced Stereo Vision Using Free-Form Surface Mirrors

Alexander Wuerz, Stefan K. Gehrig, and Fridtjof J. Stein

Research Institute DaimlerChrysler AG
{Alexander.Wuerz,Stefan.Gehrig,Fridtjof.Stein}@DaimlerChrysler.com

Abstract. In the domain of stereo vision, the presence of repetitive patterns results in multiple matching hypotheses. The choice of the wrong hypothesis leads to an incorrect distance measurement. In applications such as automotive vision-based navigation a high precision in matching and distance calculation is vital. A common approach is the use of multiple cameras. Unfortunately, in vehicle applications this is often not feasible. However, the shiny varnished body parts of the car supply a free-form surface mirror. In combination with a camera system they form virtual cameras with a different viewing direction. This additional information can be used to select the correct matching hypothesis and to increase the depth measurement accuracy. The free-form surface mirrors yield distorted pictures without a single viewpoint which prevents a purely perspective reconstruction. We will discuss the problems arising from the use of free-form surface mirrors and present solution strategies to take advantage of the information.

1 Introduction

Stereo vision using two cameras to recognize and match objects and estimate their distances is a common method for vision-based navigation (Franke et al. [5]). Objects are matched in the images and their disparity is computed. The distances are calculated from this result on the basis of triangulation knowing the baseline distance of the cameras. There are different ways to build a stereo vision system. The most obvious is to use two cameras and align them with parallel optical axes and without rotational differences in the image planes. These two conditions facilitate the matching task and increase the computational speed. However, the adjustment is difficult and time consuming. If the stereo vision system is intended for use in automotive applications with motions at higher velocities, it is also necessary to ensure a time-correlated readout of the images in the cameras.

To overcome these problems a variety of single camera systems using mirrors to produce stereo images have been proposed. Cafforio et al. [3], Goshtasby et al. [7] and Inaba et al. [8] introduce different, so-called catadioptric[1] systems

---

[1] Dioptric systems consists of lenses, catoptric systems of mirrors. For the combination of mirrors and lenses the name *catadioptric* has been established.

R. Klette, S. Peleg, G. Sommer (Eds.): Robot Vision 2001, LNCS 1998, pp. 91–98, 2001.
© Springer-Verlag Berlin Heidelberg 2001

using planar mirrors in special configurations. These applications produce stereo images with a single lens and camera, with adjustable fields of view, without undesired geometry or intensity differences and with corresponding points in the same image row. Using just one physical camera and producing the stereo image with virtual cameras yields the advantage of easy adjustment and no problems with time-correlated image readout. However, planar mirror systems have the disadvantage of small fields of view or short baseline distances of the virtual cameras. This makes it difficult to achieve a high accuracy at large distances.

Other proposals use non-planar mirrors to increase the field of view. Nayar [9] uses a rigid configuration of two specular spheres to determine depth in a large field of view and proves that this is possible with non-planar reflective surfaces. In these cases, it is not possible to calculate distances in the usual way. Nevertheless, due to the known geometry, the distance of objects can be recovered. However, all of these systems use special configurations or specifically designed mirrors and cameras to produce virtual cameras with desired intrinsic parameters.

In certain applications it is not possible to rely on special design systems to achieve an increase in accuracy or field of view. One of them is vision-based navigation in the automotive area. These systems use CCD-based stereo cameras with aligned optical axes and adjusted rotational differences in the image planes (Franke et al. [5]). Nevertheless, it is important to achieve high accuracy in matching objects and calculating their distances.

Fig. 1. A traffic scene reflected in the hood of a car

Stereo vision systems using two images always face the problem of correctly matching repetitive patterns, particularly if a part of the pattern is occluded for one of the cameras. A solution to this problem is the use of additional cameras. These cameras do not have to be physical. Virtual cameras produced by mirrors will do as well. In the case of an automotive application, such a mirror can be provided by the car body. Fig. 1 shows a traffic scene reflected in the shiny varnished surface of the hood.

These car body parts are not designed to be used as mirrors nor do they have special geometric properties such as the mirrors in the applications mentioned above. Nevertheless, they are useful to provide additional information to improve the probability of choosing the correct matching hypothesis.

In combination of the different ideas introduced above, we intend to use free-form surface reflections in the hood of a car to produce virtual cameras[2]. These will assist a common two camera stereo vision system in selecting the correct matching hypothesis. We will discuss the problems arising from the use of free-form surface mirrors in Section 2, present strategies to solve them in Section 3 and draw our conclusions in Section 4.

2   Simulation of Free-Form Surface Mirror Reflections

2.1   Description of Free-Form Surfaces

Free-form surfaces in computer aided design systems are usually described using non-rational uniform b-splines (NURBS). In our simulation we will focus on the convex central part of the hood. The other concave parts between the hood and the fender are not considered. The resolution there is poor and the significant objects for our application are not visible.

2.2   Perspective Imaging

Most algorithms in computer vision are based on perspective imaging, a mapping of the three dimensional world onto a plane through a pinhole. This way of mapping is commonly used because it is similar to human vision. In order to reconstruct a distorted reflected image into a purely perspective pinhole image it is necessary for the catadioptric system to retain a single viewpoint from where the reflected rays seem to originate. Shree K. Nayar's group at the Columbia University, New York, has extensively researched this problem. Baker et al. [2] study three criteria for catadioptric sensors: the shape of the mirrors, the resolution of the cameras, and the focus settings of the cameras. In particular, they derive the complete class of mirrors that can be used with a single camera to give a single viewpoint. The mirrors that satisfy this single-viewpoint constraint are conic sections and planes in combination with perspective or orthographic cameras. Chahl et al. [4] research the same topic, using a differential equation. They draw the same conclusions from their results.

2.3   Simulation Results

We used ray-tracing methods in the ASAP[3] application to confirm these results in our special case. Fig. 2 shows the setup of the simulation.

---

[2] patent pending
[3] Advanced Systems Analysis Program by Breault Research Organization Inc.

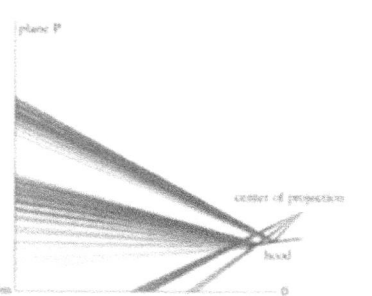

Fig. 2. Simulation of the reflection of a regular grid and intersection of the rays with a plane P, 5 m in front of the car. Not all of the simulated rays intersected with the reflective surface

The central part of the hood was used as a mirror and a camera was placed above the surface. A regular grid of points was projected through a pinhole onto the surface and the reflected rays were intersected with a plane P at a distance of 5 m, perpendicular to the ground plane. We calculated different patterns to see the impact of extrinsic camera calibration. We also determined the focal points of each bundle of reflected rays to see the distribution of viewing points. In the presented simulation (Fig. 3) the optical axis of the camera enclosed an angle of -35° with the ground plane. The results clearly show that a single viewpoint cannot be achieved and the extrinsic calibration of the camera is essential for the results. The violation of the single-viewpoint constraint signifies that a purely perspective, planar reconstruction of the image is not possible (Gluckman et al. [6]).

The distortion of the circular-shaped ray bundles, due to the circular shape of the simulated pinhole can be viewed in Fig. 3. In addition the distortion of the rectangular grid due to the curvature of the surface is also visible (Seidel aberrations).

An additional problem is due to the colored varnish of the hood. The contrast within the reflected picture is degraded. For this reason, correlation algorithms using normalized gray values are necessary.

3   Solution Strategies

Due to the results presented in paragraph 2.2 we had to refrain from a perspective reconstruction of the image in the usual sense and develop new ideas how to make use of the information contained in the reflection seen in the hood.

Nayar [9] proves that the knowledge of the geometry is sufficient to reconstruct the object direction. He uses regular spheres which make it much easier to calculate the ray directions. Nevertheless, NURBS descriptions of free-form

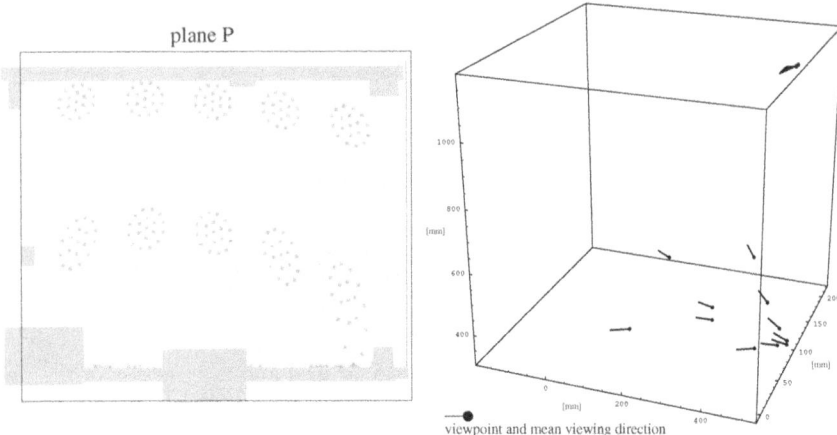

Fig. 3. Simulation results. Left: Mapping of the rectangularly distributed circular-shaped grid points with the free-form surface mirror onto the plane P (see Fig. 2). Right: Distribution of the focal points and mean viewing directions of each ray bundle shown in the left. The focal points and mean viewing directions in the upper right part of the figure show the location of the simulated camera

surfaces yield the same information but they are harder to handle mathematically.

We were able to show that the extrinsic calibration of the camera is crucial for successfully calculating the direction of the reflected ray. We use an algorithm based on the concept presented by Tsai [10]. This approach provides the intrinsic and extrinsic camera parameters so that we are able to position the camera within the coordinate system determined by the data of the free-form surface. With this information, we are able to relate a ray direction with each pixel of the reflected vision area. Fig. 4 shows the basic triangulation scheme that is applied.

With our approach it is necessary to use a matching algorithm that is capable of handling a distorted, intensity warped image and an ordinary image. One of these algorithms is the normalized mean-free cross-correlation function (CCFMF) presented by Aschwanden et al. [1]. The contrast of the reflected image is degraded but the gray value distribution form remains the same. The CCFMF is able to compensate for this effect. The intensity is compared in two two-dimensional image areas along epipolar lines. The epipolar lines can be recovered using Fermat's principle: the path of light is such that the time of propagation is minimal.

Looking at Fig. 4 it is obvious that the object point, the point of reflection on the surface, and the pinhole form a plane. The lines connecting these points are in the same plane and consequently, the points of intersection with the image

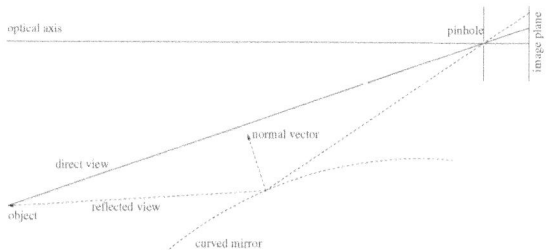

Fig. 4. Basic triangulation scheme using geometric data of the free-form surface

plane too. If the object point is moved along the line of the direct view the corresponding point of reflection will move along the convex reflective surface according to the direction of the object movement. To proof this we analyze it in two dimensions. The function $f(x)$ describes the convex segment. Without loss of generality we can choose the direction of direct view along the coordinate axis. The parameter u describes the location along this axis. This results in a distance function $s(x, u)$:

$$s(x, u) = \sqrt{x^2 + f(x)^2} + \sqrt{(x - u)^2 + f(x)^2}$$

The evaluation of the derivatives $d_u s(x, u)$ and $d_x s(x, u)$ yields the result.

The same argument can be applied to the direction perpendicular to the plane described by the three initial points depicted in Fig. 4. The path of light for these points is minimal. The point of reflection for another object location along the line of direct view cannot leave the plane. This is due to that fact that any other point outside the plane would prolong the path and therefore the time of propagation. One restriction in this argument is the use of a convex surface. The convex surface has a continuous negative curvature that produces a minimum in the action functional and therefore a stable solution. We analyzed this in three dimensions using the same assumptions. The convex surface is described by $f(x, z)$ and this results in a distance function $s(x, z, u)$:

$$s(x, z, u) = \sqrt{x^2 + f(x, z)^2} + \sqrt{(x - u)^2 + z^2 + f(x, z)^2}$$

The evaluation of the derivatives $d_u s(x, z, u)$ and $d_x s(x, z, u)$ again yields the result. The quantitative analysis of this proof with $f(x)$ being a NURBS function is not a part of this paper and a topic of our ongoing research.Nevertheless, these simple formulations and their evaluation on convex surfaces conveys the proof.

A concave surface could exhibit a different behavior. However, we have restricted ourselves to examine only the convex part of the hood. The conclusion drawn from this argument is that for a certain pixel in the direct view area of the image the corresponding pixel can be found along a straight line in the reflected view area of the image. The straight lines do not necessarily correspond to an image line or row. This behavior is similar to the epipolar geometry of two cameras in an ordinary setup. However, the disparity along this epipolar line cannot be converted to a distance using a simple reciprocal function. The structure of the epipolar distance relationship is warped by the geometry of the convex surface and can be recovered either through calculation using geometric data of the surface or through calibration of the setup.

Using these results we are able to set up an algorithm to evaluate the information in the reflected image. We match pixels along the epipolar lines and calculate, using triangulation, the distance to the matched object (see Fig. 5). This distance can be used directly to choose the correct matching hypothesis provided by the evaluation of the two direct view stereo images.

Fig. 5. Processing of the traffic scene. We evaluate correspondences along epipolar lines (here corresponding to a and b). Matching of features in A and B and the knowledge of the intrinsic geometry yields the distance to the object

4   Conclusion

We have presented a new approach to extract information contained in the reflections from free-form surfaces. The reconstruction of the image information uses the special geometric properties and a known matching algorithm to obtain an object's distance. This information is used to choose the correct match amongst different matching hypotheses provided by a stereo vision system. This additional information is crucial in resolving ambiguous cases such as partially occluded repetitive patterns. It also offers a variety of new matching possibilities

(Fig. 6). It is also possible to implement such a solution strategy in a single camera application. Furthermore, it can be used as a fallback system. In a two camera system one image might be temporarily unusable due to the windshield wiper or raindrops. This is important for real-time applications especially in the automotive field.

Fig. 6. Object matching possibilities using a free-form surface mirror (hood)

References

1. P. F. Aschwanden: Experimenteller Vergleich von Korrelationskriterien in der Bildanalyse. Hartung-Gorre Verlag Konstanz (1993). 95
2. S. Baker, S. K. Nayar: A theory of single-viewpoint catadioptric image formation. *Computer Vision* **35** (1999) 175–196. 93
3. C. Cafforio, F. Rocca: Precise stereopsis with a single video camera. *Signal Processing III: Theories and Application* (1986) 641–644. 91
4. J. S. Chahl, M. V. Srinivasan: Reflective surfaces for panoramic imaging. *Applied Optics: Optical Technology and Biomedical Optics* **36** (1997) 8275–8285. 93
5. U. Franke, D. Gavrila, S. Goerzig, F. Lindner, F. Paetzold, C. Woehlerm: Autonomous driving goes downtown. *IEEE Intelligent Systems and their Applications* **13** (1998) 40–48. 91, 92
6. J. Gluckman, S. K. Nayar, K. J. Thoresz: Real-time omnidirectional and panoramic stereo. In: Proceed. of the *DARPA Image Understanding Workshop* (1998). 94
7. A. Goshtasby, W. A. Gruver: Design of a single lens stereo camera system. *Pattern Recognition* **26** (1993) 923–937. 91
8. M. Inaba, T. Hara, H. Inoue: A stereo viewer based on a single camera with view control mechanisms. In: Proceed. of Internat. Conf. on *Intelligent Robots and Systems* (1993). 91
9. S. K. Nayar: Sphereo: determining depth using two specular spheres and a single camera. In: SPIE-Proceed. on *Optics, Illumination, and Image Sensing for Machine Vision III* **1005** (1988) 245–254. 92, 94
10. R. Y. Tsai: An efficient and accurate camera calibration technique for 3d machine vision. In: Proceed. of Internat. Conf. on *Computer Vision and Pattern Recognition* (1986) 364–374. 95

RoboCup-99: A Student's Perspective

Jacky Baltes

Center for Imaging Technology and Robotics
University of Auckland, Auckland, New Zealand
j.baltes@auckland.ac.nz

Abstract. One of the reasons for organizing robotic games is that they allow researchers to evaluate their systems and approaches on a level playing field. This evaluation is important in a quickly developing field such as robotics with few real world applications. This paper investigates through a case-study how much participating at the RoboCup-99 competition has benefited a MSc. student at the University of Auckland. Although the participation was certainly stimulating, its influence on the research was indirect. The paper makes a number of suggestions that will make it easier to quantitatively evaluate research at these competitions and thus influence research more directly.

1 Introduction

Robotic games are extremely popular. Apart from their entertainment value, the ability to evaluate research progress and compare one's own approach in a competitive environment against that of other teams is often cited as one of the reasons for organizing competitions.

However, there is only anecdotal evidence of the impact of robotic games on research programmes.

This paper describes the experiences of a student, Nicholas Hildreth, participating in the RoboCup robotic soccer competitions and how they influenced his research in path planning for mobile robots.

Section 2 gives a short introduction to the All Botz, the University of Auckland RoboCup team. The adaptive path planner is briefly introduced in section 3. Section 4 discusses how the robotic games were used to support the empirical evaluation of the adaptive path planner. Section 5 summarizes the paper and suggests some ways in which robotic games could have better supported the evaluation of our research.

2 History of the All Botz

Nick's first experience with robotics was when he took the graduate course on "Intelligent Active Vision" at the University of Auckland starting in February of 1998. The course description of this paper reads as follows:

R. Klette, S. Peleg, G. Sommer (Eds.): Robot Vision 2001, LNCS 1998, pp. 99–106, 2001.
© Springer-Verlag Berlin Heidelberg 2001

Control of autonomous mobile agents in a realistic, dynamic, and uncertain environment. It covers a variety of different areas including robotics, planning and machine learning.

The paper was offered for the first time and things were a bit chaotic. We used remote controlled toy cars for the practical work in the paper, since there was insufficient funding to purchase a "real robot." We designed and built parallel port interfaces to the remote control transmitters so that the cars could be controlled from a computer. Position and orientation information for the robots are provided by a global vision system mounted on the ceiling.

It was a real eye opener to see how difficult it was to do even simple things, such as driving a straight line, with these toy cars. However, the course turned out to be very motivating in the end. The culmination of the project was a race amongst different groups of students, the so-called Aucklandianapolis. To complete this challenge successfully, a team of students had to implement a complete mobile robot system including video processing, path planning, and path tracking control. Figure 1 shows our first mobile robot competition.

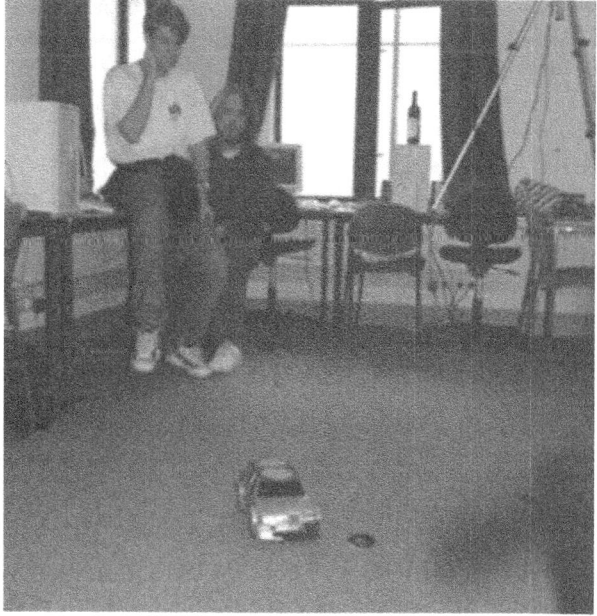

Fig. 1. The First Aucklandianapolis Competition

After the course finished in June 1998, Nick started a project on the design of an agent architecture for a mobile robot. The project resulted in some preliminary work. However, at the same time, we learned about the PRICAI-98

RoboCup competition. Since we had had good success at the Aucklandianapolis competition, we decided to extend the current architecture to create a team to enter the competition. After having implemented a striker and a goal keeper, we thought we were ready for the competition.

Two weeks before the competition, we found out that our toy cars were too large. However, with luck we managed to buy some used smaller toy cars that operated on the same frequency and could be controlled with the parallel port interfaces that we had built.

After recalibrating the system for the new cars, we managed to be ready for our first international competition. Our team consisted of only two players: a striker and a goal keeper.

At the competition, we realized that our approach to the problem was different from that of other teams. The difference in hardware became apparent. The All Botz were the only team that used non-holonomic robots and a camera that is mounted on the side. All other teams used holonomic robots and mounted the camera directly overhead. However, not all teams were following this trend and had developed their own approach. For example, the CIIPS glory team from UWA, Perth, was the only local vision team at the competition. It was interesting to see the wide variety of approaches. Some teams focuses on the mechanics on their robots, putting a lot of effort into the selection of motors and tires. Other teams focused on the embedded controller and the electronics. The main emphasis for these teams are additional sensors, such as ultrasound, infrared, or even local vision. yet other teams emphasize the cooperation amongst the individual players of a team.

At the competition, we noticed our path planner, an extension of Bicchi's non-holonomic path planner ([1]), did not perform well in the dynamic environment. Our robots would spent most of their time planning, but before the planner would finish, an object would move in the domain and the path planner hat to be restarted.

One of the disappointing aspects of the games were the amount of set plays that were used by some of the teams. Some teams spent a long time manually positioning their robots for free kicks and other stoppages in play. This was disappointing, since our main goal was to advance our research, not to take part in a game of robot chess.

3   Adaptive Path Planning

The experiences at the PRICAI-98 RoboCup competition motivated Nick to work on his Master's thesis and he chose the topic: *"the problem of path planning for car-like mobile robots in highly dynamic environments"* [3]. This section is a brief introduction into the basic idea behind adaptive path planning and the methodology used for evaluating this research.

At the PRICAI-98 competition, we used a version of Bicchi's path planner that was optimized for the RoboCup domain. This planner proved to be too slow for a soccer game. On a Pentium 200MMX running the Linux operating system, the planner would take more than one second to create a path. The level of the competition in Singapore was such that the ball would rarely sit still for more than a second, so that our path planner would spent a lot time replanning. In few cases would it actually start executing a plan.

The main motivation for the adaptive path planner that he developed in his thesis was the realization that:

- Path planning is an expensive operation, so the result of this work should be reused if possible.
- The result or output of a path planner is a path
- Assuming that changes in the domain are small between individual planning episodes, the current plan will be structurally similar to the plan for the new situation.

This motivation is similar to Hammond's case-based planning ([2]) with some important differences. Firstly, case-based planning assumes that a plan database exists with previous plans and that the most similar plan to the current situation can be found by using a similarity metric. Secondly, the database of previous plans needs to be maintained; new plans need to be added so they can be reused in the future or old plans that are not useful must be removed. So, a lot of work on case-based planning focuses on the design of suitable similarity metrics and on database policies.

In adaptive path planning, we assume the existence of a albeit slow static path planner that can be used to create an initial plan. The previous plan is the most similar one to the current situation and that therefore, there is no need to maintain a plan database.

At the heart of any path planner is the plan representation. A plan consists of a sequence of path segments. Most other path planners use a representation with different segment types such as straight lines, turns to the left, etc. This representation makes it difficult to adapt a path, since the adaptations will need to be specific for a given segment type.

To simplify the adaptation, the adaptive path planner uses a uniform representation for all path segments. Each segment contains the following information:

- start point I
- initial bearing α
- length of the segment L
- radius of the segment R
- time limit to traverse the segment T
- A possibly empty attachment A. An attachment is used to attach an object to a path segment, so that if the object moves, all attached path segments will move as well.

This representation, shown in Fig. 2, proved very useful, because plans in this representation can be easily adapted to compensate for movements of objects or goal locations in the domain.

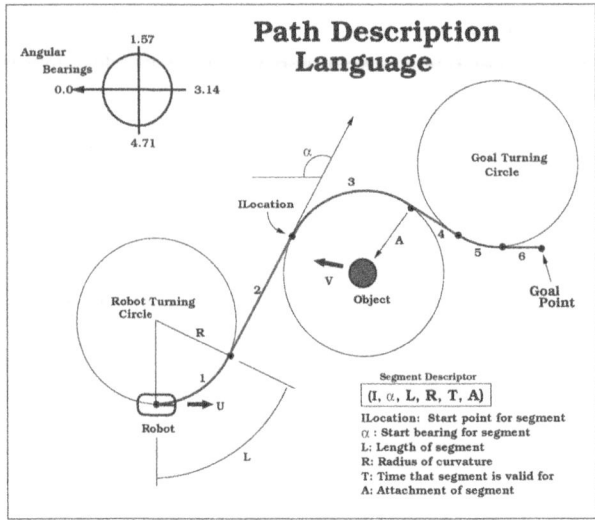

Fig. 2. Path Description Language

Objects are attached and detached from path segments dynamically. If an obstacle moves too close to a path, the path is split at the closest point to the obstacle and a new segment is inserted which is attached to the object. As the object continues to move the attached segment will move as well. Once the object is too far from the path, the object is detached from the path segment.

3.1   Evaluation Methodology

Although an intuitive argument can be made for the advantages of the adaptive path planner, to quantitatively evaluate the performance of such a path planner is difficult. For example, it is also intuitive that an adaptive path planner will perform poorly if there is no relationship between the current state and the next state of the world.

Firstly, the adaptive part of the path planner is not *complete*, that is there are situations in which the adaptive planner will find a suitable combinations of adaptations to create a valid path, but such a path exists in practice. In this case, the static path planner is called to create a new plan from scratch. This usually occurs when the world state changed dramatically from the current one. The worst case run-time of the adaptive path planner is thus worse than that of the static path planner. The motivation is that for sustained planning, the adaptive path planner can avoid calling the static path planner often.

It is true, that in principle any adaptive path planner that can delete and insert segments into a plan can be converted into a complete path planner. In the worst case, the planner would delete all segments from the path and then start creating a new path from scratch. However, if this is done, the worst case complexity would be similar to that of other static path planners. This would be missing the point of an adaptive path planner completely. One would replace one search problem with another one in a similar space. The main point is that there is a small set of highly specific repair strategies that can adapt a path to the new situation quickly. The success rate of these repair strategies must be high and they planner must be able to apply them quickly.

This argument then relies on a sequence of path planning problems that reflect the dynamics of the application environment.

Through an empirical evaluation of a simulated world and a number of case-studies, the evaluation of the adaptive planner shows that this set of adaptations is sufficient to fix a plan in most cases, yet small enough to not increase the run-time of the path planner significantly. The repair strategies have specific pre-conditions, which when met result in the new path having a high chance of being able to be completed.

However, the method used to create the random path planning problems and case studies is under control of the designer of the algorithm to be tested. Even with his best intentions, it is difficult to prevent subconscious assumptions from entering both the design of the synthetic as well as the path planner. It would be far more convincing to have a standard environment for mobile robots that could be used to compare the different path planning approaches on realistic problems.

There are few real world applications for mobile robots that are convincing and general enough to allow different robots to take part in them. The state of the art in mobile robots is currently a vacuuming robot that can maneuver through an unstructured environment.

Therefore, it was decided that the research should be evaluated at the RoboCup competition. The motivation was that a complex robotic competition is the next best thing to an unbiased environment.

4   Evaluation at RoboCup-99

A prototype version of this path planner was finished just in time for RoboCup-99. It was the intention, that we would use the competition as an opportunity to test the new path planner under real world conditions.

However, this turned out to be much more difficult than expected. The main problem is that every team has only a few games and there was no opportunity for practice games. The first few games were very important, so we used our best path planner. For a proper empirical evaluation, we needed to play two sessions against each team, one with the adaptive path planner and one with a standard path planner.

Another disappointment was that the promised video tapes of our games were not made available, so we had no possibility to view our games afterwards.

Nevertheless, the competition proved valuable, since we noticed some re-curring patterns, which the prototype path planner was not able to deal with efficiently. After returning from RoboCup-99, we redesigned part of the path planner and completed its implementation paying attention to the patterns that emerged during the competition.

Figure 3 shows a case study of a situation that was inspired by the RoboCup competitions. It shows how the adaptive path planner changes the path to compensate for a moving ball and an interfering object.

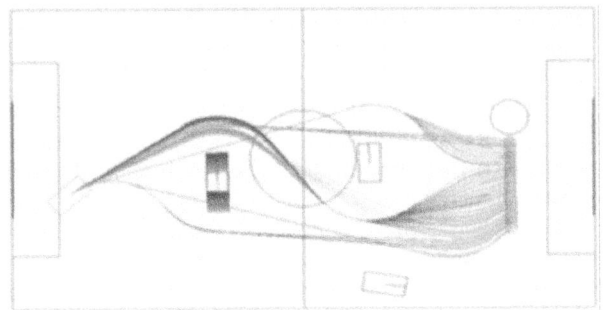

Fig. 3. Case Study Derived From Observed RoboCup Competition

Furthermore, Nick gained some important information by talking to other researchers at the competition and at the workshop. For example, Nick learned about practical potential field path planning through informal discussions with other teams. The most impressive performance was that of the Robotics team from Korea, which showed the best control. The Robotics robots moved about three times faster than the other teams.

Flexibility and robustness, obviously very important characteristics for a robot to survive in the real world, have taken a back seat to special purpose solutions. The RoboCup committee suggested that instead of every team mounting their own camera, a high quality camera would be provided by the organizing committee. All teams except the All Botz refused to even try the idea. One team forced the organizing committee to repaint all playing fields, since the center line was 2mm too wide and they could not compensate for this change in their software. Also, most teams did not switch sides at half time, since their robots performed significantly better on one half of the field than on the other. Even during the RoboCup-99 finals, teams remained on their own side.

5   Conclusion

This paper describes a student's involvement in the RoboCup initiative and the influence that the participation at the robotic competitions had on his research.

The main impact that the competition had on his research were two-fold. Firstly, experiences gained in the soccer games provided a better understanding of the path planning problem in real-world situations. Secondly, it gave Nick an excellent opportunity of discussing with researchers that were extremely familiar with the issues.

The robotic games did not provide sufficient opportunity to evaluate quantitatively the performance of the adaptive path planner. It would have been a good idea to setup a special playing field so that teams can hold short practice games.

Nick Hildreth has now completed his MSc. and is considering doing a PhD in Computer Science.

Acknowledgment

This work was supported by AURC grant XXX/9343/3414107 from the University of Auckland.

References

1. A. Bicchi, G. Casalino, C. Santilli: Planning shortest bounded-curvature paths for a class of nonholomic vehicles among obstacles. In: Proceed. of the IEEE Internat. Conf. on *Robotics and Automation*, (1995) 1349–1354. 101
2. K. J. Hammond: *Case Based Planning*. Academic Press Inc., (1989). 102
3. N. Hildreth: An adaptive path planning system for car-like mobile robots. Master's thesis, University of Auckland, (February 2000). 101

Horus: Object Orientation and Id without Additional Markers

Jacky Baltes

Center for Image Technology and Robotics
University of Auckland, Auckland, New Zealand
j.baltes@auckland.ac.nz

Abstract. This paper describes a novel approach to detecting orientation and identity of robots using a global vision system. Instead of additional markers, the original shape of the robot is used to determine an orientation using a general Hough transform. In addition the movement history as well as the command history are used to calculate the quadrant of the orientation as well as the identity of the robot. An empirical evaluation shows that the performance of the new video server is at least as good as that of a traditional approach using additional colored markers.

1 Introduction

This paper describes a new approach to image processing in the RoboCup domain, that has been implemented in the video server of the All Botz, the University of Auckland F180 team.

In the F180 league, most teams use a global vision system to control up to five robots per team in a game of robotic soccer. In order to be able to control the robots, colored markers or bar codes are put on top of the robots to simplify the vision task.

Colored markers or bar codes are an easy and robust method, but have two big disadvantages. Firstly, the calibration of sets of colors, so that they can be detected over the entire playing field and do not interfere with each other is a very time consuming task. The resulting calibration is not robust. Even small changes in the lighting conditions require a re-calibration. Secondly, these methods do not scale to large teams of robots with eleven players or more.

Therefore, the All Botz developed a new flexible and scalable approach, that uses the original shape of the robot as its sole source of vision information. In other words, the robots are unaltered except for the addition of a marker ball, which is required by the F180 RoboCup rules. A generalized Hough transform is used to infer the orientation of the robot from a sub-image. The image processing determines an exact orientation of one side of the robot (an angle between 0 and 90 degrees), but there is not sufficient information to determine the quadrant of the angle. Thus, the quadrant is determined by correlating the movement history (e.g., dx, dy) and current command (e.g., move forward) to the motion of the robot.

R. Klette, S. Peleg, G. Sommer (Eds.): Robot Vision 2001, LNCS 1998, pp. 107–114, 2001.

The most difficult vision problems in the RoboCup domain is to determine the identity of a robot. All other teams use unique features of the robots to determine their id. As the number of robots increases it becomes more difficult to find unique features that can be recognized efficiently and robustly. In our system, the identity of the robot is determined through correlating the command stream from the individual controllers to the observed behavior of the robot.

Section 2 describes the vision problems associated with the F180 league and how these problems were addressed by other teams previously. Section 3 describes the design of HORUS, the new video server of the All Botz. The results of an empirical evaluation comparing the performance of HORUS against that of a traditional video server are shown in section 5. Directions for future research and further improvements are shown in section 6.

2   Global Vision in the RoboCup

Most teams in the F-180 league of the RoboCup initiative use a global vision system to obtain information about objects in the domain, including the robots, the opponents, and the ball.

There are three important pieces of information that the global vision system must provide: position, orientation, and identification. The following subsections describe related work in obtaining the necessary information.

2.1   Position

The rules of the F-180 league require each robot to mount a colored table tennis ball in the center of the robots. Each team is assigned a color (either yellow or blue). The two goal boxes are also painted yellow and blue respectively. The yellow team shoots on the blue goal and vice versa.

The position of a robot can easily be determined by the image coordinates of this marker ball. Given the height of the robot as well as the extrinsic and intrinsic camera parameters, this location can be mapped back to real world coordinates. The All Botz use a pinhole camera model with two non-linear lens distortion parameters.

The geometry of the All Botz video camera setup makes the accuracy of the camera parameters more important and the computation of these parameters more difficult than that of other teams. However, this side view setup is general and versatile.

Briefly, a calibration pattern is used to find real world coordinates for a number of image coordinates. This mapping from image to real world coordinates is computed using an automatic iterative method. Given this set of calibration points and their real world coordinates, the Tsai camera calibration method is used to compute the parameters of the camera model [4].

2.2   Orientation

Although a single point on the robot is sufficient to determine its position, additional information (e.g., a second point or a vector) is needed to determine the orientation of a robot.

Most teams in the RoboCup competition use additional colored markers to create a second point on the robot. In the simplest case, the two points have a distinct color, which makes it easy to determine the orientation of the robot by relating it to the orientation of the line between the two points.

The distance between the two points determines the accuracy of the orientation: the further apart the two points, the better the orientation. The maximum length of the robot is limited by the rules.

The All Botz used this method previously with good success. The variance in the orientation for a static object was less than 10 degrees at the far side of the field.

2.3   Identification

One of the most difficult aspects of the vision processing is the visual identification of robots. To be able to control a robot, the control system needs to know the identity of the robot, so that the commands are sent to the correct robot (e.g., Robot 3 move forward).

So far, the only solution to this problem suggested by teams competing in the RoboCup are to use individual color markers, "bar codes" or manual tagging.

Most teams identify their robots through different colors. The major problem is that it is non-trivial to find a parameters for a color model that allows the detection of a color over the entire playing field.

Another possibility is to identify a robot using some easily distinguishable geometrical pattern. Popular methods are to identify different patterns based on their size or their aspect ratio.

A third possibility is to manually identify (tag) each robot before game starts. For example, the user may click on robot one through five in turn. The vision server then continues to track the robot until there is an occlusion or the robot is occluded. This occurs usually during a stoppage in play.

This procedure is time consuming and error prone. Assigning an identification takes about 30 seconds, but needs to be done in every stoppage in play. Also, in the heat of battle it is easy to mistake two robots. Furthermore, as the skill level increases and stoppages in play become less common, there will be fewer chances to change an erroneous assignment.

3   The Horus Videoserver

The solutions described in the previous section have severe disadvantages since they do not scale up to larger teams and to more flexible camera positions. If we do not want to use additional patterns, then what else is there? The only information left is the image of the robot itself. So the goal was to design a videoserver that uses only a single marker ball and no other patterns on the robot.

Position information in the current implementation is still based on the marker ball on top of the robot. Since the rules require this marker, it seems reasonable to use it for position information. Since the processing of the orientation (described in more detail in the next section) is computationally more expensive than simple blob detection, the position information is used to "anchor," that is to constrain the following computation to a small subimage (approximately 64 by 64 pixels).

3.1   Orientation Information Using the Generalized Hough Transform

Figure 1 contains three zoomed views of our robots from our video camera. The views correspond to the worst case (i.e., the robot is at the far end of the playing field) for our vision system. As can be seen, the most prominent features are the edges along the top of the robot. Other features (e.g., the wheels are not always visible and hard to distinguish). Therefore, we decided to use these edges as features and to infer the orientation of the robot from them.

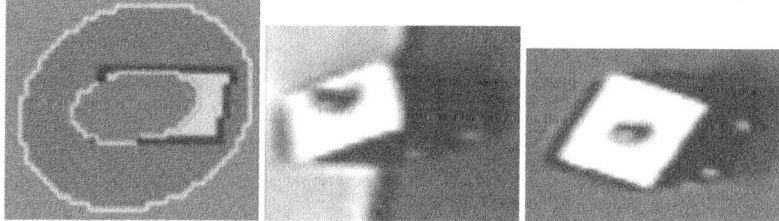

Fig. 1. Some sample images of our robots taken at the far side of the field

This idea faces an immediate problem, since the robots are almost square. This means that it is *impossible* to determine the orientation of the robot from a single image. Given the angle of the edge, there are four possible orientations for the robot, which can not be distinguished without further information.

Furthermore, since all robots have exactly the same shape, it is impossible to identify the robot. Therefore, we decided to use additional information (e.g., history of the cars, current commands, motion of the robot) available to the video server to disambiguate the orientation and to identify the robot. This part of the system is described in section 4.

Given the real world coordinates of the robot, the surrounding image corresponding to a square area of the diameter of the robot is extracted. The maximum size of this window depends on the geometry of the camera position. In most "practical" situations, the size of the window is less than 64 * 64 pixels.

All further processing is limited to this local neighborhood. The image is divided into four regions, which are shown in the Fig. 2.

- Pixels that are more than half a diameter away from the position. These can not be part of the robot and are ignored.
- Pixels that belong to the marker ball or are very close to it. These pixels are usually noisy and are ignored.
- Pixels that match the top color of the robot.
- Pixels that belong to the contour of the robot. These pixels are determined after tracing the contour of the robot using a standard edge walking algorithm.

Figure 2 shows the output for the three sample images given in Fig. 1. The contour of the robot is shown in black. As can be seen, using even a very coarse color calibration, the edges of the robot can be traced accurately even under worst case conditions.

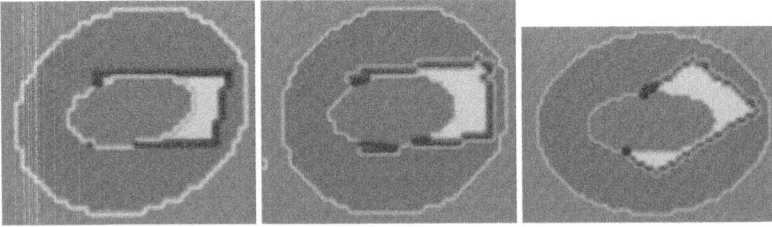

Fig. 2. The image of the robot after preprocessing. Pixels that are too far or too close are ignored. Pixels matching the color of the top of the robot and pixels on the contour

Given the position of the edge pixels, a general Hough transform is used to compute the possible orientation of the robot in the first quadrant [1].

The Hough transform is a popular method in computer vision to find lines and other shapes. The basic idea for the Hough transform is to collect evidence to support different hypothesis. The evidence for different hypotheses is accumulated and the hypothesis with the strongest support is returned as the solution.

Figure 3 shows an example of the geometry in our problem. Each edge pixel can be at most on four possible edges (E_1, E_2, E_3, E_4 in the figure). It is easy to see that

$$\alpha = sin^{-1}(w/d)$$
$$\beta = sin^{-1}(l/d)$$

Therefore, the corresponding angles for the edges can be computed as:

$$E_1 = \theta + \beta$$
$$E_2 = \theta - \beta$$
$$E_3 = \theta + \alpha$$
$$E_4 = \theta - \alpha$$

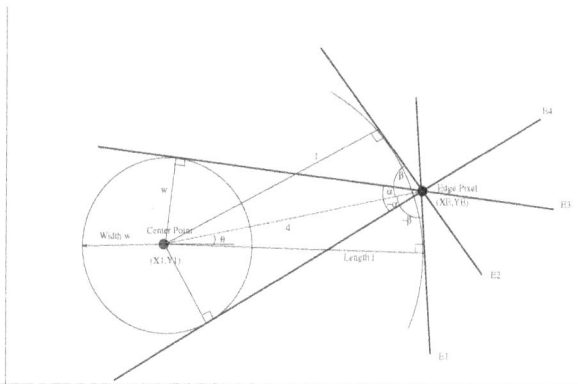

Fig. 3. The four possible edges E_1, E_2, E_3, and E_4 (and therefore, the possible orientations) for an edge pixel (X_E, Y_E)

Edges E_1 and E_2 are only possible solutions if the distance d between the center point (X_C, Y_C) and the edge pixel (X_E, Y_E) is greater than the length of the robot l. Similarly, edges E_3 and E_4 are only solutions if d is greater than the width w of the robot.

In theory, this information is sufficient to determine the orientation of the robot +/- 180 degrees. In practice, we normalize the angles to within 90 degrees, since the distance between the edge pixel and the center point is noisy. This makes little difference in resolving the disambiguity about the quadrant, since for a car-like robot two of the four possible orientations can immediately be rules out, since a car can not move sideways.

The hough space consists of a one-dimensional array with 18 entries, which gives us a resolution of 5 degree. For each edge pixel, the value in the array for that position is incremented. Finally, the angle corresponding to the maximum value is returned.

4   Identification Using Bayesian Probability

As mentioned previously, since all robots in our team look identical, the vision information is insufficient to identify them. HORUS uses two additional sources of information to determine the identity.

HORUS maintains a probability for the identity of each robot. We use a simple Bayesian model to update this probability when new evidence is encountered.

Firstly, HORUS predicts the motion of the robot and tracks it. Should the robot be found in the predicted position, its identity and its associated probability is not changed. If the robot is found in the neighborhood of the predicted position, its identity is not changed, but the probability in the identity is reduced by a factor of 0.9 or 0.7, dependent on how far the robot was found from the predicted position.

Secondly, Horus observes the motion of the robot over a number of frames and assigns it one of seven states: not moving, forward left, forward straight, forward right, backwards left, backwards straight, and backwards right. The actual steering angle is not determined, so there is no difference between, for example, full left and gently left.

Initially as well as after some errors in the assignment, a robot will have an unknown identity. If a robot has an unknown identity, Horus will assign it the first free identity that matches the observed behavior of the robot. The initial probability of this identity assignment is 0.5.

5   Evaluation

The performance of Horus was compared against the performance of our original video server, both with respect to speed and accuracy. The evaluation shows that the performance of the new videoserver is at least as good as that of our original video server. The performance of the vision processing is not a limiting factor in the overall system.

5.1   Horus' Processing Speed

The Hough transform is a compute intensive method, which as a rule should be avoided in real time applications. However, since the position of the robot is known, only a small neighborhood (64x64 pixels) of the image needs to be considered. Also, the number of edge pixels in that neighborhood is small. In the worst case, there are 256 edge pixels. Also, the Hough space is reduced, since we are not interested in the location of the line and since the possible orientations are only between 0 and 90 degrees.

These factors explain why there is no noticeable difference in the processing speed of the two videoservers. Both videoservers are able to maintain a 50 fields/second frame rate in the RoboCup domain with eleven objects.

5.2   Horus' Accuracy

Evaluating the accuracy of the orientation information is more difficult. Horus is unable to determine the orientation completely from just a single image or from a stationary object.

Knowing the orientation of static objects is rarely useful though. We are interested in moving our robots to their targets, so the accuracy of the orientation information for a dynamic object is much more important. A dynamic evaluation is more difficult than a static one, since we have no way of knowing the correct orientation for a moving object.

We tested Horus by driving a simple pattern (a circle to the left in the center of the playing field at constant speed) and by observing the orientation information. The correct information was inferred from a kinematic model of the robot. This test showed that the average error of Horus was slightly less (less than approx. 5 degrees) than that of our original videoserver (less than approx. 10 degrees).

Another factor that determines the quality of the videoserver is its interaction with the control algorithm. For example, it is unclear if the maximum or average errors are more important. Therefore, we tested the interaction of the orientation information with a non-holonomic control algorithm based on Egerstedt's controller ([3]. This control algorithm was used in time-trials on the Aucklandianapolis race track ([2]) using orientation information from the two video servers. The times were identical in both cases. The limiting factor for the speed was in this case not the accuracy of the video server information, but rather the latency in the control loop, which is at least 20ms.

6   Conclusion

This paper presents a new approach to vision in the RoboCup domain. Instead of colored markers, the system uses geometrical features of the robots to determine their orientation.

This means, that the only colored marker on the robots are marker balls, which are used to determine the position of the robot. The orientation is determined by the projection of the robot in the image.

The system uses a generalized Hough transform to find edges in the neighborhood of the position of the robot. These edges are used to determine four possible angles (offset by 90 degrees) for the orientation of the robot.

The videoserver correlates the movement of the different robots with the observed behavior to disambiguate the angles and identify each robot.

References

1. D. H. Ballard: Generalizing the Hough transform to detect arbitrary shapes. *Pattern Recognition* **13** (1981) 111–122.   111
2. J. Baltes, Y. Lin: Path-tracking control of non-holonomic car-like robots using reinforcement learning. In: Proceed. of the *IJCAI Workshop on RoboCup*, Stockholm, Sweden, (July 1999).   114
3. M. Egerstedt, X. Hu, A Stotsky: Control of a car-like robot using a dynamic model. In: IEEE-Proceed. of the Conf. on *Robotics and Automation*, Leuven, Belgium, (1998).   114
4. R. Y. Tsai: An efficient and accurate camera calibration technique for 3d machine vision. In: IEEE-Proceed. of Conf. on *Computer Vision and Pattern Recognition*, Miami Beach, FL, (1986) 364–374.   108

An Stereoscopic Vision System Guiding an Autonomous Helicopter for Overhead Power Cable Inspection

Pascual Campoy[1], Pedro J. Garcia[1], Antonio Barrientos[1], Jaime del Cerro[1],
Iñaqui Aguirre[1], Andrés Roa[2], Rafael Garcia[2], and José M. Muñoz[2]

[1] U.P.M - DISAM - División de Ingenieria de Sistemas y Automática
Universidad Politécnica de Madrid
José Gutiérrez Abascal 2, 28006 Madrid. Spain
{campoy,pjgarcia,barrient,jcerro,iaguirre}@disam.upm.es
[2] Line Maintenance and Construction Department, Red Eléctrica de España, S.A.
Po. del Conde de los Gaitanes, 177, 28109 La Moraleja, Madrid, Spain
{aroa,rafgarcia,jmmunoz}@ree.es

Abstract. The present paper describes the objectives, structure, present stage, results and future milestones of the project ELEVA. This project is aimed to control an autonomous helicopter in order to follow an overhead power cable by means of a stereo computer vision system. The helicopter is aimed to have always in sight the overhead power cable, to follow it by using it as an external visual reference guide and to record it for its ulterior visual inspection. These objectives are achieved by using a 3D computer vision system to generate the reference trajectory to be followed and by using internal sensors to control its stability and its trajectory. The paper presents the results obtained so far: visual detection and tracking of the power cable, robust under changing environments, and robust stationary control of the helicopter, now linked to a safety mechanical platform. Finally this paper describes the future challenges of the project and its temporal milestones.

1 Introduction

The ELEVA project is an application-oriented research project for the automatic inspection of overhead power cables. Some top research institutes in the world are developing unmanned helicopters able to follow predefined trajectories [3], some of these projects use computer vision for recognition of the terrain and for some kind of spatial orientation. Most of these unmanned helicopter projects use the available accuracy of GPS systems for the global trajectory guidance [13].

1.1 Motivation

The key differing feature of this fully application-motivated project, which arises directly from its main goal, is the fact that the helicopter uses the overhead power cable-layout as a permanent visual guidance for its trajectory. This fact presents

R. Klette, S. Peleg, G. Sommer (Eds.): Robot Vision 2001, LNCS 1998, pp. 115–124, 2001.

two countered aspects. In one hand it has the advantage of having a permanent physical guide for the helicopter to be used as its three dimensional reference. And on the other hand, the cable has to be always in sight in order to be recorded and inspected, i.e. the cable is not only an external reference but also the aim of the helicopter flight, which has to be continuously followed.

The overhead power cable has to be in sight and recorded and it has to be achieved continuously under a high variety of backgrounds and illuminations. This fact encourages the desired robustness of the vision algorithms in a changing outdoors environment. If the cable is instantly out of sight or if it becomes lost anyhow, the vision system has to track the cable again and the helicopter has to accomplish a recovering strategy in order not to lose any part of the cable to be inspected.

1.2   State of Art

Nowadays dangerous tasks like rescues, inspections or works in contaminated areas or military missions are performed by human piloted aircrafts. Some of these tasks could be performed using autonomous systems. Modern navigation systems like differential GPS, sophisticated inertial sensors and small powerful computers allow to control small flying aircrafts.

Several universities have his own prototype that performs different tasks in the annual UAV (Unmanned Aerial Vehicle) Competition (MIT, Stanford, CMU, Georgia Tech...). The competition consists on simulations of real dangerous situations. Typically the helicopter must find some targets on ground using computer vision and transport it to specific places marked. There are commercial helicopters used for spraying rice fields or look for fish banks in Japan [Sugeno 99], others like Scandicraft and the CAMCOPTER by Schiebel.

Visual guidance of autonomous vehicles is one of the most recent and lively research fields in Computer Vision. There are already quite a lot of research centers involved in developing of both aerial [2,13] and ground [10,11,12] (intended to work in either indoor or outdoor unstructured environments) autonomous vehicles. However, after studying the state-of-the-art it becomes that current vision-based navigation systems are mostly focused on tracking strategies of mobile ground references whose size and shape are well-known. In those cases difficulties fall on locating such ground reference (as said above, an accurate visual model is available for it), identifying its motion and, as a result of that, assisting the control and navigation system of the helicopter in order for it to be able to track that reference. In other cases, research goes into how certain obstacles can be identified and avoided and static well-known ground objects located. Such features are intended to emulate an actual unmanned aerial rescue operation.

The present prototype of vision-guided autonomous helicopter differs from those referred above in the properties of the reference that must be tracked, the power cable, since such reference, even though static, is unbounded and does not show any salient geometric features (like corners or edges) that can be modelled for cable identification to get easier. As a result of that it can stated that the present helicopter actually brings up a new and challenging research project in the field of vision-guided vehicles.

2   The ELEVA Project

The ELEVA project is a three year long project supported by the CICYT (Spanish National Research Program) and by Red Eléctrica de España, that is the owner company of the transmission network within the Spanish electrical system, and is responsible for its operation, maintenance and construction. The project is planned to be finished at the end of year 2000, and after this period the whole results of the project are to be evaluated and a work plan is to be analyzed for its future extension.

The specific objectives to be fulfilled within the project ELEVA are scheduled in three main millstones:

- Milestone 1:
  1. Simulation system for 3D image acquisition, which includes the helicopter fight and several actual backgrounds.
  2. Flight data acquisition using internal sensors when prototype is manually controlled.
  3. Cable detection and tracking in the simulated system for several illumination conditions and backgrounds.
  4. Flight control of the prototype linked to an indoor safety mechanical platform.
- Milestone 2:
  1. Dynamic 3D co-ordinates estimation using stereo computer vision and robustness analysis under changing environmental conditions.
  2. Flight control of the helicopter linked to an outdoor six degrees of freedom platform, using its internal sensors.
- Milestone 3:
  1. Autonomous unmanned flight of the prototype following a real overhead power cable in a tested field, using both internal sensors and the computer vision system.
  2. Results evaluation and conclusions about the functionality of the developed prototype and analysis of its extrapolation to bigger prototypes.

3   General Structure of the Project

Figure 1 shows the architecture of ELEVA prototype. It can be differentiated two subsystems, one on board the helicopter and the other in ground. On board system is composed by sensors, microcontrollers, cameras, and wireless communications devices.

In order to estimate the attitude and the position of the helicopter an Inertial Measurement Unit is used. It provides angular velocities and accelerations using strap down systems navigation algorithms implemented into a microcontroller.

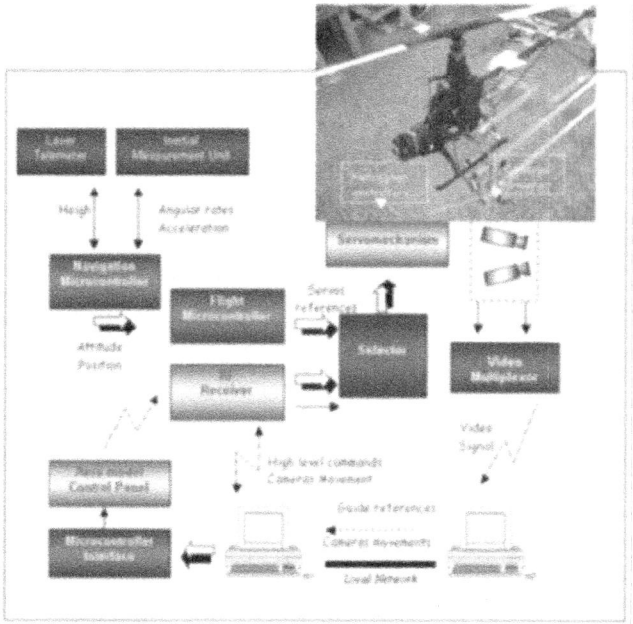

Fig. 1. Architecture of ELEVA Prototype

This information goes to the ground control computer via radio link and also to the on board control microcontroller. The use of a laser telemeter allows to determine the distance between the craft and ground. Control algorithms use this information to generate new servo references.

By other hand, using a commercial control panel for models, a man can supervise the flight. If something not expected happens, only by touching a key the human operator would take the control of the helicopter. The ground computer sends high level commands to the on board controller, generating trajectories corrections using the information from the vision system. In order to generate 3D coordinates, two video cameras are mounted on the helicopter. Signals from these cameras are multiplexed and sent to the ground image-processing computer. The ground control and vision computers are connected by a local network. Image processing computer is able to send movement commands to the cameras servos trough the control computer using a specific interface with the control panel.

3.1 Autonomous Control

The helicopter model is based on other previous works by Furuta et al. [5] , Johansson [7], Persson and Klang [9] and Maki [8]. These studies conclude that the helicopter model may be separated in two parts. The first part represents the main rotor and the second one represents the tail rotor dynamic. The mini-helicopter mathematical model is obtained in Aguirre et al. [1].

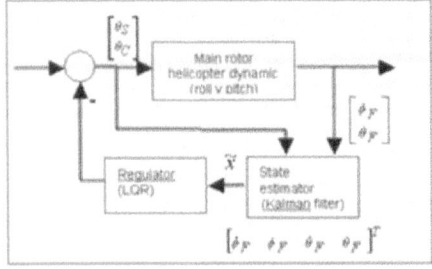 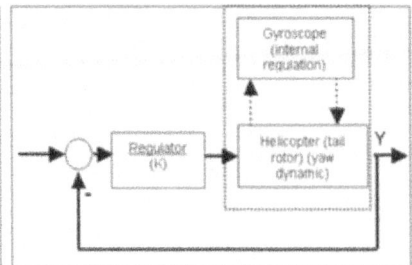

Fig. 2. Control block diagram for pitch and roll

Fig. 3. Control block diagram for yaw

Robustness is considered as the main requirement of control in miniheli-copters. The control action must give a precise answer at any time in such a manner to minimize the risk of a crash. The helicopter has a very particular and complex dynamics due to its couplings and high non-linearities. The first control goal is to attempt to control the attitude of the craft (roll, pitch and yaw).

The first step to build a controller is to model the dynamics of the helicopter near the operation point, which in our case is hovering. Once the mathematical model is obtained using the empirical measurements from the craft, the next step is to propose the type of control to be implemented. Due to the facilities of the mathematical model represented in state space, a Linear Quadratic Gaussian Control (LQG) is selected for controlling the roll and pitch of the craft in hover.

Linear Quadratic Control is utilized to calculate the controller necessary to fulfill the planned objectives. Due to the fact that the only disposed variable is position, it is recommended to estimate the other variables. To estimate variables Kalman Filter is applied. The Linear Quadratic Gaussian Control refers to an optimal control problem for a linear plant model in which Linear Quadratic Control and Kalman Filter are utilized.

Figure 2 shows the block diagram utilized for control of the roll and pitch movements of the craft. After the linearization in hover, the proposed linear model will be composed of four states and two control variables. Due to the use of potentiometers, only positions are acquired. The use of Kalman filter is required to calculate the velocity, thus Kalman filter performs the estimation of the variables. Once the state variables are obtained, the control action is calculated using the Linear Quadratic Regulator [3]

An internal regulation is done using a piezoelectric gyroscope for the yaw. A higher level regulation is obtained using a proportional control. The control scheme for the yaw control could be seen in Fig. 3.

In hover, the control maintains the craft near hover as shown in Fig. 4. The main characteristic of hover is that roll and pitch angles must be close to zero. It can be seen that the three orientation angles, corresponding to the roll, pitch and yaw angles of the minihelicopter, maintain their reference values of zero degrees

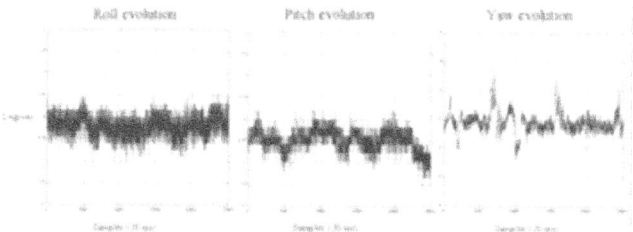

Fig. 4. Roll, pitch and yaw in hover

with a small error. The control is obtained for the three angles and keeps the helicopter in hover. The maximum error committed by the pitch and roll value with respect the reference of 0 degrees is of 1 degree but in most of the samples, the error is close to 0. For the yaw figure there are some peaks in short period of time that the error approaches 3.5 degrees and then it maintains the value close to the reference.

To evaluate the robustness of the controller, a test was carried out by disturbing the helicopter attitude while it was in hover. The disturbances were imposed upon the craft by pulling the helicopter out of the hover position with a cable connected to the test platform. An air compressor was also used to disturb the helicopter orientation. As can be seen in Fig. 5, the orientation of the helicopter returns to the reference values once the perturbation is compensated by the controls' actions. The disturbances are introduced in sample 0 and in sample 400. The outputs of the system show that the controller works very well leading the helicopter to a stationary flight. The system delays 4 seconds in returning to the reference values after the perturbations were imposed.

As seen in Fig. 5, the Linear Quadratic Gaussian Regulator calculated for the roll and pitch movements behaves very well and is robust in the presence of external disturbances. The PID control for yaw also works correctly in the presence of external disturbances.

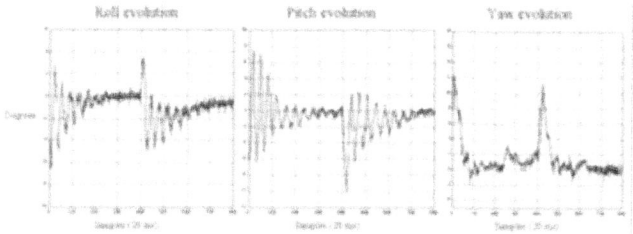

Fig. 5. Roll, pitch and yaw evolution in presence of disturbances

Fig. 6. Simulator user interface for the Vision System

3.2   Vision Guidance

In order to justify why a visual guidance is needed just bear in mind how the helicopter must fly: the reference that it has to follow is the cable itself, so global coordinates systems would be useless for navigation (unless UTM coordinates of every pylon were available, on-board navigation equipment included GPS and, furthermore, both georeferences are accurate enough). What is more, there are three conditions the flight must complies with all the time, one related to trajectory and the two others related to position, always with respect to the line:

- a safety lateral distance (of about 8 m) related to the nearest cable must be kept.
- a relative height with respect to such cable must also be guaranteed in order to improve visual guidance accuracy.
- parallelism between the cable and helicopter trajectory must not be lost.

The first two topics are related to the so-called "vision-based lateral control" [10] and the latest to the "longitudinal control" of the helicopter. These are the majors goals reached so far:

- A software simulator for the on-board helicopter image acquisition system has been built.
- Robust algorithms in order for the cable to be identified and tracked, over both real an virtual image sequences, have been developed.

The software-based image acquisition simulator allows to generate virtual inspections according to any desired line topology and flight trajectory and to obtain the sequence of images that a real inspection of such line would supply if such trajectory were followed. This tool provides an excellent testbed since it let assess the performance of differences line identification algorithms as well as that of cameras pan-tilt positioning algorithms. Furthermore, extreme situations can be simulated whose critical nature can derive from either how dangerous it may become (e.g.: the helicopter approaches the line too much) or how difficult visual tracking may get in such conditions. Figure 6 shows the user interface of the application and one of the pair of images generated over a simulation.

Image processing algorithms intended to cable identification have also been developed for both virtual (simulator-generated) and actual (coming from a real

Fig. 7. Results of cable detection under several environments

inspection of the 400 kV line Aragñn-Escatrñn) sequences. Algorithms developed so far make possible for a single-circuit line to be identified and tracked along a flight trajectory. Current work is focused on generalizing those results for double-circuit lines. It must be noticed that images used to assess the performance of these algorithms have been mainly those stemming from real inspection sequences, since virtual images are not as complex as real images are. This fact is due to the big variety of textures and types of grounds that real images bring up as well as the presence of luminous reflections on the cables landscape features and vegetation. Bear in mind that all these factors generally make line tracking more difficult. Figure 7 shows some line identification examples over a set of real images. Notice that certain landscape features, like tracks or furrows, can suppose a problem to deal with.

The image processing algorithms that have been implemented exploit advanced techniques for straight lines detection, based on the vector-gradient Hough transform [4] applied in a multiresolution strategy [6], that is to say, firstly the approximate position of the line is determined and next a high-resolution processing is carried out around that position. Once the line has been located in the first image, only the second processing stage (high-resolution) has to be run over the next images in order to track it., making use of the past history of the line position for this task to become easier. Figure 8 shows the low and high resolution Hough transforms applied on the first of the previous images.

The control structure of Fig. 9 ensures the visual tracking of the cable by both cameras. The three dimensional position of the vision system (and therefore

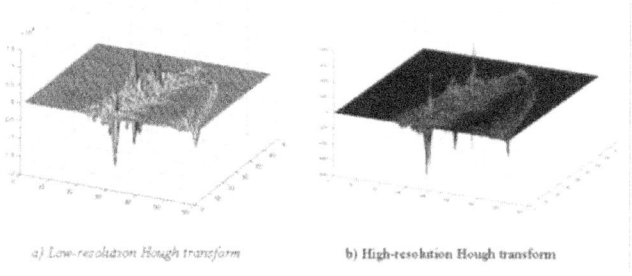

a) Low-resolution Hough transform   b) High-resolution Hough transform

Fig. 8. Hough transform of images for detection of most significant lines

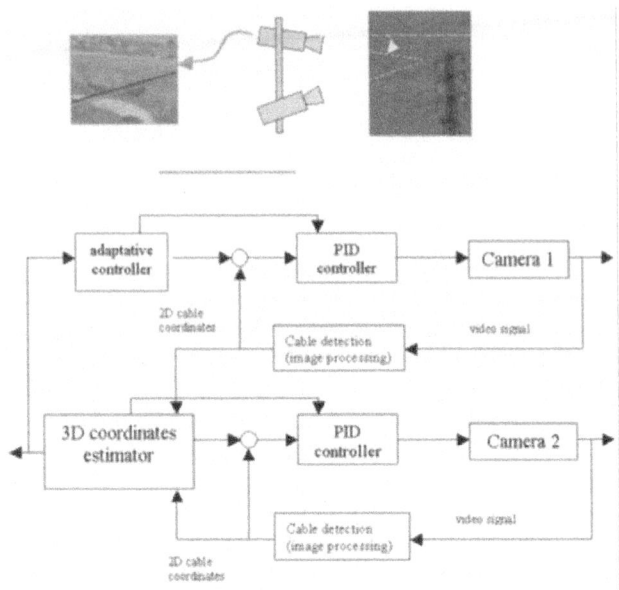

Fig. 9. Visual control structure

of the helicopter) is calculated from the orientation of the cameras tracking the line. This data is deliver to the control system and also used by the individual controllers of the cameras in order to adapt their internal parameters.

4   Future Stages and Goals

The final objective of the ELEVA's project is to develop an autonomous aerial vehicle capable to navigate by his own while it detects electrical fails in the cable lines. To obtain the final goal of the ELEVA's prototype the following propositions are made to be developed:

- Integration of the vision system on board the minihelicopter utilizing DSP's.
- Automatic detection of electrical lines visual failures.
- Developments of emergency routines on board the craft when communication is lost.
- Development of a master-slave with force reflection system to teleoperate the craft.
- Study of a bigger helicopter that could allow more payloads.

5   Conclusion

Use unmanned aerial vehicle for inspection of electrical power line allows a significant reduction of cost and risk of this task. Two main problems must be resolved in order to achieve this objective: Visual guidance system and Automatic unmanned helicopter control. In the actual stage, ELEVA project permits the hovering control of the vehicle, and the location of the power line using computer vision. In the future, other kind of fly control will be implemented and the visual guidance system will be integrated with a GPS and a remote telecontrol station in order to be able to perform power line unmanned inspection test.

Acknowledgements

The authors wish to thank the CICYT that supports the presented project through the TAP97-1092-C03-02 and R.E.E. that supports the project through the research contract with U.P.M.-Disam.

References

1. I. Aguirre, J. Del Cerro, A. Barrientos: Development of a low cost autonomous mini-helicopter for power lines inspections. *SPIE Conf. Mobile Robots XV: Intelligent systems and Smart Manufacturing*, Boston, (November 2000). 118
2. O. Amidi: An autonomous vision-guided helicopter. Ph.D. Thesis, Electrical and Computer Engineering Department, Carnegie Mellon University (1996). 116
3. F. Barbariol: Implementation of an autonomous helicopter flight controller. Masters Thesis. Linkoping Institute of Technology, Linkoping University, Sweden (1995). 115, 119
4. R. Cucchiara, F. Filicori: The vector-gradient Hough transform. *IEEE Trans. Pattern Analysis and Machine Intelligence*, **20** (1998) 746–750. 122
5. K. Furuta, Y. Ohyama, O. Yamano: The modelling and control of RC helicopter. Department of Control Engineering, Tokyo Institute of Technology, Japan (1983). 118
6. J. Illingworth, J. Kittler: The adaptive Hough transform. *IEEE Trans. Pattern Analysis and Machine Intelligence* **9** (1997), 690–697. 122
7. H. Johansson: Modellering av en RC-helicopter. Masters Thesis. Linkoping Institute of Technology, Linkoping University, Sweden (1994). 118
8. E. Maki: System Identification and Attitude Control of an R/C Helicopter. Honors Thesis for B. S. Department of Mathematics and Computer Science, Central Missouri State University. USA (1998). 118
9. M. Persson and R. Klang: Autonomous flying helicopter: attitude control. Masters Thesis. Center for Computer Architecture, Halmstad University (1997). 118
10. C. Taylor, J. Malik, J. Kosecka, R. Blasi: Vision-based lateral control of vehicles. *Conf. Intelligent Transportation Systems*, Boston (November 1997). 116, 121
11. C. Taylor, J. Malik, J. Weber: A real-time approach to stereopsis and lane-finding In: *Proceed. IEEE Intelligent Vehicles Symposium* (1996). 116
12. W. B. Thompson, H. L. Pick: Vision-based navigation. In: *Proceed. ARPA Image Understanding Workshop* (April 1993). 116
13. B. Woodley H. Jones II, E. LeMaster, E. Frew: Carrier phase GPS and computer vision for control of an autonomous helicopter. *ION GPS-96*, Kansas City, Missouri, (September 1996). 115, 116

3D Stereo Vision-Based Nursing Robot for Elderly Health Care

Wee-Soon Ching, Edward Ho, Christopher Ong, Hs Tay, and Sai-Mui Lim

School of Engineering (Electronics)
Nanyang Polytechnic, AMK Campus, Singapore 569830
CHING_Wee_Soon@NYP.gov.sg

Abstract. With the population aging in most of the developed countries, using robot for health care and related service industry is of increasing importance in recent years. This paper describes the design and development of a 3D stereo vision-based [1] nursing robot for elderly health care home. It focuses on presenting a novel image segmentation method, and an innovative and effective method of locating the facial features, which include the two nose points, for reliable mouth detection. A 3D stereo vision-based method for determining the 3D position of the mouth is also described. Experimental results are discussed.

1 Introduction

The system consists of a six-axis seven drive modular robot, a vision sub-system, and a voice recognition sub-system, as shown in Fig. 1.

The vision sub-system, which is 3D stereo vision-based, uses 2 cameras, and acts as the eyes of the nursing robot. Its main function is to locate the mouth of the person, and in determining the 3D positions, so that the robot can be calibrated to feed the patient with food, or medical pills. The IC-Async frame grabber from Imaging Technology, which is able to support up to 4 cameras, is

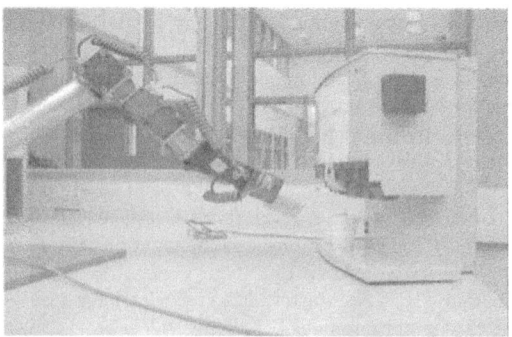

Fig. 1. 3D stereo vision-based nursing robot

R. Klette, S. Peleg, G. Sommer (Eds.): Robot Vision 2001, LNCS 1998, pp. 125–130, 2001.

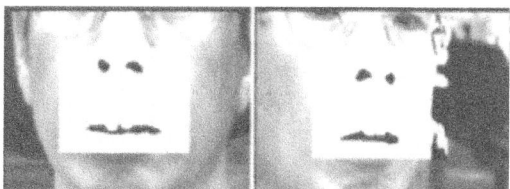

Fig. 2. Stereo views of segmented image with two nose points above the mouth

used. The voice recognition sub-system, which is built in the central processor, enables the patient to activate the radio or television through voice commands. As a safety feature, the system is equipped with touch sensors.

2   Image Segmentation for Mouth Detection

The stereo images are first captured, and then the noises are filtered out through median filter. The segmentation algorithm is then applied to detect the mouth. The image segmentation algorithm can be divided into the following major steps:

1. The threshold value, which will change adaptively according to the environmental lighting conditions, and the skin color of the person, is first set to a fixed value of 100.
2. A 1st level decision algorithm that eliminates impossible blobs that might be caused by unwanted noise and marking of the person is then executed. These blobs are usually of much smaller sizes, and can be eliminated easily.
3. A 2nd level decision algorithm that makes use of the logical relationship between the position of the nose and that of the mouth is next executed. As the noise will appear as two black holes, the algorithm can determine the nose positions easily, and uses the logical relationship between the nose position and the mouth (nose will be above the mouth by a certain known distance) to further rule out impossible matches. A typical segmented image is shown in Fig. 2.
4. A 3rd level decision algorithm that ignores those blobs that are caused by the hairs of the persons is executed next. These blobs are usually on the sides of the ROI, and thus touch the boundary, as shown in Fig. 3.
5. If a reliable mouth that has two nose points above it and within acceptable distance cannot be detected at both the stereo cameras, the threshold value will be changed to a new value. Steps 2 to 4 are then repeated. The threshold is changed adaptively, in a converging manner, until a reliable mouth is detected.

Note that the computational speed for mouth detection is slower during the system set up stage, which takes about 1 second, running on a Pentium III 450 MHz PC. Subsequent mouth detection can be determined at a much faster speed.

3   Enhancing the Robustness of Mouth Detection

The above image segmentation was found working under many situations, except for the situations where the nose shapes are sharp, and are thus more difficult to locate. This section presents two methods of enhancing the robustness of mouth detection.

3.1   Method 1: Inclusion of Eyes

In enhancing the algorithm, we first include the eyes of the patients' eyes to enhance the robustness of detection. Figure 4 shows the stereo images where mouths are reliably detected, with the inclusion of eyes. That is, the eyes and nose points act as the conditions in determining the location of mouth, as they are above the mouth by a more-or-less fixed distance ranges, and thus similar algorithms as explained in Section 2 can be used.

3.2   Method 2: Pointing the Camera Slighting Upward

The above inclusion-of-eye method works at the expenses of additional computational complexity in the detection of eyes and provided that the spectacles are coated. Figure 5 shows that if the spectacles that the patients wear are not coated, the eyes cannot be found reliably.

In view of the above limitations, we have adopted an innovative method, which is more effective, in solving the problem, with better results and with no increase in computational loads. The method is to lower the cameras to a level that is lower than the mouth's level by about 8.00 cm to 10.00 cm, and pointing the stereo cameras upward so that the nose's location can be extracted reliably, regardless of the nose shape of the patient. This is shown in Fig. 6 below.

4   Determining the 3D Position of Mouth

Once the 2D location of the mouth is determined on both cameras, the next step is to use stereo vision for finding the 3D mouth location. The 3D stereo vision systems consist of two cameras, aligning in parallel, pointing towards

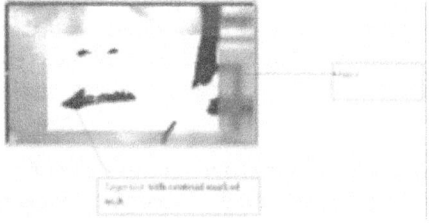

Fig. 3. Segmented image with hair that touches the side of the ROI boundary

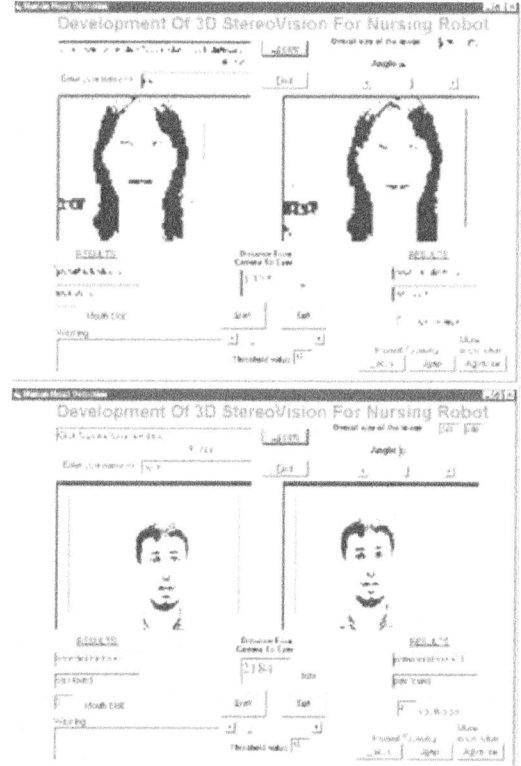

Fig. 4. Segmented stereo images with the inclusion of eye for enhancing the robustness

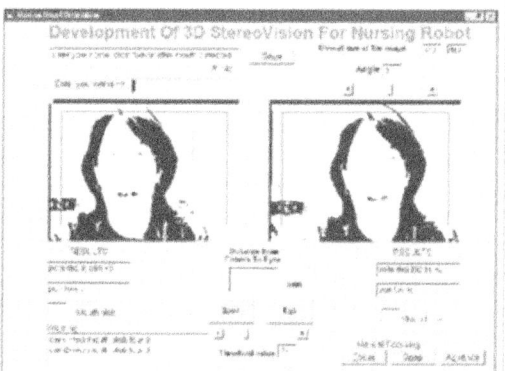

Fig. 5. Segmented stereo images with non-coated spectacles (eyes cannot be located)

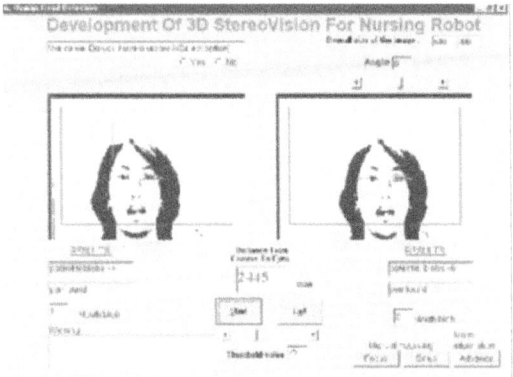

Fig. 6. Segmented stereo images with reliable mouth detection (with cameras pointing slightly upward, and with coated spectacle)

the person. Such parallel alignment of cameras is adopted for the simplicity in implementation. This in terms gives the speed needed when implemented on Pentium III PC, for greater customer satisfaction.

As the environment is controlled, both cameras can image the mouth clearly, within a selected field-of-view. As the patient, who sits on a wheel chair, will be pushed by a human nurse to a position that his/ her mouth is within the field-of-view of both cameras, no occlusion problem or strong specular-highlights problems need to be resolved. The key problems are the changing skin color and environmental lighting, which have already been tackled by the image segmentation algorithm, as described in the previous sections.

5   Experimental Results and Discussion

It was noted that the image segmentation algorithm works reliably, especially when the mouth is opened, which gives a dark spot, which is ideal for image segmentation. As the patient have to open the mouth when they want to eat their food, or taking the medical bills, the 3D stereo vision-based nursing robot works reliably in the health care nursing home.

The camera-pointing-slightly-upward method has further improved the mouth location algorithm, and it works with patients of all kind of nose shape (sharp nose, flat nose, high nose etc..), and is regardless of whether the spectacles that the patients wear are coated, or not.

The next task is to examine the 3D depth accuracy. System calibration is carried out to determine the exact 3D coordinates of the person, by taking the average and peak deviation of 50 actual and measured distances, for 4 different persons having different skin color. The measurements are taken over the measuring depth range of approximately 500 mm to 650 mm. The results obtained show that it has a maximum distance deviation from the actual distance of only

14.5 mm, which is acceptable to the present application. The distance deviation is also within the tolerance of the mouth movement of the patient.

Better accuracy can be obtained by performing sub-pixel measurement, by performing multi-file-of-view and multi-resolution image analysis, or through adaptive movement of the robotic arm, which improves the accuracy, at the expense of increased response time. As the fast response time of the nursing robot is more important for greater customer satisfaction than the accuracy in the present application, the system has not adopted these computational more expensive approaches [2].

6   Conclusion

This paper describes the design and development of a 3D stereo vision-based [1] nursing robot for elderly health care home. A novel image segmentation method, and a simple and effective method of locating the two nose points, for reliable mouth detection are presented. A 3D stereo vision-based method for determining the 3D position of the mouth is also described. Experimental results show that the 3D stereo-based vision system that we have developed for nursing robot can effectively lessen the major problem face by the elderly health care industry.

References

1. W. S. Ching, P. S. Toh, M. H. Er: Robust vergence with concurrent identification of occlusion and specular highlights. *Computer Vision and Image Processing* **62** (1995) 298–308.   125, 130
2. G. H. Sim, M. C. Yap, S. G. Ong: Development of a 3D stereo vision-based nursing robot. *FYP Project report* (December 1999).   130
3. S. Bhuvaneshwary, D. Oh: Image segmentation and 3D depth determination for nursing robot. *FYP Project report* (June 2000).

Efficient Computation of Intensity Profiles for Real-Time Vision

Ernst Dieter Dickmanns

Unibw Munich, Institut fuer Systemdynamik und Flugmechanik,
D-85577 Neubiberg, Germany
Ernst.Dickmanns@unibw-muenchen.de

Abstract. For the EMS-vision system realized on distributed general-purpose processors with a set of video cameras on an active gaze control platform, an efficient method for exploiting area-based image information has been developed (as opposed to edge features preferred in real-time vision systems up to now). It relies on the same oriented intensity gradient operators as have been used for edge localization in the past (K(C)RONOS). However, the goal achieved now is fast derivation of one-dimensional intensity profiles with piecewise linear shading models. First, regions of large intensity changes (so-called 'non-homogeneous' regions) are separated from 'homogeneous' ones containing at most moderate intensity changes (to be specified by a threshold parameter). The average intensity values and ternary mask responses in these areas yield information for a coarse linear (first order) intensity model. Then, in the homogeneous regions, the one-dimensional equivalent of a pyramid (a triangle-) representation is derived for the residues between the actual intensity values and the coarse linear model. Depending on the size of the homogeneous region and the number of intensity peaks, a certain triangle level for further processing is selected. Again, a (different) ternary mask operator is used for intensity gradient computation and for finding the zero-crossings of the gradient. This information is sufficient for determining the fine structure of regions with linear shading models. Examples are given for road and vehicle detection and recognition.

1 Introduction

In the 4-D approach to dynamic machine vision, it has been proven that modeling over time with dynamical models for motion representation can lead to advantages in image sequence understanding, if the image evaluation frequency can be kept sufficiently high for exploiting temporal continuity conditions (Wuensche 1987; Dickmanns and Graefe 88). This is an extension of the well-known Extended Kalman Filter (EKF) to perspective mapping. In order to achieve fast image evaluation, image processing had been confined to intelligently controlled edge feature extraction in certain areas based on expectations derived from spatio-temporal predictions.

Most of the applications have been confined to well structured environments like vehicle guidance on highways or landmark navigation for low flying aircraft

R. Klette, S. Peleg, G. Sommer (Eds.): Robot Vision 2001, LNCS 1998, pp. 131–139, 2001.

(Schell 1992; Werner 1997). The general method developed has led to third generation systems based on Commercial-off-the-shelf (COTS) hard- and software components (Gregor et al. 2000); it has been dubbed 'Expectation-based, Multifocal, Saccadic'- (EMS-) Vision and makes use of a number of features developed in biological vertebrate vision systems like: 1. Foveal-peripheral joint evaluation of parallel visual data streams based on, 2. full spatio-temporal models for active agents; 3. similar models for visual/inertial data integration. For the own body, inertial measurements allow pose predictions almost without time delays; with delay times in vision of typically several hundred milliseconds, this considerably alleviates visual dynamic scene understanding under strong perturbations.

EMS-vision has been realized in object-oriented programming with the language C++. The 3-D environment is represented by a scene-tree, the edges of which are homogeneous coordinate transformations. The actual best estimates for the states of all objects or components are collected in a dynamic object database (DOB) common to all interacting distributed devices. With this general framework in place, visual recognition has to be improved in order to lift the performance level further. However, processing power still is not yet sufficient to fully digest several video data streams in parallel in real time. This has led to the concept of intelligently controlled 1-D 'cuts' through image regions for determining intensity profiles; proper selection of these cuts allows a wide range of interpretations:

1. Equally spaced vertical stripes allow to understand the static environment by sampling and interpolation;
2. at the same time, moving objects can be detected: they will receive special attention by tracking groups of features with crosswise centering of search stripes on them in future frames.
3. Looking almost parallel to the ground, horizontal lines in the images directly correspond to distances on the ground. Linking this to features far away (near the horizon) allows determining pitching motion of the own body (redundant to inertial pitch estimation).
4. Specially oriented search stripes may allow tracking of important features, for example, broken lane markings in perspective projection for vehicle guidance; the corners of the lane markings allow distance and speed estimation not available from solid lines.

2   Edge Localization with Ternary Masks

Oriented ternary mask evaluations in search stripes have shown to be very efficient because of multiple use of intermediate computational steps (Kuhnert 1988; Mysliwetz 1990; Dirk Dickmanns 1997). The expected edge orientation is used first for collapsing all pixel within the mask width into a single (directionally low-pass filtered) vector. The actual mask response is then computed using combinations of these data; for example, a commonly used mask sums up n pixel values to each side of a central region (e.g.: — -1 -1 -1 — 0 — +1 +1 +1 —). The extremum value of this mask response yields the position of the strongest gradient in intensity; it may be determined to the precision of a small fraction of a pixel.

The average intensity value in one half-field of the mask may allow associating some area-based information with the object part yielding the edge. This has proven useful in the past for tracking objects in more complex scenes. Based on this experience, the current approach has been developed for systematic appreciation of intensity profiles in connection with spatio-temporal models of objects. Homogeneous regions are to be separated from non-homogeneous ones, and the homogeneous ones are to be represented by a few significant numbers for scene interpretation like average intensity level, shading effects or some measure of texture on different geometrical scales.

3   Image Stripes for 1-D Intensity Profiles

Of course, full image interpretation would require a 2-D analysis of the image data; however, it is well known that for certain application areas with known continuity conditions, crosswise 1-D sub-sampling will provide most of the information available in 2-D at a small fraction of the cost. Depending on the homogeneity of the scene and on the geometrical scale to be resolved, different stripe widths and orientations may be selected. This point is not elaborated here; it may vary according to the field of application and the point of interest. We start here from a stripe vector of width 1, into which the original stripe has been condensed taking preferred directions into account. For this vector, piecewise linear approximations to the intensity profile have to be found. These analytically represented parameters and ranges are used as additional features for object recognition and tracking in the recursive estimation approach underlying EMS-vision ('4-D approach' (Dickmanns, Wuensche 1999)). This embedding assures both spatial (in 3-D) and temporal continuity conditions, respectively points of discontinuity.

4   Separation of Homogeneous and Non-homogeneous Regions

Discontinuities (like sharp edges) have to be preserved since they, usually, correspond to boundaries between objects or regions of different qualities. Therefore, the first step is to isolate regions with relatively strong intensity gradients. In order to achieve this, first the response to the following mask over the entire vector is evaluated: — -1 -1 — 0 — +1 +1 —; this operator, shown as heavy line in Fig. 1, has the advantage of twofold smoothing. First, it averages two values to each side (empty squares = image intensity at pixel position) before, second, it forms the central difference known to be a second order approximation to the slope. One sixth of the mask response is a good approximation to the slope d(intensity)/d(pixel) of image intensity in vector direction. Figure 2 gives a typical response of this operator to a step change in image intensity. It shows

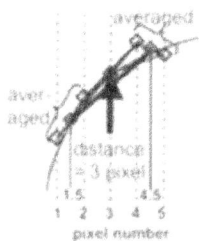

Fig. 1. Visualization of mask response as slope of a function

twice the amplitude of the simple central difference (lower triangles), has a wider range of response, but the same location of the peak. In many tests with typical image data, this operator seemed to yield the best results in the average; it is computationally not too expensive, and yet has a good smoothing effect. During its computation, statistical data of the intensity distribution in the stripe are also evaluated like: minimum, mean and maximum values of the intensity and of the mask response (correlation values or intensity gradients).

A threshold factor for multiplying the difference between the maximum and minimum correlation value is selected in order to arrive at an absolute threshold value for separating homogeneous from non-homogeneous regions; factors between 5 and 30 More experience has to be accumulated in order to resolve the question, how an optimal value can be selected automatically.

5   Linear Coarse (Low Spatial Frequency) Models for Homogeneous Regions

With a threshold value fixed, all regions showing larger mask responses in magnitude, or short ranges of smaller mask responses of extension smaller than a minimal segment length 'segmin' are collected as 'non-homogeneous segments' in a separate statistic (CharNonHom). These represent regions in the image with rapid intensity changes at rather large spatial frequencies. The other segments with limited gradient values are represented with their statistical data in a table (CharHom). These tables contain beginning and end of the segments, the

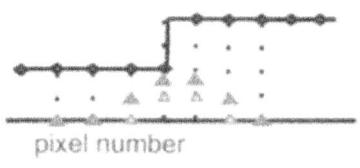

Fig. 2. Mask response (lower triangles) to step input (upper squares)

minimal, maximal and mean values of image intensity and mask response, location of appearance and other values of potential interest. Keeping track of the mean values of both the signed and the unsigned mask responses (absolute values) allows some insight into the average 'texture' of the segment. Note, that there are only two parameters to be specified beforehand, first the threshold factor for multiplying the difference between the actually measured maximum and minimum correlation value, and second the minimal segment length.

In EMS-vision, the new photometric data are used for grouping regions with similar features in neighboring or orthogonal stripes into candidates for objects. In this way, the 2-D extension of objects in the image can be explored with relatively little effort (a few 1-D cuts). These regions are then tracked over time (3 to 5 cycles); after arrival at a consistent feature set, a jump to a 4-D object hypothesis (or several ones in parallel) is made encompassing three essential ingredients: (3-D) shape, aspect conditions and dynamical model. This big jump has been a standard step in the 4-D approach for over a decade and has proven to be viable; the additional photometric aspects becoming available with the present extension are used for arriving at more robust recognition processes.

The mean values of segment intensity and of mask response (the latter one divided by 6 for obtaining the average single pixel intensity gradient (dI/dpel)), yield a coarse straight-line approximation $(I = a * x + b)$ to the intensity function over segment length (from pixel k to pixel m, say). This allows to easily compute the end points of this straight line. When the magnitude of 'a' is small, a large image segment may be characterized by the intensity value 'I' from pixel k to pixel m. The average value of the magnitude of mask response (relative to 'a') allows to judge the goodness of fit. The residues as differences between pixel intensity measured and the value corresponding to the coarse-line- model are taken as input to a second, finer approximation step. However, in order to be efficient, a multiple scale representation like the 'pyramid' in 2-D (Burt 1981) has been chosen for this vector, yielding higher levels with half the pixel count each level up. In correspondence, this is called 'triangle-representation'; it is computed by averaging the intensity of two neighboring pixels on the lower level into one on the next higher triangle level. This is repeated until the base vector is represented by a small number of pixels (8 to 16 pixels have shown to be convenient); this defines the index 'TriaLevMax' up to which triangle levels are determined. During this process, statistical information is collected on all triangle levels, including the number of zero crossings; this information is stored in a table like Table 1. It can be seen how maximal and minimal values of the intensity in the stripe approach the mean value with increase in triangle level (rows 1, 3 and 5 from left to right), and how dynamic range is reduced due to averaging (row 7). The mean magnitude of the intensity in relation to the intensity range on the triangle level may increase (row 8). The location of occurrence of the extremum values may jump (a factor of two downward from left to right would be normal); the location of the maximum mask response (row 12) shows this behavior. Depending on the grouping of pixel, the gradient values on the triangle levels may increase considerably (rows 9, 11, 13 and 14). This is an indication for

Table 1. Statistical data on image intensity and mask response — - 0 + — in a homogeneous segment for the residues relative to the coarse linear model on all computed triangle levels 1 to TriaLevMax

ipelmax = [177 88 44 22 11]; number of pixels on triangle level

TriaYChar

row	1	2	3	4	5	triangle level (TriaLev)
1	-96.5	-96.1	-94.6	-83.3	-64.5	minimum intensity Imin
2	143.0	72	36.0	18.0	9.0	minimum intensity location/pel
3	97.2	92.5	91.4	90.4	87.9	maximum intensity Imax
4	93.0	47	23.0	12.0	6.0	maximum intensity location/pel
5	3.56	3.55	3.55	3.55	3.55	mean intensity
6	41.4	41.5	41.3	41.0	40.3	mean —intensity—
7	194.0	188.0	186.0	173.0	152.0	intensity range Imax - Imin
8	0.214	0.220	0.222	0.236	0.265	mean —intensity—/(Imax-Imin)
9	-27.6	-37.7	-51.3	-88.4	-115.3	minimum mask response
10	11.0	6.0	3.0	2.0	1.0	minimum mask response location/pel
11	18.4	27.8	34.1	44.0	72.9	maximum mask response
12	69.0	87.0	43.0	8.0	4.0	maximum mast response location/pel
13	-0.14	-0.45	-1.25	-2.21	-5.24	mean mask response
14	6.60	11.4	19.6	33.9	50.6	mean —mask response—
15	-0.021	-0.040	-0.064	-0.065	-0.103	ratio —mask response—
16	37.0	13.0	5.0	3.0	3.0	number of zerocrossings

regions with rather smooth local brightness changes on a high-spatial frequency scale, but large intensity changes on a low-spatial-frequency scale. This spatial-frequency-dependent texture is reflected in the values of row 15.

The last line of the table shows the decrease of the number of zero crossings of the mask response with triangle level. Zero crossings in the gradient function correspond to extremum values in intensity. On the lower levels, many of these occur due to noise corruption and minor local extremum values not of interest here; a drastic reduction in zero crossings can usually be observed when moving up the triangle level. For capturing essential information on the finer but not too local structure of the intensity signal, a linear representation between extremum intensity values on a higher triangle level might be sufficient. This, however, corresponds to a line segment interpolation in the region between two zero crossings of the gradient mask response. This is done in a similar way as for coarse segmentation.

6   Derivation of Medium-Spatial-Frequency Intensity Structure from Suitable Triangle Level

Looking for at most half a dozen to a dozen piecewise straight-line approximations in each homogeneous segment (for efficiency reasons), the following

Fig. 3. Road scene analyzed with stripes selected; counting starts from top to bottom (hor.) and left to right (vert.)

approach, based on the discussion above, has been selected. From the last row shown in table 1, select that level 'TriaLevFine' for further fine approximation of the residues with linear sub-segments, for which the number of zero crossings is below a limit value for the first time. In the example given, a value of 10 has been selected as the limit, and TriaLevFine = 3 results. The ensuing interpolation is done as follows: For a sub-segment to be accepted as a homogeneous one, a minimal number of pixels 'SegMinTriaFine' to the next zero crossing on the level TriaLevFine is requested. This is equivalent to the number of pixels with the same sign of the slope dI/dpel. Because of the magnification factor relative to the original pixel base, a value around 3 has shown to be a good compromise. The straight-line approximation again is obtained from the mean value of the intensity and the mean value of the mask response, both localized at the center of the sub-segment (all of this is done in original pixel values). If the condition SegMinTriaFine is not met (the so-called non-homogeneous sub-case), two cases may occur: A single such element is just ignored and no interpolation is done. For several such cases occurring in a row (corresponding to a textured area on this scale), the same interpolation is done as for the homogeneous case; however, it may be specially marked. This approach has tried to avoid multiplications and divisions for fast computation in real time; the results are surprisingly good as may be seen from the next section.

7   Experimental Results

The algorithm (developed with Matlab) has been tested on a large number of road scenes like the one given in Fig. 3. The grid shown by white lines are the stripes for which the intensity profiles have been determined. The results for the horizontal lines 8 to 10 are shown in Fig. 4. The horizontal scale units are image pixels (1 to 768). Vertically, the image intensity is shown, basically, however with a bias for showing the results of multiple stripes in a single image. The thin lines are the result of coarse segmentation; it can be seen that for stripe row 10 the shadow underneath and to the left side of the truck as well as the one underneath the left car are clearly separated from the rest of the scene mapped. However,

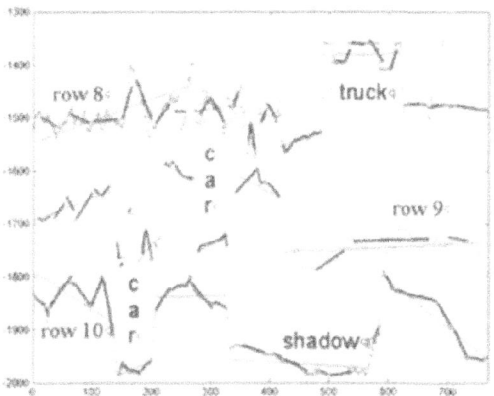

Fig. 4. Intensity profiles from three horizontal cuts (stripes 8 to 10 in Fig. 3, spread vertically)

local shading effects are represented only poorly. The fine segmentation, however, indicated by solid lines, yields sufficiently good information on local shading. In the central part of Fig. 4 (horizontal stripe row 9) a cut through the lower end of the truck body in front right (a typical non-homogeneous region from fine-segmentation) is shown to the right; average intensity values are improved as compared to the coarse segmentation result. The initial neighboring shading effect to the left side on the truck body is represented well; then, however, the part with constant lower shade bound has local intensity dynamics prohibiting a line fit. [In EMS-vision, this situation would request attention by one of the tele-cameras with improved resolution (factors 3 and 10).] The two cars to the left with dark paint are clearly indicated as discontinuities from coarse segmentation. Since this cut happens to go through the rear light group and the license plate, the vehicle appears quite non-homogeneous on this level. Note, however, that in our overall approach, each moving objects will be tracked by a separate tracker (with its own representation in the scene tree and its own set of feature extractors centering on the object by crosswise adjustment of search stripes from the 4-D model).

The upper stripe (row 8) clearly shows the white body of the truck on the left. The second truck in front and the two cars to the left, however, are cut through the lower edge, respectively the upper part of the body where intensity changes happen to be strong and inconclusive. This confirms previous experience that boundary regions of vehicles have to be interpreted with special care; one should concentrate on central regions and entire intensity profiles for robust recognition. [Here, limited space does not allow discussion of the results derived from the vertical stripes.]

8 Conclusion

Vision is considered to be an animation process in the 'mental' world of internal representations built from these knowledge elements on the one side and feature collections derived from image sequence processing on the other side. Area based features allow generating more secure object hypotheses and more robust dynamic scene understanding. In addition to the frame-oriented stripes shown for recognition of the general background, sets of local, cross-wise oriented short stripes are scheduled for tracking each object in smooth pursuit over time by prediction error feedback. The algorithm developed and its exploitation for dynamic scene understanding allow a next step towards full area-based image sequence analysis in real-time (several 2-D signals in parallel at video rate) with computing power actually available at moderate costs. Linear shading elements on a coarse and a fine scale are derivable with moderate computing effort from ternary mask responses hitherto used for edge localization only. These additional features are used for more robust object hypothesis generation and testing in the 4-D approach. Only a joint interpretation of intelligently selected horizontal and vertical sets of stripes with edge positions will yield sufficiently robust feature sets for starting and running 4-D estimation processes

References

1. P. J. Burt, T. H. Hong, A. Rosenfeld: Segmentation and estimation of image region properties through cooperative hierarchical computation. *IEEE Trans. Syst. Man and Cybern.* **12** (1981) 802–805.
2. Dickmanns, Dirk: Rahmensystem für visuelle Wahrnehmung veränderlicher Szenen durch Computer. Dissertation, UniBwM, Informatik,(1997).
3. E. D. Dickmanns, V. Graefe: a) Dynamic monocular machine vision. b) Application of dynamic monocular machine vision. *J. Machine Vision Application*, Springer-Int, (1988) 223–261.
4. E. D. Dickmanns and H.-J. Wünsche: Dynamic vision for Perception and Control of Motion. *In Jaehne(ed.): Handbook of Computer Vision and Applications, Vol.3: Systems and Applications* Academic Press (1999) 569–620.
5. R. Gregor, M. Lützeler, M. Pellkofer, K. H. Siedersberger, E. D. Dickmanns: EMS-Vision: A Perceptual System for Autonomous Vehicles. *Proc. Int. Symposium on Intelligent Vehicles (IV'2000), Dearborn, (MI), October 4–5, 2000.*
6. K.-D. Kuhnert: Zur Echtzeit-Bildfolgenanalyse mit Vorwissen. Dissertation UniBw Munich, LRT, (1988).
7. B. Mysliwetz: Parallelrechner-basierte Bildfolgen-Interpretation zur autonomen Fahrzeugsteuerung. Dissertation UniBw Munich, LRT, (10.8.1990).
8. F. R. Schell: Bordautonomer automatischer Landeanflug aufgrund bildhafter und inertialer Messdatenauswertung. Dissertation UniBw Munich, LRT, (23.3.1992).
9. S. Werner: Maschinelle Wahrnehmung für den bordautonomen automatischen Hubschrauberflug. Dissertation, UniBw Munich, LRT, (17.7.1997).
10. H.-J. Wuensche: Erfassung und Steuerung von Bewegungen durch Rechnersehen. UniBw Munich LRT (1987).

Subpixel Flow Detection by the Hough Transform

Atsushi Imiya and Keisuke Iwawaki

Computer Science Division,
Dept. of Information and Image Sciences, Chiba University
1-33 Yayoi-cho, Inage-ku, 263-8522, Chiba, Japan
`imiya@ics.tj.chiba-u.ac.jp`

Abstract. In this paper, we show that randomized sampling and voting processes allow to treat linear flow field detection as a model-fitting problem. If we use an appropriate number of images from a sequence of images, it is possible to detect subpixel motion in this sequence. We use the accumulator space for the unification of these flow vectors which are computed from different time intervals.

1 Introduction

In this paper, we deal with a random sampling and voting process for linear flow detection. In a series of papers [1,2], the author introduced the random sampling and voting method for problems in machine vision. The method is an extension of the randomized Hough transform which was first introduced by a Finnish school for planar image analysis [3]. Later they applied the method to planar motion analysis [4] and shape reconstruction from detected flow fields [5]. These results indicate that the inference of parameters by voting solves the least-squares problem in machine vision without assuming the predetermination of point correspondences between image frames. We show that the randomized sampling and voting process detects linear flow fields. We introduce a new idea to solve the least-square model-fitting problem using a mathematical property for the construction of a pseudoinverse of a matrix. If we use an appropriate number of images from a sequence of images, it is possible to detect subpixel motion in this sequence. In this paper, we use the accumulator space for the unification of flow vectors detected from many time intervals.

2 Random Sampling and Voting Process

For a system of equations

$$\xi_\alpha^\top a = 0, \ \alpha = 1, 2, \cdots m, \tag{1}$$

setting

$$\Xi = (\xi_1, \xi_2, \cdots, \xi_m)^\top, \tag{2}$$

R. Klette, S. Peleg, G. Sommer (Eds.): Robot Vision 2001, LNCS 1998, pp. 140–147, 2001.

the rank of matrix $\boldsymbol{\Xi}$ is n if vector $\boldsymbol{x}_\alpha$ is an element of $\mathbf{R}^n$. Therefore, all $n \times n$ square submatrces $\boldsymbol{N}$ of $\boldsymbol{\Xi}$ are nonsingular. Setting N_{ij} to be the ij-th adjacent of matrix $\boldsymbol{N}$, we have the equality

$$f_{\alpha 1} N_{11} + f_{\alpha 2} N_{21} + \cdots + f_{\alpha n} N_{n1} = 0, \tag{3}$$

if the first column of $\boldsymbol{N}$ is $\boldsymbol{\xi} = (\boldsymbol{x}_\alpha^\top, 1)^\top$. Therefore, the solution of this system of equations is

$$\boldsymbol{a} = (n_{11}, n_{21}, \cdots, n_{n1})^\top, n_{i1} = \frac{N_{i1}}{\sqrt{\sum_{j=1}^n N_{j1}^2}}, \tag{4}$$

that is, the solutions distribute on the positive semisphere. If the dimension of the parameter of the model is 3, we have the solution

$$\boldsymbol{a}_{\alpha\beta} = \frac{\boldsymbol{\xi}_\alpha \times \boldsymbol{\xi}_\beta}{|\boldsymbol{\xi}_\alpha \times \boldsymbol{\xi}_\beta|}, \tag{5}$$

for a pair of randomly selected vectors.

A generalization of this property is based on the following proposition.

Proposition 1 *[7] Assuming that matrices $\boldsymbol{P}_k$ and $\boldsymbol{O}$ are a $k \times k$ permutation matrix and the $(m - n) \times n$ null matrix, respectively, for an $m \times n$ matrix $\boldsymbol{A}$ such that $m > n$ and $\mathrm{rank}\boldsymbol{A} = n$. Let vector $\boldsymbol{c}$ such that*

$$\boldsymbol{b} = \left(\boldsymbol{A}_n^{-1} \ \boldsymbol{O} \right) \boldsymbol{c} \tag{6}$$

for

$$\boldsymbol{A}_n = (\boldsymbol{P}_n \ \boldsymbol{O}) \, \boldsymbol{P}_m \boldsymbol{A} . \tag{7}$$

Then it holds that vector $\boldsymbol{c}$ minimizes the criterion $|\boldsymbol{c} - \boldsymbol{A}\boldsymbol{x}|^2$.

There are ambiguities for the selection of $\boldsymbol{P}_n$ and $\boldsymbol{P}_m$. The proposition implies that if an $m \times n$ system matrix is column full-rank,

1. selecting n equations from the system of equations, and
2. solving this nonsingular equation,

we obtain a solution of the least-squares optimization. If we randomly select column vectors, this method also derives an extension of the randomized Hough transform.

3   Flow Field Detection

Setting $f(x, y, t)$ to be a time dependent gray-scale image, the linear optical flow $\boldsymbol{u} = (u, v, 1)^\top$ is the solution of the linear equation

$$\boldsymbol{f}^\top \boldsymbol{u} = 0, \quad \frac{df(x, y, t)}{dt} = \boldsymbol{f}^\top \boldsymbol{u}, \tag{8}$$

where

$$\boldsymbol{f} = \left(\frac{\partial f(x,y,t)}{\partial x}, \frac{\partial f(x,y,t)}{\partial y}, \frac{\partial f(x,y,t)}{\partial t} \right)^{\top}. \tag{9}$$

Assuming that the flow vector $\boldsymbol{u}$ is constant in an area S, the flow vector in an area is the solution of a system of equations

$$\boldsymbol{f}_{\alpha}^{\top} \boldsymbol{u} = 0, \ \alpha = 1, 2, \cdots, N, \tag{10}$$

where

$$\boldsymbol{f}_{\alpha} = \left(\frac{\partial f(x,y,t)}{\partial x}, \frac{\partial f(x,y,t)}{\partial y}, \frac{\partial f(x,y,t)}{\partial t} \right)^{\top} \Bigg|_{x=x_{\alpha}, y=y_{\alpha}} \tag{11}$$

for a sample point $(x_{\alpha}, y_{\alpha})^{\top}$ in an area S, see [8].

For a system of equations defined in eq. (10) in a windowed area, we have

$$\boldsymbol{a} = \frac{\boldsymbol{f}_{\alpha} \times \boldsymbol{f}_{\beta}}{|\boldsymbol{f}_{\alpha} \times \boldsymbol{f}_{\beta}|}, \ \boldsymbol{a} = (A, B, C)^{\top}, \tag{12}$$

if we assume that $|\boldsymbol{a}| = 1$. Furthermore, since $\boldsymbol{u} = (u, v, 1)^{\top}$, that is, assuming that $\frac{\partial f(x,y,t)}{\partial t} = 0$, setting

$$\boldsymbol{u} = \lambda(\boldsymbol{f}_{\alpha} \times \boldsymbol{f}_{\beta}), \ \boldsymbol{\alpha} = (\alpha, \beta, \gamma)^{\top} \tag{13}$$

for a nonzero real constant λ, we have $\boldsymbol{u} = (\frac{\alpha}{\gamma}, \frac{\beta}{\gamma}, 1)^{\top}$.

4   Computation of Subpixel Motion

For a sequence of images

$$S_m = \langle f(x, y, -m), f(x, y, -m+1), f(x, y, -m+2), \cdots, f(x, y, 0) \rangle, \tag{14}$$

setting $\boldsymbol{f}_{(k)}$, which is computed from $f(x, y, -k)$ and $f(x, y, 0)$, to be the spatiotemporal gradient between k-frames, we define the k-th flow vector $\boldsymbol{u}_{(k)}$ as the solution of a system of equations

$$\boldsymbol{f}_{(k)\alpha}^{\top} \boldsymbol{u}_{(k)} = 0, \ \alpha = 1, 2, \cdots, m \tag{15}$$

for each windowed area. From a sequence of images S_m, we can obtain flow vectors $\boldsymbol{u}_{(1)}, \boldsymbol{u}_{(2)}, \cdots, \boldsymbol{u}_{(m)}$, For this example, if we assume the size of a window is $a \times a$, we have $_{(a \times a)}C_2 \times m$ constraints among m frames.

Setting $s = kt$, we have the equation,

$$f_x \frac{dx}{ds} + f_y \frac{dy}{ds} + f_s \frac{ds}{dt} = 0. \tag{16}$$

Since $\frac{ds}{dt} = k$, this constraint between the flow vector and the spatiotemporal gradient of an image derives the expression $\boldsymbol{u}_{(k)} = (\frac{dx}{ds}, \frac{dy}{ds}, k)^{\top}$ for the flow vector detected from a pair of images $f(x, y, (-k + 1))$ and $f(x, y, 0)$. If the speed of an object in a sequence is $1/k$-pixel/frame, the object moves 1 pixel in sequence $S_{(k-1)}$. Therefore, in the spatiotemporal domain, we can estimate the average motion of this point between a pair of frames during the unit time as

$$\overline{u_k} = (\frac{1}{k}u_k, \frac{1}{k}v_k, 1)^\top. \tag{17}$$

form vector $u_{(k)} = (u_k, v_k, k)^\top$. We vote 1 to point $\overline{u_k}$ on the accumulator space for the detection of subpixel flow vectors from a long sequence of images. Therefore, we can estimate the motion of this object from $\{u_{(k)}\}_{k=1}^{m}$ which is computed from $f(x, y, 1)$ and $f(x, y, m)$. For the unification of vector field $\overline{u_{(k)}}$, we use the accumulator space.

In the accumulator space, we vote $w(k)$ for $u_{(k)}$ for a monotonically decreasing function $w(k)$, such that $w(1) = m$ and $w(m) = 1$. In this paper, we adopt $w(k) = \{(m + 1) - k\}$. This weight of voting means that we define large weight and small weight for short-time motions and long-time motions, respectively.

5   Numerical Examples

We have evaluated the performance of the random sampling and voting process for solving LMS in the flow-vector detection.

5.1   Evaluation for Synthetic Data

For this evaluation, we detected the flow vectors in synthetic images, the "**Translating tree**" and "**Divergent tree**." We used frames 7, 8, 9, 10, and 11. In these examples, the total number of combinations of linear constraints is 600. In these examples, we detected flow vectors whose votes are more than $50\% \times 600$, and we have evaluated the performance of the random sampling and voting process for presmoothed images, because the gray levels of these images are not so smooth for the detection of linear flow fields which are given as a collection of linear constraints in windowed areas. In Tab. 1, we have listed the statistical data for the performance analysis of the traditional method and our method.

In this evaluation, we have computed the angle ϕ

$$\phi(x, y) = \cos^{-1} \left(\frac{v_E^\top v_R}{|v_E| \cdot |v_R|} \right) \tag{18}$$

for each pixel where $v_E = (u_E, v_E, 1)^\top$ and $v_R = (u_R, v_R, 1)^\top$ are the expected flow vector and the estimated flow vector of each pixel (x, y), respectively, for a synthetic image sequence. Furthermore, we have used the averages and densities of $\phi(x, y)$ over the domain. This table shows that the performance of our method for obtaining smooth imagesequences with stationary motion is good compared to the traditional method.

Table 1. Statistics of Motion Detection

Methods	Avarage Errors [deg]	Density[%]
Proposed Method 1[1]	1.75	25.8
Proposed Method 2[2]	4.04	54.4
Horn and Schunck [3] $\lambda = 0.5$	11.16	100.0
Lucas and Kanade [4] $\lambda \geq 0.5$	1.09	51.5
Nagel	41.76	100.0
Anadan	4.54	100.0

1. threshold=99%×600, with presmoothing.
2. threshold=90%×600, with presmoothing.
3. with presmoothing.
4. with presmoothing.
The definition of λ is based on Barron *et.al* [8]

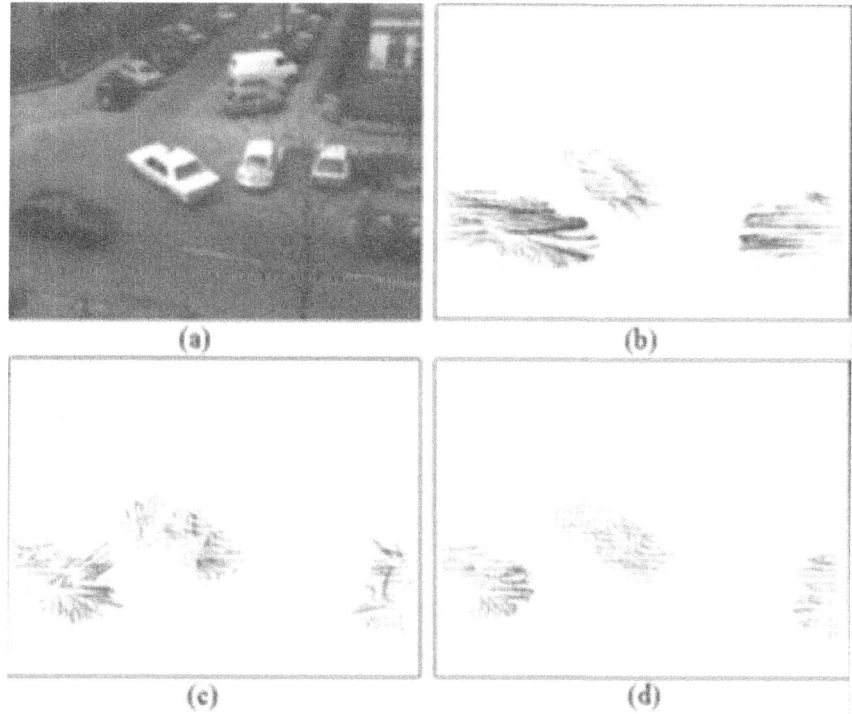

Fig. 1. Detection of Subpixel Motion

5.2   Detection of Subpixel Motion

The next evaluation is for the performance analysis of the multiframe effect for the detection of slow motion. In this examples, we set the size of the windowed area to be 5×5 for Fig. 1(b) and Fig. 1(c), and 7×7 for Fig. 1(d). Furthermore, we have utilized frames 1, 2, 3, 4, 5, and 6. Without presmoothing, we detected the median from the voting of all combinations for the selection of linear constraints.

Figures 1(a), 1(b), 1(c), and 1(d) show the original image for frame 1, the flow field detected using frames 1 and 2, the flow field detected using frames 1 and 5, and the flow field detected using frames 1, 2, 3, 4, 5 and 6, and using weighted voting in the accumulator space. In Fig. 1(b), we could not detect the motion of a pedestrian because the speed of the pedestrian was 0.5 frame/pixel. In Fig. 1(c), we can detect the motion of this pedestrian because, during 5 frames, the pedestrian moves 1.2 pixels. However, the flow vectors for three cars contain large amount of error because during 5 frames these cars move 9 pixels. Therefore, the lengths of flow vectors on each car are longer than the length of the edge of windows. In Fig. 1(d), we could see good performance for the detection of all moving objects in the scan. Furthermore, there is no error in the background of the scan, although there exists errors on the walls of houses in the results of "**Lucas and Kanade**", as shown in Fig. 2.

In Fig. 2, (a) shows the flow field detected by "Lucus and Kanade" method, and (b) the field detected by our method.

We compared the results for same images using **Lucas and Kanade** with preprocessing. The preprocessing is summarized as follows [8].

- Smoothing using an isotropic spatiotemporal Gaussian filter with a standard deviation of 1.5 pixels-frames.
- Derive the 4-point central difference with mask coefficients $\frac{1}{12}(-1, 8, 0, 8, 1)$.
- The spatial neighborhood is 5 × 5 pixels.
- The window function is separable in vertical and horizontal directions, and isotropic. The mask coefficients are $(0.00625, 0.25, 0.375, 0.25, 0.00625)$.
- The temporal support is 15 frames.

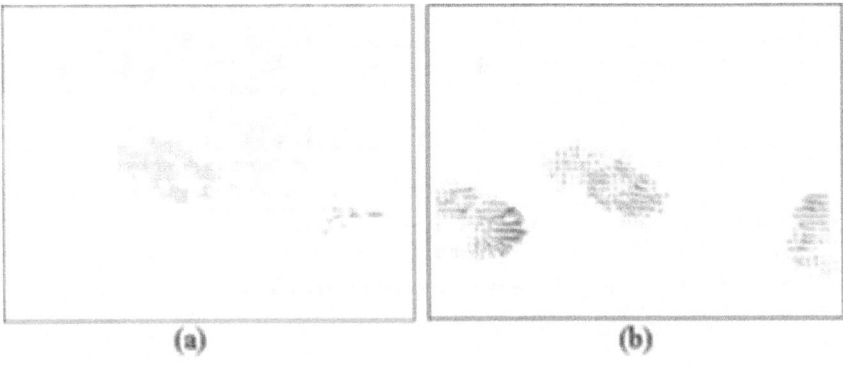

(a)   (b)

Fig. 2. Comparision of (a) **Lucus and Kanade**-method and (b) Voting Method

However, our method does not require any preprocessing. Therefore, we detect the flow field from at least two images.

For "**Hamburg Taxi**," our method detects all motions of a taxi which is turning in the center of this scan and two cars which are crossing the scan in opposite directions. Furthermore, the method detects the subpixel motion of a pedestrian using five frames without presmoothing. However, we could not detect a walking pedestrian using **Lucas and Kanade** method even if we used the 15-frame support.

From these numerical results, the performance of our new method without preprocessing is of the same level as **Lucas and Kanade**, which is a very stable method. For the detection of flow vectors, we selected 50% combinations of equations from all possible combinations of a pair of linear equations in the windowed area. The weight for voting is considered as filtering. Therefore, our method involves postprocessing for the detection of motion in a long sequence of images.

For the detection of flow vectors at time $t = 0$, traditional methods require the past observations

$$P_m = \{f(x, y, -m), f(x, y, -m + 1), f(x, y, -m + 2), \cdots, f(x, y, 1)\}, \qquad (19)$$

the present observation $N = \{f(x, y, 0)\}$, and the future observations,

$$F_m = \{f(x, y, 1), f(x, y, 2), \cdots, f(x, y, m)\}, \qquad (20)$$

if methods involves spatiotemporal smoothing. Therefore, the traditional methods involve a process which causes timedelay with respect to the length of the support of a smoothing filter with respect to the time axes.

Our method detects flow vectors of time $t = 0$ using m images $f(x, y, -m + 1)$, $f(x, y, -m + 2)$, $\cdots$, $f(x, y, 0)$, which are obtained for $t \leq 0$, that is, the we are only required data from past. As we will show our method does not require any spatiotemporal preprocessing for this sequence. Our method permits the computetation of flow vectors from past and present data, although the traditional methods with spatiotemporal presmoothing require future data for the computation flow vector. In this sense, our method satisfies the causality of events. Therefore, our method computes flow vectors at time $t = 0$, just after observing image $f(x, y, 0)$. This is one of the advantages of our method. Furthermore, in traditional method, oversmoothing delete slow motions in a sequence of images. Our method preserves slow motions in a sequence of images since the method does not require presmoothing. This is the second advantage of our method.

6   Conclusions

In this paper, we showed that the random sampling and voting process detects a linear flow field. We introduced a new method of solving the least-squares model-fitting problem using a mathematical property for the construction of a pseudoinverse of a matrix. The greatest advantage of the proposed method is simplicity because we can use the same engine for solving multiconstraint problem with the Hough transform for the planar line detection. Our method for the detection of flow vectors is simple because it requires two accumulator spaces for a window, one of which is performed by a dynamic tree, and usually it does not require any preprocessing. Furthermore, the second accumulator space is used for the unification of the flow fields detected from different frame intervals. These properties are advantagous for the fast and accurate computation of the flow field.

References

1. A. Imiya, I. Fermin: Voting method for planarity and motion detection. *Image and Vision Computing*, **17** (1999) 867–879. 140
2. A. Imiya, I. Fermin: Motion analysis by random sampling and voting process. *Computer Vision and Image Understanding*, **73** (1999) 309–328. 140
3. E. Oja, L. Xu, P. Kultanen: Curve detection by an extended self-organization map and related RHT method. Proceed. Internat. on *Neural Network Conf.*, **1** (1990) 27–30. 140
4. H. Kälviäinen, E. Oja, L. Xu: Randomized Hough transform applied to translation and rotation motion analysis. 11th IAPR Proceed. of Internat. Conf. on *Pattern Recognition* (1992) 672–675. 140
5. J. Heikkonen: Recovering 3-D motion parameters from optical flow field using randomized Hough transform. *Pattern Recognition Letters*, **15** (1995) 971–978. 140
6. K. Kanatani: Statistical optimization and geometric inference in computer vision, *Philosopical Transactions of the Royal Society of London, Series A*, **356** (1997) 1308–1320.
7. C. R. Rao, S. K. Mitra: *Generalized Inverse of Matrices and its Applications*. John Wiley & Sons, New York, (1971). (Japanese Edition: Tokyo Tosho, Tokyo (1973)). 141
8. J. L. Barron, D. J. Fleet, S. S. Beauchemin: Performance of optical flow techniques. Report No. 299, Department of Computer Science, The University of Western Ontario, London (1992). 142, 145

Tracking of Moving Heads in Cluttered Scenes from Stereo Vision

Ruijiang Luo and Yan Guo

RWCP (Real World Computing Partnership)
Multi-Modal Functions KRDL (Kent Ridge Digital Labs) Lab
21, Heng Mui Keng Terrace, Singapore 119613, Republic of Singapore
{rjluo,yguo}@krdl.org.sg

Abstract. Tracking a number of persons moving in cluttered scenes is an important issue in computer vision. It is the first step of automatic video-based surveillance systems. In this paper we present a binocular vision system using stereo information for moving head detection and tracking. After background subtraction, the remained foreground disparity image is used as a mask to delete background clutter as well as to reduce the search space, which greatly improve the tracking performance when occlusion happens. With a local sampling method together with the stereo information obtained, we are now able to reliably detect and track people in cluttered natural environments at about 5 Hz on standard PC hardware.

1 Introduction

Tracking multiple persons moving in cluttered scenes is a key problem in video surveillance. It is also a challenging research topic in computer vision. The difficulty includes high-dimensional parameter space and complicated situations such as occlusion.

Over the years numerous algorithms for person tracking have been proposed, such as methods that use color information [1] and sets of point features [2,3], and systems that use deformable template [4] and curved outlines [5], but all the methods have difficulties in tracking multiple moving objects. That is because they use only one attribute for global tracking. As the estimation progresses over many frames, the posterior state density estimation may increasingly bias towards objects with dominant likelihood. One solution is to keep multiple hypotheses about the object locations in image [6,7]. This allows a wider range of motion to be supported by keeping enough information to recover from locally bad situations. An obvious drawback of keeping multiple hypotheses is that as the problem size increases, it rapidly becomes impractical to perform such an exhaustive search for the object in real time. Another problem is it is unable to adequately detect occlusions.

With the fast increasing processing speed of microcomputers, real-time stereoscopic analysis becomes a reality. This paper presents an approach to near

R. Klette, S. Peleg, G. Sommer (Eds.): Robot Vision 2001, LNCS 1998, pp. 148–156, 2001.

real-time person tracking in crowded environments with stereo information obtained from a binocular vision system. Doing background subtraction with range information makes the detection and tracking more robust to shadows and lighting changes [8]. Besides, occluding surfaces can also be found and dealt with. This is one of the most important properties of stereo.

In the next sections of this paper we will introduce our stereo based tracking system. Section 2 is an overview of our system; Section 3 and Section 4 introduce how the stereo information is used in person detection and tracking; In Section 5 we present experimental results of the system, and Section 6 is the conclusion.

2   System Overview

The system architecture is shown in Fig. 1. The stereo system has two different cameras, one color, one monochrome. The color camera is used for color based person detection and recognition, which is not in the scope of this paper. For the stereo based detection and tracking, we do not use the color information right now. We first change the color image into 8 bit gray one, then the 2 gray images are fed into the stereo unit, which performs efficient area correlation algorithm to extract disparity information [9]. By a simple background subtraction, the moving objects, whose disparities are different from the learned background, will be isolated into different layers. Therefore, the influence of the cluttered background is eliminated.

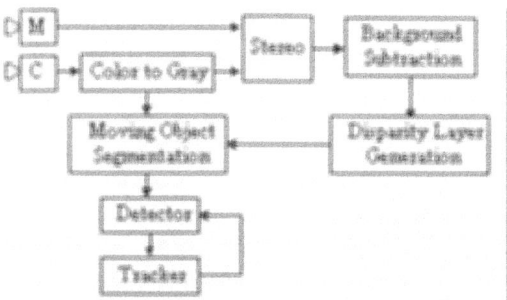

Fig. 1. Flowchart of the system

In detection, we use standard head-shoulder contours as our initial pattern for each person. First we adjust each contour to several possible sizes according to the disparity layer the person is in, then match them with edges in the masked intensity image. By some adjustment of contours' shape according to the moving head they match, we create initial contour model set of each person in the scene. All the obtained contour sets are sent to the tracking module, and will be updated in each step of tracking operation. We use a local sampling method to track moving heads with these contour models, which greatly improves the system performance. If a person is no longer recognized in the scene, the system will delete him by eliminating his contour models from model sets. If there are any unmatched moving objects in the scene, we treat them as new objects and will call detector for them. The result will be added to the contour sets if they are new heads. In next two sections we give detailed descriptions of the algorithms.

3   Person Detection by Head Contour and Disparity Info

The purpose of person detection is to get initial head-shoulder contours of each person, which can be used in subsequent steps for fast and robust tracking. To deal with background noise and occluded heads, we incorporate range information from stereo vision since range information is less sensitive to lighting conditions or extraneous motion.

3.1   Background Subtraction On Disparity Map

A pair of stereo images is taken from the binocular stereo vision system. A disparity image is obtained from each stereo pair, using the area correlation method. For an ideal pinhole camera with the optical axis being set to parallel to the ground, the disparity is inversely related to the depth according to the formula:

$$d = bf/z \, , \tag{1}$$

where d is the disparity, z is the normal distance from the image plane to the object, b is the baseline of the binocular stereo system, f is camera's focal length. The larger d is, the closer the object is to the camera plane. Observing that there are more vertical feature lines than horizontal ones in an indoor environment, we have set our cameras aligned vertically to get better distance resolution. This setup will also help to accommodate people of different heights, which may provide another useful feature for human recognition in the future.

We first apply Laplacian of Gaussian (LoG) filter on the input images. This filtering enhances the image features as well as removing the effect of intensity variations on images due to ambient light, difference of camera gains, etc. Sample images before and after LoG filter are shown in Fig. 2.a and Fig. 2.b. Next, we implement a correlation-based stereo algorithm following the approach taken by Fua [9]. The algorithm computes similarity scores for every pixel in the image by taking a fixed window in one image and shifting it along the epipolar line in the other image. The scores are determined using the normalized mean-squared difference of gray levels:

$$s = max(0, 1 - c) \, , \quad \text{with} \tag{2}$$

$$c = \frac{\Sigma((I_u(x+i, y+j) - \bar{I}_u) - (I_i(x+dx+i, y+dy+j) - \bar{I}_i))^2}{\sqrt{\Sigma(I_u(x+i, y+j) - \bar{I}_u)^2 - \Sigma(I_i(x+dx+i, y+dy+j) - \bar{I}_i)^2}} \, .$$

Fig. 2. Example images: (a) input image, (b) after LoG filter, (c) disparity image

I_u, I_i are intensities of the upper and lower image respectively. $\bar{I}_u, \bar{I}_i$ are their average values over the correlation window. and represent the disparity of x and y along the epipolar line, the pair that maximizes s is the best answer. In our system, the size of captured image is $384(W) \times 288(H)$ pixels. After calibration, the obtained epipolar line is almost vertical. Along the epipolar line from top to bottom, the maximum dx is 5 pixels. That is, for a disparity searching range of 60 pixels, the difference of dx is only 1 pixel. Therefore, we set $dx = 0$ to make calculation simpler without much error. A sample result is in Fig. 2.c. After background subtraction, the obtained foreground disparity image is separated into different layers. Each layer is a connected region contains only pixels of similar disparities. For example, in Fig. 2.c there are two layers, one is brighter, the other is darker.

After separating foreground objects into layers, we come to the next step: find heads of people.

3.2   Head Detection

People detection is a computation intensive operation, involving searching over the image and matching against the model. The problem is quite complicated since a person may appear in different sizes (scale change), and facing different directions.

To deal with different head orientations, we use a template of head-shoulder contour, which is insensitive to head rotation. The initial model C_0 is obtained manually beforehand in the following way: First, take pictures of different persons facing the camera. Next, draw points along their head and shoulder edges, connect the points with B-Spline curves to get initial contours. Put these contours together we get C_0. To handle the scale change, the detection is carried on in each layer, which can be divided into following steps (presume there are L layers. $i = n = 0$):

1. Select the layer that is closest to cameras, labeled as Layer(d) (d is the disparity)
2. Use Layer(d) as a mask to get the corresponding region in intensity image as I_d, do edge detection in I_d to get the edge image E_d.

Fig. 3. Samples of contour detection

3. Change contours in to suitable sizes with affine shape transformation by the limit of disparity d to get candidate contours $C_{ij}, j = 1, ..., N$.
4. For $j = 1, .., N$, match candidate contours C_{ij} to the upper part of image E_d. If the fitness of any C_{ij} is higher than a threshold, we consider there is a person in this layer, therefore create a candidate contour set $C_{n,k}, k = 1, ..., M$ for him with best M contours in $C_{ij}, j = 1, ..., N$.
5. Delete Layer(d) from current frame; $n = n + 1$; $i = i + 1$; if $i < L$, go to step 1.

After head detection, the obtained contour set $C_{n,k}$ of each person is sent to the tracking module, which is used as his head model for subsequent tracking operations. The detailed operation of step 3 is explained as follows: Consider a person with width w standing a distance z from the camera, he should project to a width w in the image plane, so by similar triangles we get:

$$\frac{z}{w} = \frac{f}{w^1} \Rightarrow w^1 = \frac{fw}{z}, \tag{3}$$

combine with equation (1) we can get

$$w^1 = d \cdot w/b \tag{4}$$

where w/b is a constant that can be measured in calibration step. When processing the layer with disparity d, we change the size of contours C_0 in by planar affine shape transformation to make their width around w^1. This transformation is summarized as:

$$C_{ij} = WX_j + C_0 \tag{5}$$

where

$$W = \begin{pmatrix} 1 & 0 & C_0^x & 0 & 0 & C_0^y \\ 0 & 1 & 0 & C_0^y & C_0^x & 0 \end{pmatrix} \tag{6}$$

is a 6 degree shape matrix and

$$X = \begin{pmatrix} a\,b\,c\,d\,e\,f \end{pmatrix}^T \tag{7}$$

is a shape-space vector. For details, please refer to [10]. In detection, we use 25 predefined shape-space vectors to change each contour in C_0 to get the initial

Fig. 4. Occluded head detection with contour: (a) without stereo information, (b) with the help of stereo information

sample contours. Figure 3 shows the robustness of our contour match algorithm. See the left person, no matter how much he turns his head around, we can always locate his head correctly.

Figure 4 gives a comparison between detection with and without stereo information when occlusion happens. In left image, the contours are attracted by the edges of the person in the behind, so the obtained contours are much larger than the actual size of the person's head in the front, and locate with deviation from their right position. In the right image, because we can get proper contour size from disparity, we locate the person in the front with suitable contours around his head. The improvement is quite obvious.

4   People Tracking

In the process of tracking, we consider that at any time instance t, the state of an object $x_t \in X$ is only influenced by its previous state x_{t-1}. Its observation z_t is independent of other previous object states or observations. This is summarized as

$$P(x_1, ..., x_n; z_1, ..., z_n) = P(x_1)P(z_1 \mid x_1)\Pi_{t-2}^n[P(x_t \mid x_{t-1})P(z_t \mid x_t)] \qquad (8)$$

The tracking problem can be termed as the computation of the a posteriori distribution $P(x_t \mid Z_t)$ for given observations $Z_t = \{z_1, z_2, \ldots, z_t\}$. To increase speed, we use a sampling algorithm to estimate the a posteriori densities. In our test, we use image coordinate system (x, y, d) to represent a 3D position with x axis parallel to the ground and $x - y$ plane vertical to the floor; d is disparity. The position in world coordinate system (X, Y, Z) can be obtained from projection transform.

Global sampling method such as the factored sampling algorithm used in [11] has a disadvantage that many samples are required to get accurate estimation, which is very time consuming. Therefore, we use a local sampling method instead. Presume that each person moves no more than a speed limit in front of the cameras, our algorithm is summarized as follows (after background subtraction, L persons in scene, $i = 0$):

1. For person i, estimate his possible head centroid position

$$(x_{min}, x_{max}), (y_{min}, y_{max})$$

   and (d_{min}, d_{max}) for current frame from its last position (x^1, y^1, d^1).
2. Threshold foreground disparity map between $[d_{min}, d_{max}]$ to generate a disparity layer, named Layer(d), which covers area within $[x_{min}, x_{max}]$ and $[y_{min}, y_{max}]$. If no layer is found, delete person $i, i = i + 1$. if $i < L$, go to step 1; else, stop.
3. Map Layer(d) as a mask back to the input image to get the masked intensity image I_d. Do edge detection in I_d to get the edge image E_d.
4. Create M samples $C_{ij}, j = 1, \ldots, M$ from C_i obtained in last step [11], change each sample contour C_{ij} to suitable size with affine shape transformation by the limit of disparity d to get new candidate contours $C_{ij,k}, k = 1, \ldots, N$.
5. Match all $N \times M$ candidate contours $C_{ij,k}$ to the upper part of image E_d. Keep best M contours in C_i. If the maximum fitness of newly obtained contours is less than a threshold, delete person i. Else, save person average location (x, y, d).
6. Set $i = i + 1$. If $i < L$, go to step 1; else, stop

The step 4 in tracking is similar to the step 3 in detection. The difference is, in detection, because the contours in C_0 are different from persons in scene, we need to do more changes to find the best contours for each person. So we use a larger N. In our test, each model contour is changed for 25 shapes. But in tracking, because we take samples from contours obtained in last circle of the same person, we only need to do small changes of the shape, so a smaller N will be fine. In our test, we use $N = 3$, which greatly improves the tracking speed.

After all candidate persons are distinguished, if there are still some disparity areas left, then we consider that there are new objects enter the scene. The system will call person detection module automatically for these areas. Any newly matched persons will be added to the candidate set automatically.

5   Experimental Results

Our binocular stereo vision system is connected to a standard PC with Intel Pentium III processor at 700 MHz. For stereo, we use 2 CCD cameras aligned vertically, one is SONY XC-73 CCD Monochromic Video Camera, the other is SONY XC-003 Color Video Camera. The baseline of the system is 15cm, the maximum disparity search is 60 pixels in a 384×288 image.

To test the system, we place it inside an office room facing the door, as shown in figures above. We catch three video sequences for our test, one contains one person walking inside the scene, another contains two persons walking without occlusion, and the last one contains two persons move with occlusion. We judge the tracking performance as follows: If the obtained contour locates at the head position of the person with a suitable size, we take it as "good", otherwise it is "not good". By this criterion, we define two parameters: Tracking Rate

Table 1. Experiment result of our system. TR: tracking rate, FM: false match

Persons in scene	Has occlusion	With stereo		Without stereo	
		TR	FM	TR	FM
1	no	93%	0%	90%	9%
2	no	92%	2%	86%	7%
2	yes	82%	6%	55%	20%

(TR),which is the percentage of person number that the best contour of him is good to total person number appeared in all frames. False Match (FM) is the frame percentage that has wrong match of contour to a place without a person. The system performance can be seen in table 1. The result is obtained with a preset of $M = 10$. To make compare, we also test the tracking result without stereo information, which is also added into Tab. 1. We can find that the system performance is greatly improved when occlusion happens.

6   Conclusion

In this paper we introduced our near real-time stereo based human head tracking system. With the help of the range information obtained from stereo system, we can greatly decrease the influence of occlusion. By using a stereo based local sampling algorithm, we achieve a tracking speed of 5 frames per second with images of 384×288 in size. We can also handle the situation such as newly added person easily, which is very important for a robust surveillance system.

However our system still needs further improvement. For example, in Fig. 4, when occlusion happens, it is quite clear that their are two persons in the scene, and the one in the back is clear enough for us to recognize who he is. But for our system, we lost him because the remained part of his head is quite different from our head-shoulder contour model. We are currently working on combining skin color model together with our contour model for a better detection and tracking system.

References

1. C. Wren, A. Azarbayejani, T. Darrell, A. Pentland: Pfinder: real-time tracking of the human body. *In Proc. 2nd Int. Conf. on Automatic Face and Gesture Recognition* (1996) 51–56. 148
2. J. Costeira, T. Kanade: A multi-body factorization method for motion analysis. *In Proc. 5th Int. Conf. on Computer Vision* (1995) 1071–1076. 148
3. P. H. S. Torr: An assessment of information criteria for motion model selection. *In Proc. Conf. Computer Vision and Pattern Recognition* (1997) 148
4. G. Hager, K. Toyama: X vision: combining image warping and geometric constraints for fast visual tracking. *In Proc. 4th European Conf. Computer Vision,* **Vol. 12,** (1996) 507–517. 148

5. A. Blake, M. Isard, D. Reynard: Learning to track the visual motion of contours. *J. Artificial Intelligence* **78**, (1995), 101–134. 148

6. M. Isard. A. Blake: Contour tracking by stochastic propagation of conditional density. *In Proc. 4th European Conf. Computer Vision, Cambridge, England,* (April 1996) 343–356. 148

7. C. Rasmussen, G. D. Hager: Joint probabilistic techniques for tracking multi-part objects. *In Proc. IEEE Conf. on Computer Vision and Pattern Recognition, Santa Barbara, CA* (1998) 16–21. 148

8. C. Eveland, K. Konolige, R. C. Bolles: Background modeling for segmentation of video-rate stereo sequences. *In Proceedings IEEE Conf. on Computer Vision and Pattern Recognition* (June 1998) 266–271. 149

9. P. Fua: A parallel stereo algorithm that produces dense depth maps and preservers image features. *Machine Vision and Applications* **6** (1993), 35–49. 149, 150

10. A. Blake, M. Isard: "Active Contours". *Berlin, New York: Springer,* (1998), 74–79. 152

11. M. Isard, A. Blake: CONDENSATION - conditional density propagation for visual tracking. *Int. J. Computer Vision,* **29** (1998) 5–28. 153, 154

Servoing Mechanisms for Peg-In-Hole Assembly Operations

Josef Pauli, Arne Schmidt, and Gerald Sommer

Christian-Albrechts-Universität zu Kiel
Institut für Informatik und Praktische Mathematik
Preußerstraße 1–9, D-24105 Kiel, Germany
jpa@ks.informatik.uni-kiel.de

Abstract. Image-based effector servoing is a process of perception-action cycles for handling a robot effector under continual visual feedback. Apart from the primary goal of manipulating objects we apply servoing mechanisms also for determining camera features, e.g. the optical axes of cameras, and for actively changing the view, e.g. for inspecting the object shape. A peg-in-hole application is treated by a 6-DOF manipulator and a stereo camera head. The two robot components are mounted on separate platforms and can be steered independently. In the first phase (inspection phase), the robot hand carries an object into the field of view of one camera, then approaches the object along the optical axis to the camera, rotates the object for reaching an optimal view, and finally inspects the object shape in detail. In the second phase (insertion phase), the system localizes a board containing holes of different shapes, determines the relevant hole based on the extracted object shape, then approaches the object, and finally inserts it into the hole. At present, the robot system has the competence to handle cylindrical and cuboid pegs. For treating more complicated objects the system must be extended with more sophisticated strategies for the inspection and/or insertion phase.

1 Introduction

Image-based robot servoing (visual servoing) is the backbone of Robot Vision systems. The book edited by Hashimoto [3] collects various approaches of automatic control of mechanical systems using visual sensory feedback. A tutorial introduction to visual servo control of robotic manipulators has been published by Hutchinson et al. [5]. Quite recently, a special issue of the International Journal on Computer Vision has been devoted to image-based robot servoing [4].

Our work demonstrates the usefulness of servoing for treating all sub-tasks involved in an overall robotic application. The novelty is to consider servoing as a universal mechanism for *camera-robot calibration, active viewing, shape inspection,* and *object manipulation.* Furthermore, we consider minimalism principles by extracting just the necessary image information and *avoiding 3D reconstruction,* which leads to real-time usage. Related to the particular application of

R. Klette, S. Peleg, G. Sommer (Eds.): Robot Vision 2001, LNCS 1998, pp. 157–166, 2001.
© Springer-Verlag Berlin Heidelberg 2001

peg-in-hole assembly operations it is favorable to integrate video and force information [6]. Due to limited paper size, we concentrate on servoing mechanisms for the vision-related sub-tasks of the overall peg-in-hole application.

As a survey, we describe (in Section 2) the components of the robot system, present (3) the general measurement-based control mechanism, use (4) servoing for determining the optical camera axis, apply (5) the servoing mechanism for optimally viewing and inspecting the object (Fig. 1), and use (in Sections 6 and 7) servoing to suitably approach the object to the relevant hole (Fig. 2).

2   System Description

The computer system consists of a Sun Enterprise (E4000 with 4 UltraSparc processors) for image processing and of special processors for computing the inverse kinematics and motor signals. The robot system is composed of a robot manipulator including a hand with parallel jaw fingers and a robot head including two monochrome, stereo cameras. Based on six rotational joints of the manipulator the robot hand can be moved in arbitrary position and orientation within a certain working space. Additionally, there is a linear joint at the robot hand for opening and closing the two fingers. The tool center point is defined at the position of the hand tip, which is fixed in the middle point between the two finger tips. The robot head is mounted on a mobile platform and is equipped with motorized pan, tilt, and vergence degrees-of-freedom (DOF). Additionally, the stereo camera has motorized zooming and focusing facilities.

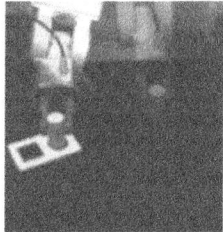

Fig. 1. Robot head, manipulator; approach an object towards a camera for shape inspection

Fig. 2. Vision-based approaching a cylindrical peg to a circular hole

3   Mechanism of Measurement-Based Control

The robot system is characterized by a *fixed state vector* S^c which is inherent constant in the system, and by a *variable state vector* $S^v(t)$ which can be changed through a *vector of control signals* $C(t)$ at time t. State and control vector are specified in the manipulator coordinate system. A subsequent state vector $S^v(t+1)$ is obtained by a transition function f^{ts}, e.g. addition of $S^v(t)$ and $C(t)$.

$$S^v(t+1) := f^{ts}(S^v(t), C(t)) \tag{1}$$

A *control function* f^{ct} is responsible for generating the control vector $C(t)$. It is based on the current state vector $S^v(t)$, a *current measurement vector* $Q(t)$ and a *desired measurement vector* Q^*.

$$C(t) := f^{ct}(S^v(t), Q(t), Q^*) \tag{2}$$

A *measurement function* f^{ms} is responsible for taking and analyzing images, and thereof generating the current and desired measurement vectors $Q(t)$ and Q^*. They are represented in the coordinate systems of the cameras.

$$Q(t) := f^{ms}(S^v(t), S^c) \tag{3}$$

Control function f^{ct} describes the relation between changes in different coordinate systems, e.g. $S^v(t)$ in the manipulator and $Q(t)$ in the image coordinate system. For defining this function, the Jacobian will be computed for a projection matrix $\mathcal{M}$ which linearly approximates (in projective spaces) the relation between the manipulator coordinate system and the image coordinate system.

$$\mathcal{M} := \begin{pmatrix} \mathcal{M}_1^v \\ \mathcal{M}_2^v \\ \mathcal{M}_3^v \end{pmatrix} \; ; \quad with \quad \begin{aligned} \mathcal{M}_1^v &:= (m_{11}, m_{12}, m_{13}, m_{14}) \\ \mathcal{M}_2^v &:= (m_{21}, m_{22}, m_{23}, m_{24}) \\ \mathcal{M}_3^v &:= (m_{31}, m_{32}, m_{33}, m_{34}) \end{aligned} \tag{4}$$

The usage of the projection matrix is specified according to [2, pp. 55-58]. Given a point in homogeneous manipulator coordinates $P := (X, Y, Z, 1)^T$, the position in homogeneous image coordinates $p := (x, y, 1)^T$ can be obtained.

$$p := f^{pr}(P) := \begin{pmatrix} f_1^{pr}(P) \\ f_2^{pr}(P) \\ f_3^{pr}(P) \end{pmatrix} := \frac{1}{\xi} \cdot \mathcal{M} \cdot P \; ; \quad with \quad \xi := \mathcal{M}_3^v \cdot P \tag{5}$$

Matrix $\mathcal{M}$ is determined with simple linear methods by considering the training samples of corresponding 3D points and 2D points. The scalar parameters m_{ij} represent a combination of extrinsic and intrinsic camera parameters which we leave implicit. The specific definition of normalizing factor ξ in equation (5) guarantees that function $f_3^{pr}(P)$ is constant 1, i.e. the homogeneous image coordinates of position p are in normalized form. The *Jacobian* $\mathcal{J}$ for the transformation f^{pr} in equation (5), i.e. for projection matrix $\mathcal{M}$, is computed as follows.

$$\mathcal{J}(P) := \begin{pmatrix} \frac{\partial f_1^{pr}}{\partial X}(P) & \frac{\partial f_1^{pr}}{\partial Y}(P) & \frac{\partial f_1^{pr}}{\partial Z}(P) \\ \frac{\partial f_2^{pr}}{\partial X}(P) & \frac{\partial f_2^{pr}}{\partial Y}(P) & \frac{\partial f_2^{pr}}{\partial Z}(P) \end{pmatrix} := \begin{pmatrix} \frac{m_{11} \cdot \mathcal{M}_3^v \cdot P - m_{31} \cdot \mathcal{M}_1^v \cdot P}{(\mathcal{M}_3^v \cdot P) \cdot (\mathcal{M}_3^v \cdot P)} \cdots \\ \vdots \quad \ddots \end{pmatrix} \tag{6}$$

Control function f^{ct} is based on deviations between current and desired image measurements and should generate changes in manipulator coordinates. For this purpose, the pseudo-inverse of the Jacobian is relevant (see following sections).

4   Servoing for Estimating Optical Axes

For estimating the *optical axis of a camera* relative to the basis coordinate system
of the manipulator we apply image-based hand-effector servoing. The optical axis
intersects the image plane approximately at the center. By servoing the hand-
effector such that the two-dimensional projection of the hand tip reaches the
image center, we finally obtain a 3D position which is a point on the optical axis,
approximately. By applying this procedure at two different distances from the
camera one obtains two distinct points located (approximately) on the optical
axis which are used for its estimation. Two virtual planes are specified which
are parallel to the $(\boldsymbol{Y}, \boldsymbol{Z})$ plane with constant offsets X_1 and X_2 on the $\boldsymbol{X}$-
axis. The movement of the hand-effector is restricted just to these planes (see
Fig. 3). Accordingly, the generic definition of the Jacobian $\mathcal{J}$ in equation (6) can
be restricted to the second and third columns, because the coordinates on the
$\boldsymbol{X}$-axis are fixed. A quadratic Jacobian matrix is obtained (with two rows and
columns) which must be inverted, i.e. $\mathcal{J}^{\dagger}(P) := \mathcal{J}^{-1}(P)$.

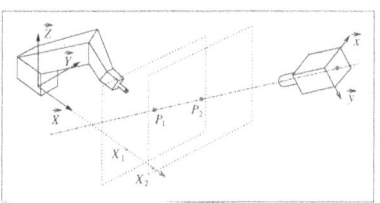

Fig. 3. Hand-effector servoing for estimating
the optical axis of a camera

Fig. 4. Course of detected
hand tip towards image center

The current measurement vector $Q(t)$ is defined as the 2D image location of
the hand tip and the desired measurement vector Q^* as the image center point.
Hough transformation and normalized cross correlation are used in combination
for detecting the hand tip in the images. The variable state vector $S^v(t)$ consists
of the two variable coordinates of the tool center point in the selected plane
$(X_1, \boldsymbol{Y}, \boldsymbol{Z})$ or $(X_2, \boldsymbol{Y}, \boldsymbol{Z})$. With these redefinitions of the Jacobian we can apply
the following control function.

$$C(t) := \begin{cases} s \cdot \mathcal{J}^{\dagger}(S^v(t)) \cdot (Q^* - Q(t)) : \|Q^* - Q(t)\| > \eta \\ \qquad\qquad 0 \qquad\qquad\quad : \qquad else \end{cases} \qquad (7)$$

A proportional control law is defined, i.e. the change is proportional to the deviation between desired and current measurements. Servoing factor s influences
the step-size of approaching the goal place. The hand position is changed by a
non-null vector $C(t)$ if desired and current positions in the image deviate more
than a threshold η. According to our strategy, first the hand tip is servoed to

the intersection point P_1 of the unknown optical axis with plane $(X_1, \boldsymbol{Y}, \boldsymbol{Z})$, and second to the intersection point P_2 with plane $(X_2, \boldsymbol{Y}, \boldsymbol{Z})$. Figure 4 shows for the hand servoing on one plane the succession of the hand tip extracted in the image, and the final point is located at the image center. The two resulting positions define a straight line in the manipulator coordinate system which is located near to the optical axis. The estimated line will be used to approach an object towards the camera for detailed inspection.

5   Servoing for Shape Inspection

Prior to the inspection phase of the peg-in-hole application the object must be grasped with the parallel jaw fingers of the robot hand [7, pp. 127-129, 210-217]. The grasped object is carried to a specific pose in the viewing space of one camera. Concretely, the specific position is the intersection point of the optical axis and the bottom rectangle of the pyramid viewing space, and the specific orientation of the fingers is orthogonal to the optical axis. As an example, the orientation of a grasped cylinder is such that the camera has an orthogonal view from the top or bottom, circular cylinder face. Due to the large distance from the camera (most distant viewing position), the depiction of the circular face is small. In order to inspect the shape of an object face it is desirable to have the face depicted in the image as large as possible. For this purpose, the robot hand must be servoed along the optical axis towards the camera, which is illustrated in Fig. 1 for one step of movement.

For this servoing process it is convenient to take as image measurements the appearance of the robot fingers. Due to their well-known shape the fingers can be extracted much easier (e.g. through Hough transformation) than the unknown shape of the object. We take the width of a robot finger (number of pixels) for defining current and desired measurement scalars (instead of vectors) $Q(t)$ and Q^*. Just as the measurements, also the control vector $C(t)$ is a scalar. With this definitions the control function of equation (7) can be applied for reaching the optimal viewing distance. The Jacobian may be simply defined by constant value 1, because servoing factor s can be used anyway for affecting the step-size. After having finished the approaching process we obtain an *acceptable size of the depicted object*, like in the first image of Fig. 5.

The inspection of the object shape is based on extracting the relevant region in the image. Especially, the regions of the robot fingers must be suppressed. This task can be simplified by first applying once again hand servoing. It is intended to obtain a standardized (i.e. vertical) appearance of the robot fingers, as shown in the second image of Fig. 5. For this purpose, the robot hand must rotate around the optical axis with the tip of the robot hand taken as the rotation center. For the servoing process we take as image measurement the finger tilt relative to the vertical image axis. Just as the measurements, also the control vector $C(t)$ is scalar. Therefore, a simple control procedure can be applied.

The usefulness of the *standardized finger appearance* is to be able to apply simple pattern matching techniques. We use the second image of Fig. 5, then

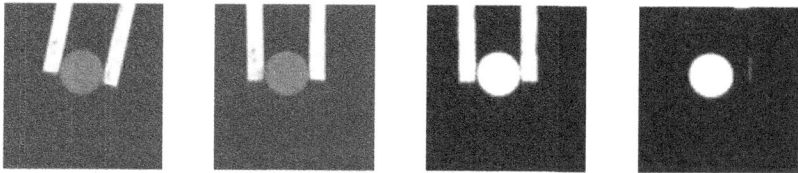

Fig. 5. (a) Appropriate size of depicted situation; (b) Vertical appearance of robot fingers; (c) Binary image with object and fingers; (d) Extraction of grasped object

move the robot hand outside the viewing space and take an image from the background. The subtraction of both images reveals a binary image containing only the fingers and the object (third image of Fig. 5). The suppression of the finger regions is reached with given finger patterns which were acquired in an offline phase under similar viewing conditions. Actually, it is this matching process which can be performed efficiently due to the standardized finger appearance. The right image in Fig. 5 is obtained which contains just the relevant object region. Undesired noisy effects (isolated white pixels) can be suppressed by applying simple morphological operations.

Our approach for describing the shape of the region is based on the autoregressive model proposed by Dubois [1]. It results in a characterizing vector of features which is invariant under region translation and rotation. The same approach is applied as well for describing the holes of the board which results in a characterizing vector for each hole, respectively. Based on euclidean metric one determines the hole whose shape is most similar to the shape of the peg. This concludes the inspection phase of the peg-in-hole application. The second phase consists in approaching the peg appropriately to the relevant hole.

6   Servoing for Object Assembly

The two cameras take images continually for the visual feedback control of approaching an object to a goal place. In each stereo image both the object and the goal place must be visible for determining the distance between current and desired measurement vectors, respectively. The critical issue is to extract the relevant features from the stereo images. For example, let us assume a *cylindrical object and a circular goal place* as shown in the first image of Fig. 6.

The binarization based on thresholding the gradient magnitudes is shown in the second image of Fig. 6. In the next step, a specific type of Hough transformation is applied for approximating and extracting half ellipses (third image in Fig. 6). This specific shape is expected to occur at the goal place and at the top and bottom faces of the object. Instead of full ellipses we prefer half ellipses, concretely the lower part of full ellipses, because due to the specific camera arrangement this feature is visible throughout the complete process. From the

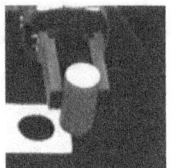 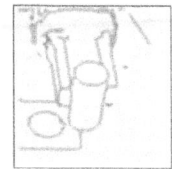 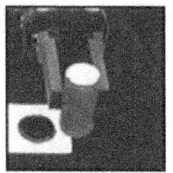 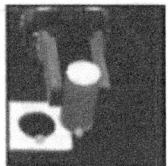

Fig. 6. (a) Cylindrical object, circular goal place; (b) Thresholded gradient magnitudes; (c) Extracted half ellipses; (d) Specific point on half ellipses of object and goal place

bottom face of the object only the specific half ellipse is visible. The process of approaching the object to the goal place is organized such that the lower part of the goal ellipse remains visible, but the upper part may become occluded more and more by the object. The distance measurement between object and goal place just takes the half ellipse of the goal place and that from the bottom face of the object into account. For computing a kind of distance between the two relevant half ellipses we extract from each a specific point and based on this we can take any metric between 2D positions as distance measurement. The fourth image in Fig. 6 shows these two points, indicated by gray disks, on the object and the goal place.

The critical aspect of extracting points from a stereo pair of images is that *reasonable correspondences* must exist. A point of the first stereo image is in correspondence with a point of the second stereo image, if both originate from the same 3D point. In our application, the half ellipses extracted from the stereo images are the basis for determining corresponding points. However, this is by no means a trivial task, because the middle point of the contour of the half ellipse is not appropriate. The left picture of Fig. 7 can be used for explanation. A virtual scene consists of a circle which is contained in a square (top part of left picture). Each of the two cameras produces a specific image, in which an ellipse is contained in a quadrangle (bottom part of left picture). The two dotted curves near the circle indicate that different parts of the circle are depicted as lower part of the ellipse in each image. Consequently, the middle points p_1 and p_2 on the lower part of the two ellipses originate from different points P_1 and P_2 in the scene, i.e. points p_1 and p_2 do not correspond. Instead, the right picture of Fig. 7 illustrates an approach for determining corresponding points.

We make use of a specific geometric relation which is invariant under geometric projection. Virtually, the bottom line of the square is translated to the circle which results in the tangent point P. This procedure is done as well in both stereo images, i.e. translating the bottom line of the quadrangle parallel towards the ellipse to reach the tangent points p_1 and p_2. Due to different perspectives the two bottom lines have different orientations and therefore the resulting tangent points are different from those extracted previously (compare bottom parts in left and right picture of Fig. 7). It is observed easily that the new tangent

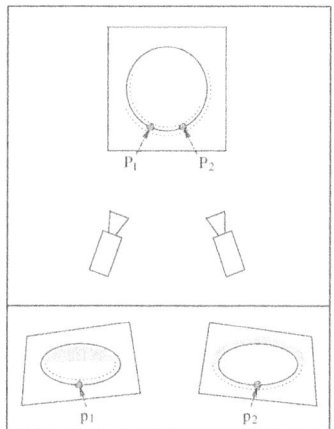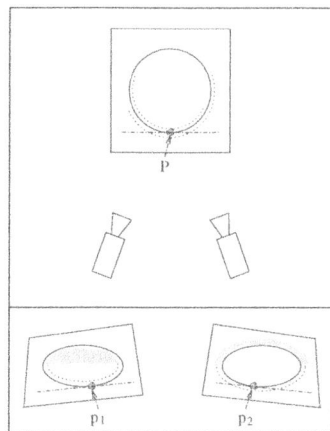

Fig. 7. (a) Extracted image points p_1, p_2 originate from different scene points P_1, P_2; (b) Extracted image points correspond, i.e. originate from one scene point P

points p_1 and p_2 correspond, i.e. originate from the same scene point P. This kind of projective compatibility can be exploited for our peg-in-hole application. The boundary of the board, which contains the holes, can be used as supporting context for stereo matching. Accordingly, both the board and the object must be fully included in the viewing space of both stereo cameras, respectively. For each stereo image the orientation of the bottom boundary line can be used for determining relevant tangent points at the relevant ellipse, i.e. virtually move the lines to the ellipses and keep orientation. Tangent points must be extracted at the half ellipse of the goal place and at the half ellipse of the bottom face of the object. These points have already been shown in the fourth image of Fig. 6.

For defining the control vector we must describe the relationship between displacements of the robot hand and the resulting displacements in the two stereo images taken by the stereo cameras. For this purpose we introduce two Jacobians $\mathcal{J}_1(P)$ and $\mathcal{J}_2(P)$ which depend on the current position P of the hand tip. If we would multiply the Jacobian $\mathcal{J}_1(P)$ (respectively Jacobian $\mathcal{J}_2(P)$) with a displacement vector of the hand position, then the product would reveal the displacement vector in the left image (respectively in the right image). The two Jacobians are joined together which results in a (4×3) matrix depending on P.

$$\mathcal{J}(P) := \begin{pmatrix} \mathcal{J}_1(P) \\ \mathcal{J}_2(P) \end{pmatrix} \tag{8}$$

In order to transform a desired change from stereo image coordinates into manipulator coordinates the pseudo inverse $J^\dagger(P)$ is computed.

$$\mathcal{J}^\dagger(P) := \left(\mathcal{J}^T(P) \cdot \mathcal{J}(P) \right)^{-1} \cdot \mathcal{J}^T(P) \tag{9}$$

The current position $P(t)$ of the hand tip defines the variable state vector $S^v(t)$. The desired measurement vector Q^* is a 4D vector comprising the 2D positions of a certain point of the goal place in the stereo images. The current measurement vector $Q(t)$ represents the stereo 2D positions of a relevant point on the object.

$$Q^* := \begin{pmatrix} p_1^* \\ p_2^* \end{pmatrix}; \quad Q(t) := \begin{pmatrix} p_1(t) \\ p_2(t) \end{pmatrix} \tag{10}$$

With these new definitions we can apply control function of equation (7).

7   Peg-In-Hole Application for other Objects

The basic assumption behind the presented technique is that the peg can be inserted successfully by taking only the shape of the bottom object face into account. Accordingly, the object surface must be composed of a top and a bottom face, which are parallel and of equal shape, and the other faces must be orthogonal to them. Apart from cylinders this constraint also holds for cuboids, whose treatment will be mentioned briefly (Fig. 8). The procedures involved in the inspection phase can be applied without any change. However, in the insertion phase we must consider that the object is not rotation-symmetric. In addition to the positions of hole and object, also the orientations have to be taken into account. Hough transformation and strategies for line organization are applied for extracting the boundaries of object and hole, respectively [7, pp. 29-98]. Based on the hole boundary and the top face boundary of the object we determine hole and object orientation. Furthermore, we take the middle point of two appropriately selected boundary lines of hole and object to determine their positions. Altogether, the current measurement vector $Q(t)$ consists of 6 components with 3 for each stereo image. These are composed of one scalar for the orientation and 2 scalars for the position of the object. Similarly, the desired measurement vector Q^* is defined for the hole. The control vector $C(t)$ consists of 4 components, i.e. three for the position and one for the horizontal orientation of the robot hand. Based on these definitions we determine the Jacobian and apply the control function of equation (7). Figure 8 shows the peg-in-hole application for the cuboid object, which includes in the second image the object boundary and the selected point for defining the measurement vector.

8   Summary

For peg-in-hole applications we used a two-component robot system which consists of a robot manipulator (including a parallel jaw gripper) and a robot head (including monochrome stereo cameras). The usefulness of image-based hand-effector servoing was demonstrated for characterizing the camera-manipulator relation, for optimal viewing and inspecting the object, and for appropriately approaching the object to the relevant hole. In our current implementation, one servoing cycle for inserting the cylindrical peg requires about 0.5 seconds. Generally, the velocity depends on the complexity of the object shape.

Fig. 8. (a) Grasped cuboid object; (b) Set of object boundary lines, selected point specifying object position in the image; (c) Insertion of cuboid into rectangular hole

References

1. S. Dubois, F. Glanz: An autoregressive model approach to two-dimensional shape classification. *IEEE Trans. on Patt. Anal. and Mach. Intel.* **8** (1986) 55–66. 162
2. O. Faugeras: *Three-Dimensional Computer Vision.* The MIT Press (1993). 159
3. K. Hashimoto: *Visual Servoing.* World Scientific Publishing (1993). 157
4. R. Horaud, F. Chaumette (eds.): Internat. J. of Computer Vision, special issue on *Image-based Robot Servoing* **37** (2000). 157
5. S. Hutchinson, G. Hager, P. Corke: A tutorial on visual servo control. *IEEE Trans. on Robotics and Automation* **12** (1996) 651–670. 157
6. M. Lanzetta,G. Dini: An integrated vision-force system for peg-in-hole assembly operations. *Intel. Comp. in Manufact. Eng.* (1998) 615–621. 158
7. J. Pauli: Development of camera-equipped robot systems. Christian-Albrechts-Universität zu Kiel, Institut für Informatik, Technical Report **9904** (2000). 161, 165

Robot Localization Using Omnidirectional Color Images

David C. K. Yuen* and Bruce A. MacDonald

The Department of Electrical and Electronic Engineering
The University of Auckland, Private Bag 92019, Auckland, New Zealand
{d.yuen,b.macdonald}@auckland.ac.nz

Abstract. We describe a vision-based indoor mobile robot localization algorithm that does not require historical position estimates. The method assumes the presence of an *a priori* map and a reference omnidirectional view of the workspace. The current omnidirectional image of the environment is captured whenever the robot needs to relocalise. A modified hue profile is generated for each of the incoming images and compared with that of the reference image to find the correspondence. The current position of the robot can then be determined using triangulation as both the reference position and the map of the workspace are available. The method was tested by mounting the camera system at a number of random positions positions in a 11.0m × 8.5 m room. The average localization error was 0.45 m. No mismatch of features between the reference and incoming image was found amongst the testing cases.

1   Introduction

Under the traditional deliberative motion control architecture, a robot needs to know its own position in the environment before making a navigation plan. If the robot is first switched on or wants to re-position itself after getting lost, no reliable previous position estimates will be available for the localization stage. Many common localization methods, notably dead-reckoning using extended Kalman filtering [4], cannot cope with such a condition.

In this paper, we describe a passive, vision-based localization technique that does not involve the use of historical position estimates, and takes advantage of the richer information in an image. An omnidirectional imaging system is introduced to provide color and textual information to the system. The distinctive features from an incoming image are extracted using a region segmentation method. The extracted features are then matched with those from a reference image to generate matched landmarks. The placement of artificial landmarks in the environment is unnecessary.

In section 2, we review previous work in vision-based localization methods that do not require historical position estimates. Section 3 outlines our localization approach. It also describes the image segmentation and triangulation

---

* This work was supported in part by the Foundation for Research, Science and Technology, New Zealand, with a Top Achiever Doctoral Scholarship.

R. Klette, S. Peleg, G. Sommer (Eds.): Robot Vision 2001, LNCS 1998, pp. 167–175, 2001.
© Springer-Verlag Berlin Heidelberg 2001

techniques adopted in the system. The test results are discussed in section 4 before summarizing the paper in section 5.

2   Maps and Landmarks

Map matching can usually be carried out without the use of an image. A local map is first generated for the area around the robot, using the measurements from a laser or ultrasonic range finder [2,7]. The local map is then matched against different regions of a global map, at different orientations. Since the map matching uses a local distance map, the localization process can be confounded if objects with similar shapes are present in the environment. Also, the correlation operation requires considerable computation.

Many industrial robots are guided by bar codes [5], reflective tape [3, p313–317], ceiling light patterns [3, p472–477] or other artificial landmarks. A global positioning system (GPS) is a notable example of an artificial emitter in an outdoor navigation. While the landmark recognition step is usually quite simple, the cost of laying out and maintaining the well calibrated landmarks can be very expensive, and impractical in some environments.

Visual images usually have high spatial resolution and can provide details such as the color and texture of the object being observed. With the extra information provided by visual sensors, the robot can have a better understanding of the complex surroundings. In many cases, natural landmarks can be extracted from the incoming images.

Using the concept of a "view field" [1], tiny visual features may be extracted from an image together with their relative spatial relations, to form a landmark. The memory requirement for the storage of typical indoor scenes is thus reduced to about 16000 bytes per m^2. Both Lin and Zhang [6,8] process the sparsely sampled omnidirectional image with neural networks to extract landmarks for localization, in which 120 and 1600 bytes were retained respectively for each image frame. While the storage of these landmarks requires only modest amount of memory, the image capturing stage involves a lot of preparative work and makes the localization system quite inflexible.

3   Vision-Based Localization System

Algorithm 1 shows the overall process of localization. Our method assumes an *a priori* map for the environment. An omnidirectional image is used to simplify camera motion; panning control is not required.

To locate the robot, a vertically central strip of an omnidirectional image is segmented into regions by analyzing the horizontal hue profile, then matched against region boundaries in a reference image, and triangulation is used to calculate the new robot position.

The imaging system comprises two Sony EVI-D31 cameras and two OMT SEQ-P1S frame grabber cards with a Pentium based controller, to be mounted on

Algorithm 1 Localize

1: On first invocation, call Initialize()
2: CurrentImage = ObtainImage()
3: Create all tokens of 3 consecutive region MHI median values for ReferenceImage
4: Create all tokens of 3 consecutive region MHI median values for CurrentImage
5: Find longest token match between ReferenceImage and CurrentImage
6: **for** each of the first, middle and last matching boundary pairs: **do**
7:     Triangulate position from the map position of the boundary pair
8: **end for**
9: return the average of the three position estimates

ObtainImage:

1: Take 8 images at 45° increments, link them together to one image
2: Extract the 30 pixel high central strip
3: Calculate the MHI for each pixel in the strip
4: **for** each 10-pixel wide band **do**
5:     Calculate the band MHI median
6: **end for**
7: Find region boundaries by differentiating the band median sequence
8: **for** each region between boundaries **do**
9:     calculate the region MHI median
10: **end for**
11: return the sequence of region MHI values

Initialize:

1: ReferenceImage = ObtainImage()
2: Load the environment map
3: Calculate the map positions of boundaries in ReferenceImage

our mobile robot as a multi-purpose flexible vision system. To ensure controllable images for testing the current development stage, a single camera is mounted on a tripod. The images captured for this study have a resolution of 320×240 pixels and a color depth of 24 bits. To facilitate comparing results, the zoom control of the camera was adjusted for a view angle of 45° (horizontal) $\times$ 34° (vertical) at 84cm above the floor. At each location, 8 images were taken in 45° increments. At present the camera head should face the same direction when taking the first image amongst each series; the purpose is to discover the robot position and later we expect to remove this constraint and also discover the orientation. The 8 images were linked together to form a panoramic view of the environment, shown in Fig. 1. A horizontal strip of 2560×30 pixels is then cut from the center of the omnidirectional image and used for the rest of the processing.

The representation of the image may be further simplified by extracting the hue channel of an HSV model. For humans, color discontinuity often represents separation between objects. While the hue channel is relatively immune to variations in illumination, some hue values have little meaning and are sensitive to minor changes, notable values near white, gray and black. The modified hue index (MHI) is then defined:

Fig. 1. Omnidirectional view of the workspace: a) the original panoramic image. b) The horizontal strip cut from original view, which is marked by the white box shown in image a). (The view shown in b) has been stretched vertically for better display.)

$$MHI = \begin{cases} -2/3 * \pi & S >= 0.15 \text{ and } V >= 10 & \text{(black)} \\ -1/4 * \pi & S < 0.15 \quad \text{and } V >= 90 & \text{(gray)} \\ -1/3 * \pi & S < 0.15 \quad \text{and } V > 90 & \text{(white)} \\ H & \text{otherwise} & \text{(other colours)} \end{cases} \quad (1)$$

where H,S,V represents the hue $[0, 2\pi)$, saturation $[0, 1]$ and value $[0, 100]$.

The image is divided into 10-pixel wide vertical bands and the median MHI is computed for each band. Most of the smaller uncharted objects, e.g. network cable ducts, electric switches etc, are removed by band median filtering.

When viewing a large object, we may find regions with relatively constant values in the MHI band median profile, as illustrated in Fig. 2. The regional boundaries may represent object edges or distinctive changes in the surface fea-

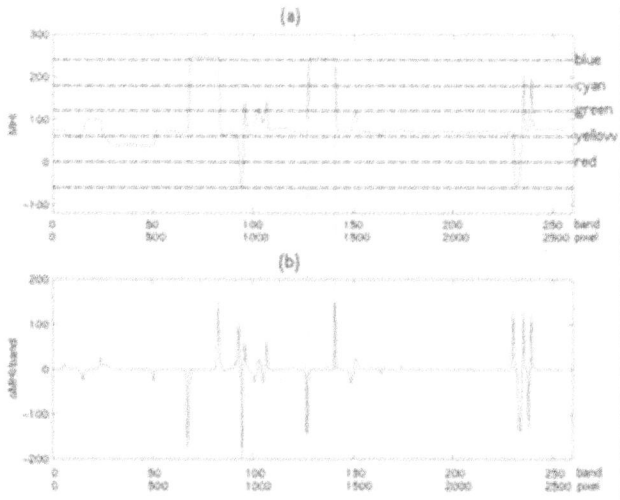

Fig. 2. (a) The modified hue index profile. (b) the differentiation of the MHI profile

tures of objects. We can locate potential regional boundary lines by thresholding the differentiated MHI band median profile. To facilitate the later matching operation, a "region median" is calculated for each detected region by calculating the median MHI of all the bands within the region boundaries.

3.1   Preparations for the Map and Reference Image

Since the band median filtering method removes minor features, the level of detail required in the map is not high, and maps should not be difficult to maintain. The complexity of the environment determines the minimum number of reference images that needs to be taken. If the visibility of different parts of the workspace to the reference point is blocked, more reference points are required. In this study, a simpler environment was considered where only one reference point was sufficient. The exact position of the reference point was determined by surveying before taking the first image.

The viewing angle from the reference point to the edges of the large objects can be calculated from the coordinates of the regional boundaries on the omnidirectional image. The map position of these objects can then be estimated by extending the line-of-sight at the given viewing angle until an intersection is formed on the map, as depicted in Fig. 3.

3.2   Localization System

An omnidirectional snapshot of the environment is taken whenever the robot needs to re-locate itself, and the MHI is calculated to identify regions. Since

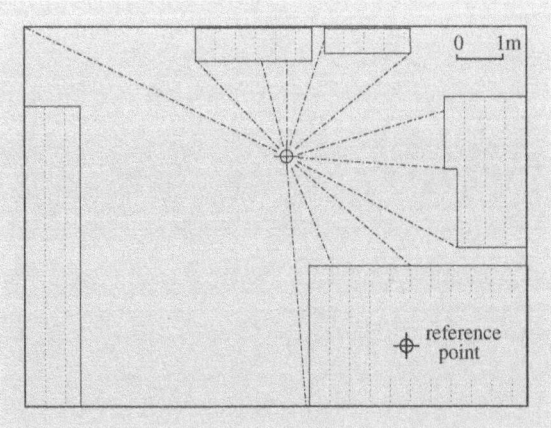

Fig. 3. Mapping of the observed feature for the reference image. The map position of an observed feature can be found by extending the line-of-sight at the given viewing angle until an intersection is formed

the positions of large objects are known, the current position of the robot can be identified using triangulation once enough matches have been established between boundary lines in the reference and current images, that represent features in the map.

The feature matching process is crucial to the performance of the localization stage. When the robot moves to different parts of the room, the relative size of the regions on the MHI profile may change. Some features may become too small and be left unaccounted for. Due to the presence of uncharted objects, some unexpected features may appear while some expected ones may be occluded. Also changes in reflectance of object surfaces may appear as features after MHI processing. The proposed matching algorithm should be tolerant to these defects.

Omnidirectional images have the important property that the *sequence* of modified hue regions remains the same, providing all the objects are still visible to the observer. A sequence of triples is formed for the reference image by grouping the region median values of three consecutive regions (that is for regions $\{(1,2,3),(2,3,4),(3,4,5),\ldots\}$) into "tokens." The list of region median values for the current image is then searched to locate the possible matches for each of the reference tokens. A match is declared if the region medians for each of the three consecutive regions of current image are within a certain tolerance from the respected regions of the reference token. The tolerance level was set to $\frac{5}{36}\pi$ radians in this study. Ideally, we can obtain a token sequence match from the incoming image that contains as many regions as the reference. In practice the longest token is taken as the best match.

The location and orientation of the robot (x, y, ϕ) can be found by solving the following non-linear simultaneous equations:

$$\tan(2 * \pi - \phi - \theta_i) = \frac{y_i - y}{x_i - x} \tag{2}$$

where x_i, y_i, represent the x, y coordinates of the ith object edge on the map, and θ_i represents the observed angle of the ith object edge from the robot. See Fig. 4 for further explanation.

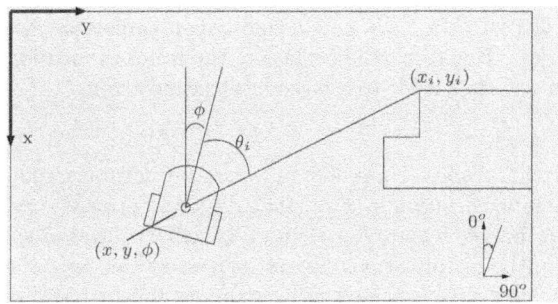

Fig. 4. Geometric conventions

In this study, the camera head was aligned to a fixed direction before taking the first image. The localization module thus needs to solve for only the two position variables (x, y), So a minimum of two matched features are required.

As an initial investigation, the average is taken of three sets of position estimates, which are generated by taking the observed angles of the first, last and the middle regional boundaries of the longest token match from equation 2.

4   Results and Discussions

The vision-based localization method was tested in an 11.0m × 8.5m laboratory. As shown in Fig. 5, nine random testing positions were generated. The test results are shown in Table 1. The average localization is 0.45 m with a standard deviation of 0.22 m. No mismatch was found between the reference and current image when examined the longest token match for each testing case.

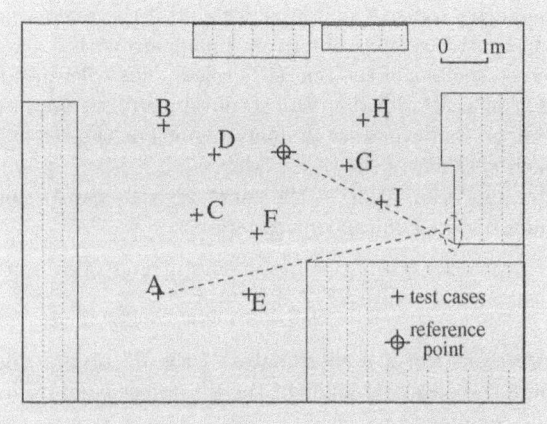

Fig. 5. The testing environment for the localization algorithm. The influence of partial occlusion is demonstrated. The dotted lines from location A and the reference point represent their line-of-sight when supposedly viewing the same edge of an object. Due to partial occlusion, the robot at location A is not really the true edge and thus leads to a large localization error

Although the proposed method may not be accurate enough for the use in a standalone localization system, that does not poise a serious problem. In this study, we intend to develop a vision-based localization system that does not depend on the historical position estimates. In this way, the relative rough position estimates can be refined using more established localization methods, such as extended Kalman filtering.

Table 1. Localization error of the testing cases

Position	A	B	C	D	E	F	G	H	I
x-coordinate (m)	5.57	2.11	3.93	2.70	5.58	4.53	2.95	1.98	3.69
y-coordinate (m)	2.98	3.10	3.82	4.14	4.91	5.13	7.02	7.43	7.79
localization error (m)	0.66	0.91	0.50	0.38	0.25	0.45	0.23	0.27	0.44

The test samples that give large localization error are located far away from the reference point. The view can be quite different from that captured at the reference point. For example, only a fraction of the partition can be visualized at location A. As a result, the observed boundary at location A is not really the true edge of the partition (circled with dots in Fig. 5) and thus leads to a large error.

In the current system, the robot position was calculated using only three of the matched features with the rest being discarded. These other matches could potentially be used to improve the accuracy and robustness of the technique. In addition, range sensors can be introduced to the system to reduce the ambiguities arisen during various stage of the operation.

5   Conclusion

A vision-based robot localization system is proposed that does not involve the use of historical position estimates. A modified hue profile is generated for each of the incoming omnidirectional images. The extracted hue regions are matched with that of the reference image to find corresponding region boundaries. As the reference image, exact location of the reference point and the map of the workspace are available, the current position of the robot can be determined by triangulation.

The method was tested by placing the camera set-up at a number of different random positions in a 11.0m × 8.5m room. The average localization error was 0.45 m. No mismatch of features between the reference and incoming image was found. While the proposed localization method may not be sufficiently accurate if used alone, it provides a good initial position estimate for the use of other more established localization methods, such as extended Kalman filtering.

References

1. Ch. Balkenius: Spatial learning with perceptually grounded representations. *Robotics and Autonomous Systems*, **25** (1998) 165–175.   168
2. A. Elfes: Sonar-based real-world mapping and navigation. *IEEE Journal of Robotics and Automation*, **RA-3** (1987) 249–265.   168
3. H. R. Everett: *Sensors for Mobile Robots: Theory and Application.* A. K. Peters Ltd., (1995).   168

4. L. Jetto, S. Longhi, G. Venturini: Development and experimental validation of an adaptive extended kalman filter for the localization of mobile robots. *IEEE Transactions on Robotics and Automation*, **15** (1999) 219–229. 167

5. J. J. Leonard, H. F. Durrant-Whyte: Mobile robot localization by tracking geometric beacons. *IEEE Transactions on Robotics and Automation*, **7** (1991) 376–382. 168

6. L.-J. Lin, Th.R. Hancock, J. S. Judd: A robust landmark-based system for vehicle location using low-bandwidth vision. *Robotics and Autonomous Systems*, **25** (1998) 19–32. 168

7. B. Yamauchi: Mobile robot localization in dynamic environment using dead reckoning and evidence grids. In: Proceed. of the IEEE Internat. Conf. on *Robotics and Automation*, Minneapolis, Minnesota, (April 1996) 1401–1406. 168

8. J. Zhang, A. Knoll, V. Schwert: Situated neuro-fuzzy control for vision-based robot localisation. *Robotics and Autonomous Systems*, **28** (1999) 71–82. 168

The Background Subtraction Problem for Video Surveillance Systems

Alan McIvor[2], Qi Zang[1], and Reinhard Klette[1]

[1] CITR, University of Auckland
Tamaki Campus, Building 731, Auckland, New Zealand
r.klette@auckland.ac.nz
[2] Reveal Ltd.
Level 1, Tudor Mall, 333 Remuera Rd., Auckland, New Zealand
alan.mcivor@reveal.co.nz

Abstract. This paper reviews papers on tracking people in a video surveillance system, and it presents a new system designed for being able to cope with shadows in a real-time application for counting people which is one of the remaining main problems in adaptive background subtraction in such video surveillance systems.

1 Introduction

Video surveillance systems seek to automatically identify events of interest in a variety of situations. Example applications include intrusion detection, activity monitoring, and pedestrian counting. The capability of extracting moving objects from a video sequence is a fundamental and crucial problem of these vision systems. For systems using static cameras, background subtraction is the method typically used to segment moving regions in the image sequences, by comparing each new frame to a model of the scene background, see, e.g., [6,13].

1.1 Evaluation Criteria

When the above methods are applied to a video surveillance problem, there are a number of key attributes and scenarios that must be handled. These are:

- The choice of models for key components.
- Initialization of the model parameters when the system is first started.
- Distinguishing objects of interest from illumination artifacts such as shadows and highlights.
- Handling uncontrollable illumination level changes, such as occur in outdoor scenes.
- Adapting to changes in the scene, such as when a new chair is introduced into a restaurant.
- Detecting when something is wrong with the processing, and re-initialization in response.

R. Klette, S. Peleg, G. Sommer (Eds.): Robot Vision 2001, LNCS 1998, pp. 176–183, 2001.

2   Review of Existing Methods

In this section, we discuss existing methods with respect to the choice of a background model and how it is maintained in the face of illumination changes, how the foreground is differentiated from the background, and what strategies are used to correct classification errors.

2.1   Background Models

The simplest and most common model for the background is to use a point estimate of the color at each pixel location, e.g., [12]. This point estimate is usually taken to be the mean of a Gaussian distribution. In some systems, e.g., [17], the variance of the intensity is also modelled.

To cope with variation in the illumination, the background estimate $\mathbf{B}_t$ has to be updated with each new image $\mathbf{I}_t$. In [12], this is updated using

$$\mathbf{B}_{t+1} = \alpha \mathbf{I}_t + (1 - \alpha)\mathbf{B}_t \tag{1}$$

In [5], the background is maintained as the temporal median of the last N frames, with typical values of N ranging from 50 to 200. The updating of the background estimate is often restricted to pixels which have been classified as background. In [9,10,11], three parameters are estimated at each pixel: $\mathbf{M}$, $\mathbf{N}$, and $\mathbf{D}$, which represent the minimum, maximum, and largest interframe absolute difference observable in the background scene.

Such simple background estimates fail to cope with scenes which contain regularly changing intensities at a pixel, such as occurs with flashing lights and swaying branches. Several more complex background models have been developed to handle such scenarios. In [7,20,14], each pixel is separately modeled by a mixture of K Gaussians

$$P(\mathbf{I}_t) = \sum_{i=1}^{K} \omega_{i,t} \cdot \eta(\mathbf{I}_t; \mu_{i,t}, \Sigma_{i,t}) \tag{2}$$

where $K = 4$ in [14] and $K = 3 \ldots 5$ in [20]. In [7,20], it is assumed that $\Sigma_{i,t} = \sigma_{i,t}^2 \cdot \mathbf{I}$. The background is updated before the foreground is detected. In [4], three background models are simultaneously kept, a primary, a secondary, and an old background. They are updated as follows:

1. The primary background is updated as in (1). At pixels marked as foreground a much reduced value of α is used.
2. The secondary background is updated as in (1), but only at pixels where the incoming image is not significantly different from the current value of the secondary background.
3. The old background is a copy of the incoming image from 9000 to 18000 frames ago. It is not updated.

In [21], two background estimates are used which are based on a linear predictive model:

$$\mathbf{B}_t = -\sum_{k=1}^{p} a_k \mathbf{I}_{t-k} \quad \text{and} \quad \hat{\mathbf{B}}_t = -\sum_{k=1}^{p} a_k \mathbf{B}_{t-k} \tag{3}$$

where the coefficients a_k are reestimated after each frame is received so as to minimize the prediction error. The second estimator $\hat{\mathbf{B}}_t$ is introduced because the primary estimate $\mathbf{B}_t$ can become corrupted if a part of the foreground covers a pixel for a significant period of time. As well as the above, an estimate of the prediction error variance is also maintained:

$$\mathcal{E}(e_t^2) = \mathcal{E}(\mathbf{I}_t^2) + \sum_{k=1}^{p} a_k \mathcal{E}(\mathbf{I}_t \mathbf{I}_{t-k}) \, . \tag{4}$$

2.2   Foreground Detection

The method used for detecting foreground pixels is highly dependent on the background model. But, in almost all cases, pixels are classified as foreground if the observed intensity in the new image is substantially different from the background model, e.g., varies bymore than a predefined threshold, or by a factor dependent on the variance estimate also maintained within the background model. In systems that maintain multiple models for the background, then a pixel must be substantially different from all background values to be classified as foreground. In [4], an adaptive thresholding with hysteresis scheme is used.

In the mixture of Gaussians approach, the foreground is detected as follows. All components in the mixture are sorted into the order of decreasing $\omega_{i,t}/\|\Sigma_{i,t}\|$ and pixels matching the first couple are marked as background.

In [21], which uses two predictions for the background value, a pixel is marked as foreground if the observed value in the new image differs from both estimates by more than a threshold $\tau = 4\sqrt{\mathcal{E}(e_t^2)}$, where $\mathcal{E}(e_t^2)$ is calculated in (4).

2.3   Error Recovery Strategies

The section above on background models describes how the various methods are designed to handle gradual changes in illumination levels. Sudden changes in illumination must be detected and the model parameters changed in response. In [21], the strategy used is to measure the proportion of the image that is classified as foreground and if this is more than 70%, then the current model parameters are abandoned and a re-initialization phase is invoked. In [12], if a pixel is marked as foreground for most of the last couple of frames (the values of 'most' and 'couple' are not given), then the background is updated as $\mathbf{B}_{t+1} = \mathbf{I}_t$. This correction is designed to compensate for sudden illumination changes and the appearance of static new objects.

Another common problem are regular changes in illumination level, such as that from moving foliage. Multiple model schemes such as the mixture of

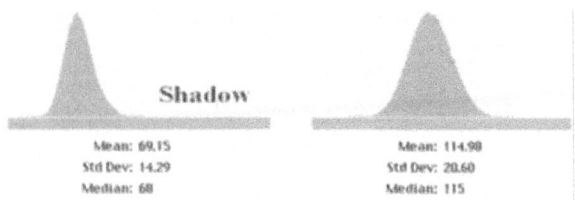

Mean: 69.15
Std Dev: 14.29
Median: 68

Mean: 114.98
Std Dev: 20.60
Median: 115

Fig. 1. Value distribution in the Blue channel of a region of the background: left - with shadow, right - without. This shows a typical situation: shadow means reduced intensity and reduced 'dynamics'

Gaussians explicitly handle such problems. However, they present a problem for single model schemes. In [12], to compensate for these problems, a pixel is masked out from inclusion in the foreground if it changes state from foreground to background frequently.

3 An Approach Incorporating Foreground Classification

The purpose of foreground classification is to distinguish objects of interest such as pedestrians from illumination artifacts such as shadows and highlights. This is usually accomplished by evaluating the color distribution within a foreground region [13]:

A *shadow* region has similar hue and saturation to the background but a lower intensity (see Fig. 3).

A *highlight* region has similar hue and saturation to the background but a higher intensity.

and, consequently, an object region has a different hue and saturation to the background.

3.1 The Background Model

The background observed at each pixel is modelled by a multidimensional Gaussian distribution in RGB space. The estimates of the mean and variance are updated with each new frame at pixels that have been classified as background, as follows:

$$\mu = (1 - \alpha)\mu + \alpha Ii$$
$$\sigma^2 = (1 - \alpha)\sigma^2 + \alpha(Ii - \mu)^2$$

Here, α is the predefined learning rate. Initially the distributions of the background are not known. The mean is initialised with the value in the first frame. The variance and brightness distortion are initialised with arbitrary large values.

3.2   Pixel Classification

Each pixel in a new image is classified as one of background, object, shadow, or highlight. Pixels are classified as background if the observed color value is within the ellipsoid of probability concentration (the region of minimum volume, centered around the background mean, that contains 99% of the probability mass for the Gaussian distribution).

The distinction between objects, shadows, and highlights among the pixels not classified as background is made using the brightness distortion [13]. This detects shifts in the RGB space between a current pixel and the corresponding background pixel. Let E_i represent the expected background pixel's RGB value, I_i represents the current pixel's RGB value in the current frame, σ_i represents its standard deviation. The brightness distortion α_i is obtained by minimizing $\phi(\alpha_i) = (I_i - \alpha_i E_i)^2$. If the current pixel's intensity is greater than the background intensity and the current pixel's brightness distortion value is less than the corresponding background pixel's brightness distortion value, this pixel will be marked as highlight. If the current pixel's intensity is less than the background intensity and the current pixel's brightness distortion value is less than the corresponding background pixel's brightness distortion value, this pixel will be marked as shadow. Otherwise the pixel is marked as an object of interest.

4   Our Results

Figure 2 shows that the algorithm works accurately in a uniformly lighted scene: The moving person is marked as light green, the shadow is marked as dark red successfully. Figure 3 shows how the algorithm is able to cope with global illumination changes after adding two extra lights. Figure 4 shows a result after initialization by using static frames without moving people. We obtain good foreground subraction and shadow detection results in real-time. Figure 5 shows that after static initialization, the background model can be updated adaptively. The algorithm updates a new background model while frames with moving people are captured. It also illustrates the robustness of background updating. Even when we capture a very different frame with moving people, this algorithm still can refine the background model and we obtain good subtraction results.

5   Conclusion and Further Work

This paper presents and discusses a background subtraction approach which can detect moving people on a background while also allowing for shadows and highlights. The background is adaptively updated. The approach has been tested in RGB space. The method is efficient with respect to computation time and storage. It only requires to store a background model and brightness distortion values. This allows real-time video processing.

The method has a number of limitations that will be addressed in future work. A static background without moving people is needed during the initialization

Fig. 2. Backgrond subtraction for uniform lighting: left - the original frame, right - the result after background subtraction

Fig. 3. Backgrond subtraction for lighting coming from window plus two extra lights: left - the original frame, right - the result after background subtraction

Fig. 4. Here, all frames before the current frame have been static (i.e. without moving people): left - the current frame, right - the result after background subtraction

Fig. 5. Here, starting with static frames (i.e. without moving people) we continue with 5 frames showing moving people until the current frame: left - the current frame, right - the result after background subtraction

phase, otherwise it takes a substantial amount of time to reliably classify pixels as foreground. Very dark parts of people are still classified as either background, or shadows. Another limitation is that shadows on very dark backgrounds or several shadows added together will not be detected effectively. Bright highlights are also a problem because of saturation in the camera sensor.

References

1. Y. Bar-Shalom, Th. E. Fortmann: *Tracking and Data Association*. Academic Press, New York (1988).
2. D. Beymer, Ph. McLauchlan, B. Coifman, J. Malik: A real-time computer vision system for measuring traffic parameters. *CVPR'97* (1997) 495–501.
3. A. Blake, M. Isard: *Active Contours*. Springer, Berlin (1998).
4. T. E. Boult, R. Micheals, X. Gao, P. Lewis, C. Power, W. Yin, A. Erkan: Frame-rate omnidirectional surveillance and tracking of camouflaged and occluded targets. *Second IEEE Workshop on Visual Surveillance.* (1999) 48–55. 177, 178
5. R. Cutler, L. Davis: View-based detection and analysis of periodic motion. *Internat. Conf. Pattern Recognition* (1998) 495–500. 177
6. A. Elgammal, D. Harwood, L. A. Davis: Non-parametric model for background subtraction. *ICCV'99* (1999). 176
7. W. E. L. Grimson, C. Stauffer, R. Romano, L. Lee: Using adaptive tracking to classify and monitor activities in a site. *Computer Vision and Pattern Recognition* (1998) 1–8. 177

8. W. E. L. Grimson, C. Stauffer: Adaptive background mixture models for real-time tracking. *CVPR'99* (1999).
9. I. Haritaoglu, D. Harwood, L. S. Davis: W^4: Who? When? Where? What? A real time system for detecting and tracking people. *3rd Face and Gesture Recognition Conf.* (1998) 222–227. 177
10. I. Haritaoglu, R. Cutler, D. Harwood, L. S. Davis: Backpack: Detection of people carrying objects using silhouettes. *Internat. Conf. Computer Vision* (1999) 102–107. 177
11. I. Haritaoglu, D. Harwood, L. S. Davis: Hydra: Multiple people detection and tracking using silhouettes. *2nd IEEE Workshop on Visual Surveillance* (1999) 6–13. 177
12. J. Heikkila, O. Silven: A real-time system for monitoring of cyclists and pedestrians. *2nd IEEE Workshop on Visual Surveillance* (1999) 74–81. 177, 178, 179
13. T. Horprasert, D. Harwood, L. A. Davis: A statistical approach for real-time robust background subtraction and shadow detection. *ICCV'99 Frame Rate Workshop* (1999) 1–19. 176, 179, 180
14. T. Y. Ivanov, C. Stauffer, A. Bobick, W. E. L. Grimson Video Surveillance of Interactions. *2nd IEEE Workshop on Visual Surveillance* (1999) 82–90. 177
15. M.-S. Lee: Detecting people in cluttered indoor scenes. *CVPR'00* (2000).
16. S. J. McKenna, Y. Raja, S. Gong: Tracking color objects using adaptive mixture models. *Image and Vision Computing* (1999) 780–785.
17. J. Orwell, P. Remagnino, G. A. Jones: Multi-camera color tracking. *2nd IEEE Workshop on Visual Surveillance* (1999) 14–24. 177
18. R. Rosales, S. Sclaroff: 3D trajectory recovery for tracking multiple objects and trajectory guided recognition of actions. *Computer Vision and Pattern Recognition* (1999) 117–123.
19. P. L. Rosin: Thresholding for change detection. *ICCV'98* (1998) 274–279.
20. C. Stauffer, W. E. L. Grimson, Adaptive background mixture models for real-time tracking. *Computer Vision and Pattern Recognition* (1999) 246–252. 177
21. K. Toyama, J. Krumm, B. Brumitt, B. Meyers: Wallflower: Principles and practice of background maintenance. *Internat. Conf. Computer Vision* (1999) 255–261. 178
22. C. Wren, A. Azabayejani, T. Darrell, A. Pentland: Pfinder: Real-time tracking of the human body. *IEEE Trans. Pattern Analysis and Machine Intelligence* (1997) 780–785.

Stable Monotonic Matching for Stereoscopic Vision

Radim Šára*

Center for Machine Perception, Czech Technical University
Faculty of Electrical Engineering, Prague, Czech Republic
sara@cmp.felk.cvut.cz

Abstract. This paper deals with stable monotonic matching (SMM), which is a generalization of stable matching that includes ordering constraint. The matching algorithm is fast, does not optimize any explicit cost functional, processes one epipolar line at a time, and requires only two parameters for disparity search range.

A designed experiment demonstrates that SMM has no occluding boundary artifacts, that it detects half-occluded regions reliably even if they are wide, and that it rarely misses thin objects in the foreground, unless the ordering is violated. On the other hand, the resulting disparity map is often not dense, especially in weakly textured areas.

1 Introduction

Stereo matching is one of the inverse problems well known in computer vision. Because of its ill-posedness much work has been devoted to posing the problem such that a prior continuity model could be incorporated while keeping the computational complexity of the matching algorithm low. Since the classical heuristic solution due to Pollard, Mayhew and Frisby [7] many authors tried to pose the problem in the framework of statistical estimation theory. The difficulty was that discontinuities and half-occlusions must be accounted for. The ML estimators of early 90's, based on dynamic programming, were later put on a more sound basis by Belhumeur [1] (assuming continuity along epipolar lines) and by Robert and Deriche [8] (assuming isotropic continuity prior). Recent disparity component matching [2] and network flow [6,9] formulations of the matching task also include isotropic prior model but their computational complexity is lower.

Standard stereo matching algorithms that incorporate a prior continuity model often produce mismatches when the image SNR is low. They fail when the prior model becomes locally stronger than the constraints given by the input data, which results in interpolation of the solution within the boundaries surrounding the weakly-textured region. But such interpolation is quite often

* This research was supported by the Czech Ministry of Education under Programme J04/98:212300013 and in part by the Grant Agency of the Czech Republic under Project GAČR 102/00/1679. Results are presented on images from the VASC Image Database at Carnegie Mellon University, the Birch Tree data was provided by SRI.

R. Klette, S. Peleg, G. Sommer (Eds.): Robot Vision 2001, LNCS 1998, pp. 184–192, 2001.
© Springer-Verlag Berlin Heidelberg 2001

erroneous, especially in scenes of deep range and relatively thin objects (like trees scene in a park). A low match error rate becomes important when point models obtained from partial views by stereo matching are to be interpreted by a high-level geometric model.

This led us to the formulation of stereo matching problem that relies just on the evidence available in input data and that does not involve any prior models. Rather than interpolating the textureless areas it rejects all matches there but does not require any prior match rejection threshold to be chosen. The approach is based on the notion of *stability* which turns out to be a powerful constraint for stereo matching. Like the known ordering constraint it only requires certain condition to hold. But unlike the ordering constraint it provides a necessary and sufficient condition for the existence of a unique matching. Since no optimization of a criterion function is directly involved, the algorithm is very fast (and simple).

Stability is a notion naturally capturing the competition among candidate matches for being selected. Roughly speaking, stable matching is one in which each selected match dominates its potential competitors. A precise definition will be given in the next section.

The Stable Matching Problem is also known as the Stable Marriage Problem [5] and has been studied intensively since Gale and Shapley published their classical paper [4]. In this paper we define stable matching for the special case when all potential matches can be globally ranked. This is often the case in area-based stereo using similarity measure like the sum of squared differences or normalized cross-correlation.

The main result of this paper deals with stable monotonic matching, which is a generalization of stable matching that includes ordering constraint. We propose a fast $O(m\,n^2 \log n)$ matching algorithm for $m \times n$ images and show how it can be used in binocular stereo. We will not deal with polynocular stereo in this paper, we refer the interested reader to our report [10].

2   Theory

Let $I, J = \{1, 2, \ldots, n\}$ be two sets indexing pixels on the left-image and the right-image epipolar lines, respectively and let $P \subseteq I \times J$. The element $p = (i, j) \in P$ will be called a *pair*. Let $c: P \to \mathbb{R}$ be a function assigning each pair p a value. In the stereo correspondence problem the $c(i, j)$ is a measure of how much binocular measurements at respective positions i and j correspond to the images of the neighborhood of the same scene point. There is an array of functions that can serve as the correlation measure but we will not distinguish among them here. For the purpose of this paper we convert all values $c(i, j)$ to respective *ranks,* so that the pair of the highest correlation value has the rank of $|P|$. We obtain a function $r: P \to \{1, 2, \ldots, |P|\}$. For given P and r, the tuple (P, r) will be called the *matching problem.* It can be arranged in a *correlation table* which is visualized by a diagram as in Fig. 1 in which each crossing (circled or not) represents a pair from P.

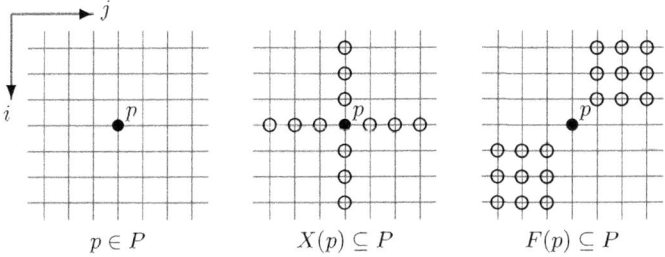

Fig. 1. A pair p in correlation table P (*left*), the X-zone $X(p)$ (*center, empty circles*), and the F-zone $F(p)$ (*right*)

The *X-zone* $X(p) \subseteq P$ of a pair $p = (i,j) \in P$ consists of all pairs $q \in P$ with the same row index i or column index j, except for the pair p itself, as shown in Fig. 1. Similarly, the *F-zone* $F(p) \subseteq P$ of a pair $p \in P$ is formed by two opposite quadrants around (i,j). Two pairs p, q such that $p \in F(q)$ will be called *discordant*. The union of $X(p)$ and $F(p)$ will be called the *FX*-zone, $FX(p) = X(p) \cup F(p)$.

A *bipartite matching* M is a subset of P in which each $i \in I$ and each $j \in J$ is represented at most once. For the sake of brevity, we omit the word 'bipartite' in the following text. The *cardinality* of the matching is $|M|$. A maximum cardinality matching has the greatest number of pairs possible.

2.1   Stable Complete Matching

Definition 1. *A matching $M \subseteq P$ is stable iff for each pair $p \in M$ and every pair $q \in X(p)$ such that $r(q) > r(p)$ there is a pair $s \in M \cap X(q)$ such that $r(s) > r(q)$.*

In other words, a pair p is matched if for each stronger pair q that competes with p there is a still stronger candidate s over-competing q.

Here we consider the maximum cardinality stable matching. To avoid the words 'maximum cardinality stable matching' we use a simpler term *stable complete matching*. Where we want to stress a matching is complete (maximum cardinality), we put a bar above the letter, e.g., $\bar{M}$.

An efficient algorithm exists because of the decomposability of the stable complete matching problem into a series of nested subproblems as follows:

Theorem 1. *Let (P_1, r) be a matching problem. Let a matching subproblem (P_2, r) be constructed such that $P_2 = P_1 \setminus (X(p_1) \cup \{p_1\})$, where p_1 is the highest-rank pair in P_1. Then the following holds:*

1. *If $\bar{M}_1$ is a stable complete matching for (P_1, r) then $\bar{M}_2 = \bar{M}_1 \setminus \{p_1\}$ is a stable complete matching for (P_2, r).*
2. *If $\bar{N}_2$ is a stable complete matching for (P_2, r) then $\bar{N}_1 = \bar{N}_2 \cup \{p_1\}$ is a stable complete matching for (P_1, r).*

For proof see [10]. As a corollary we have that for a given problem (P, r) there is always a *unique* stable complete matching which may be found as follows: The complexity of Alg. 1 is $O(n^2 \log n)$ time and $O(n^2)$ space. Each of the fixed-

---

Algorithm 1 Stable Complete Matching

---

1. (a) For each column $j = 1, 2, \ldots, n$ of the correlation table $c(i, j)$ create a max-rooted heap $H(j)$ of pairs $p(k, j)$ with keys $c(k, j)$, $k = 1, 2, \ldots, n$.
   (b) Create a max-rooted heap L of heaps $H(j)$, in which the key for each $H(j)$ is the key of its root and $j = 1, 2, \ldots, n$.
   (c) Create a flag array C of length n and initialize it to zeros.
   (d) Initialize M to an empty set.
2. Let H be the root of L and let $p = (i, j)$ be the root of H.
   (a) If $C(i) = 1$, remove p from H and update the key of the root in L.
   (b) Otherwise, add p to M, set $C(i) := 1$, and remove H from L.
3. If L is empty, terminate and return stable complete matching M. Otherwise go to Step 2.

---

size heaps in Step 1a can be created in linear time [11], which sums to $O(n^2)$ time. Step 2a is executed $O(n^2)$ times and each execution requires $O(\log n)$ time. Step 2b is executed $O(n)$ times and each execution requires $O(\log n)$ time.

The stable complete matching itself is not very useful for stereoscopic vision: it ignores half-occlusions and assigns a match to every pixel. Its importance lies in the fact that *any* stable matching must be a subset of stable *complete* matching as stated by the following:

Theorem 2. *For a given matching problem (P, r), any stable matching N on (P, r) is a subset of stable complete matching M on (P, r).*

We refer the reader to our report [10] for proof. As will be seen shortly, the practical importance of this result is that it enables the existence of a fast $O(n^2 \log n)$ algorithm for stable monotonic matching which is in the main focus of this paper.

2.2   Stable Monotonic Matching of Maximum Cardinality

A matching M is monotonic iff for each two pairs $p, q \in M$, $p = (i, j)$, $q = (k, l)$ such that $k > i$ it holds that $l > j$. We may observe that the necessary and sufficient condition for monotonicity is that no pair $p \in M$ is in the forbidden zone of a pair $q \in M$: For each $p, q \in M$ it must hold that $p \notin F(q)$. To include the monotonicity constraint we generalize the notion of stability as follows:

Definition 2. *A matching $M \subseteq P$ is FX-stable iff for each pair $p \in M$ and every pair $q \in FX(p)$ of higher rank $r(q) > r(p)$ there is a pair $s \in M \cap FX(q)$ such that $r(s) > r(q)$.*

Note the definition is very similar to Def. 1, the only formal difference is in what subset of P are the competing pairs drawn from. As a corollary of Theorem 2 we see that a unique FX-stable matching always exists for a given problem (P, r) and is a monotonic subset of stable complete matching.

The FX-stable matching for a given matching problem will be called *stable monotonic matching*. The computational problem of finding a stable monotonic matching of maximum cardinality is more difficult than that of stable complete matching. As we saw, it is possible to first find the stable complete matching and then select the largest subset that is monotonic and remains stable. Note however that not every monotonic subset is stable in the sense of Def. 2. Let $\bar{M}$ be a stable complete matching for a given problem (P, r). Consider three pairs p, q, and s from P, such that $r(p) > r(q) > r(s)$, $q \in X(p) \cap X(s)$, $s \in F(p)$, and $p, s \in \bar{M}$. According to Def. 2, if p is removed from $\bar{M}$, the matching becomes unstable because of the pair q, $r(q) > r(s)$. To restore stability, the pair s must be removed as well. We say the (deleting of the) pair $s \in \bar{M}$ is *conditioned* by the pair $p \in \bar{M}$, which we call the *conditioning pair*. If a matching is to remain stable after a conditioning pair p is removed, the corresponding conditioned pair s has to be removed too. This may be implemented by extending the F-zone of s to $\bar{F}(s) = F(s) \cup F(p)$. With this extension of the discordance relation it is now possible to find the largest stable monotonic matching as follows:

Algorithm 2 Maximum Cardinality Stable Monotonic Matching

---

1. Solve the stable complete matching problem on (P, r) by Alg. 1 to get matching M.
2. Initialize M^* and C to empty sets. Sort elements of M in the order of their decreasing rank.
3. If M is empty, terminate. The set M^* is a stable monotonic matching.
4. Let p be the first element in M. Move it from M to C and do the following: If there is no pair in C that conditions p, add p to M^* and remove all $q \in F(p)$ from M.
5. Go to Step 3.

The proof of correctness of Alg. 2 is based on the idea of extension of F-zones for conditioned pairs and on the equivalent of Theorem 1 that exist for FX-stable matchings. We omit the details of the proof for lack of space.

Alg. 2 requires $O(n^2 \log n)$ time and space, which is the complexity of the first step. Step 4 is executed n times and performs at most n conditioning tests. Step 2 requires no overhead since M is already ordered as a side-effect of executing Alg. 1.

3   Experiments

Figure 2 shows what results on standard test sets may be expected from Alg. 2. Modified normalized cross-correlation was used

$$c(i,j) = \frac{\text{cov}(W_i, W_j)}{\text{var}(W_i) + \text{var}(W_j)} \,,$$

where W_i, W_j represent image values in 5×5 matching windows in the left and right image, respectively. Note the percentage of mismatches is very low although the algorithm searches large disparity range. Note also that individual leaves are distinguished in the foreground bush in the Parking Meter disparity map. Holes (shown as white) appear in extremely low-contrast areas (like in the car back in the lower-right image corner in the Parking Meter set or in the featureless ground area in the Birch Tree set) or in ambiguous areas (like in the west wing of the building in the Pentagon set, which is imaged as a very regularly repeated pattern). In the Birch Tree disparity map, small gaps in the foreground tree trunk are present because ordering is locally violated: the solution is switching between the foreground and the background depending on the relative texture strength. Note that the occluding boundaries are crisp even if the matching was done for each epipolar line independently.

Fig. 2. Results of Alg. 2 (SMM) on three standard stereo pairs. Disparity is coded by grey-level (*bars*); small disparity is black, large disparity bright, holes are white. The Parking Meter set (*left*), frame 2 and 14 were used with disparity range $\langle 0, 30 \rangle$. The Pentagon set (*center*), range $\langle -10, 10 \rangle$. The Birch Tree set (*right*), range $\langle 0, 55 \rangle$

Fig. 3 compares SMM with maximum likelihood monotonic matching with minimum horizontal discontinuities (MLMH) implemented by dynamic programming minimizing SSD error functional with disparity discontinuity penalty of α [3]. The MLMH algorithm was modified to use a 5×5 matching window to improve its behavior in terms of mismatches and to make it comparable with the SMM which also uses a 5×5 matching window. A real scene consisting of a triple of long thin stripes (object) in front of a flat wall (background) was used. The object-background distance was chosen as the largest possible such that the ordering constraint still holds. The stripes are 14 pixels wide. The scene was illuminated by texture projector of varying intensity, which simulated varying texture contrast. The contrast was measured as the mean value of the left image. The smallest-contrast texture was invisible to the human eye but still detectable (contrast value of 3.0) and the largest-contrast texture image is shown upper left in Fig. 3 (contrast value of 74.0). The upper left plot shows the relative frequency of missing an object pixel in disparity map computed as the sum of false negatives and error in disparity larger than 1. Relative frequency is plotted with respect to the number of object pixels in ground-truth disparity map (shown lower left). The MLMH estimate used the value of $\alpha = 500$ selected as one that minimizes the error for the maximum texture contrast. We can see that the error generally decreases with increasing intensity. The SMM has a significantly better error rate when detecting a narrow object.

The upper right plot shows the relative frequency of false positive in half-occluded area with respect to the size of the area. The SMM gives small error even under very low texture contrast and performs by the order of magnitude better than the MLMH method.

The lower left plot shows the relative frequency of holes in the disparity map, which measures the disparity map sparsity. We can see that SMM gives much sparser disparity maps than MLMH method when texture contrast is low.

The lower right plot shows the total error in disparity map (the sum of false positives and pixels of disparity error larger than 1) with respect to the matched area and captures thus the mismatch rate in the resulting set of all assigned matches. We can see that SMM performs consistently better than MLMH.

4   Conclusions

The class of stable matchings was studied from which the stable monotonic matching (SMM) is the most useful for computational stereo. The SMM algorithm (1) does not explicitly optimize any criterion function, (2) can handle sparse matching problems, (3) processes one epipolar line at a time, and (4) requires only two parameters for disparity search range (the parameters are usually set to infinity). The fact it is parameter-free makes the algorithm useful for mobile robotics since the large variability of outdoor/indoor scenes usually does not allow to choose a globally optimum value for such crucial parameters as the match rejection threshold or the continuity term weight.

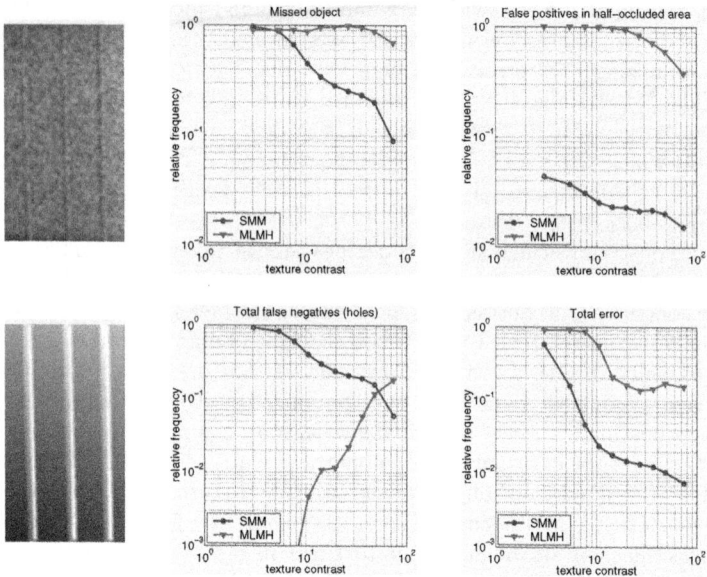

Fig. 3. Matching error probability for SMM and MLMH algorithms (*plots*). The target is a triple of thin objects in front of a planar background (*top left*). The ground-truth disparity map includes half-occluded regions (*bottom left, white*)

As long as ordering holds, SMM rarely misses thin foreground objects in front of a distant background. As opposed to functional-based matching algorithms, SMM does not suffer from the 'streak' effect. Moreover, SMM is suitable when low percentage of false positives is required in disparity map. The price is that the map is usually sparser, especially in the areas of weak and/or ambiguous texture. But the result can be made denser by taking another look and fusing the binocular maps, at which SMM is very efficient, especially for large polynocular stereo sets. More on polynocular stereo and disparity map fusion can be found in [10].

References

1. P. N. Belhumeur: A Bayesian approach to binocular stereopsis. *Int. J. Computer Vision*, **19** (1996) 237–260. 184
2. Y. Boykov, O. Veksler, R. Zabih: Disparity component matching for visual correspondence. In: Proceed. Conf. *Computer Vision and Pattern Recognition*, (1997) 470–475. 184
3. I. J. Cox, S. L. Higorani, S. B. Rao, B. M. Maggs: A maximum likelihood stereo algorithm. *Computer Vision and Image Understanding*, **63** (1996) 542–567. 190
4. D. Gale, L. S. Shapley: College admissions and the stability of marriage. *American Mathematical Monthly*, **69** (1962) 9–15. 185

5. D. Gusfield, R. W. Irving: *The Stable Marriage Problem: Structure and Algorithms.* The MIT Press, Cambridge, London, (1989). 185

6. H. Ishikawa and D. Geiger. Occlusions, discontinuities, and epipolar lines in stereo. In: Proceed. *European Conference on Computer Vision*, vol. 1 (1998) 232–248. 184

7. S. B. Pollard, J. E. W. Mayhew, J. P. Frisby: PMF: A stereo correspondence algorithm using a disparity gradient limit. *Perception*, **14** (1985) 449–470. 184

8. L. Robert, R. Deriche: Dense depth map reconstruction: A minimization and regularization approach which preserves discontinuities. In: Proc. Int. Conf. *Image Processing*, (1992) 123–127. 184

9. S. Roy, I. J. Cox: A maximum-flow formulation of the *n*-camera stereo correspondence problem. In: Internat. Conf. *Proc. Computer Vision*, (1998) 492–499. 184

10. R. Šára: The class of stable matchings for computational stereo. Research Report CTU–CMP–1999–22, Center for Machine Perception, Czech Technical University, Prague, Czech Republic, (1999). 185, 187, 191

11. R. Sedgewick: *Algorithms.* Addison-Wesley, 2nd edition, (1988). 187

Random Sampling and Voting Method for Three-Dimensional Reconstruction

Kazuhiko Kawamoto and Atsushi Imiya

Computer Science Division, Department of Information and Image Sciences,
Chiba University, 1-33 Yayoi-cho, Inage-ku, 263-8522, Chiba, Japan
kazu@icsd7.tj.chiba-u.ac.jp
imiya@ics.tj.chiba-u.ac.jp

Abstract. In the series of papers, we proposed a method for three-dimensional reconstruction from an image sequence without predetecting feature correspondences. In the method, we first collect all images and sample data, and second apply the reconstruction procedure. Therefore, the method is categorized into an off-line algorithm. In this paper, we deal with an on-line algorithm for three-dimensional reconstruction, if we sequentially measure images. Our method is based on the property that points and lines in space are uniquely computed from their projections between two images and among three images, respectively, if a camera system is calibrated. Using these property, our method determines both feature correspondences and three-dimensional positions of points and lines on an object.

1 Introduction

In the series of papers [1,2], we proposed a method for three-dimensional reconstruction from an image sequence without predetecting feature correspondences. For searching feature correspondences and estimating three-dimensional positions of rigid objects, we introduced random sampling and voting process. In the method, we first collect all images and sample data, and second apply the reconstruction procedure. Therefore, the method is categorized into an off-line algorithm. However, since the method based on random sampling and voting process needs not to predetermine the numbers of images and sample data, the method is applicable to time-varying data. In this paper, we deal with an on-line algorithm for three-dimensional reconstruction, if we sequentially observe images.

In 1962, Hough introduced voting process for line detection, called the Hough transform. Subsequently, the Hough transform was extended to conic detection and arbitrary shape detection. Comprehensive surveys of the Hough transform can be found in references [3,4]. In the last decade, for the reduction of the computational complexity of the Hough transform, the technique of random sampling has been introduced to the Hough transform. The Hough transform with random sampling is called the probabilistic or randomized Hough transform [5].In

R. Klette, S. Peleg, G. Sommer (Eds.): Robot Vision 2001, LNCS 1998, pp. 193–200, 2001.
© Springer-Verlag Berlin Heidelberg 2001

recent years, the two techniques of random sampling and voting process have been introduced in the field of motion analysis [6,7].

A typical Hough transform detects planar lines and conics. These two curves are expressed by $ax + by + c = 0$ and $ax^2 + 2bxy + cy^2 + 2dx + 2ey + f = 0$, respectively. Therefore, these two curves have the form

$$\xi^\top a = 0, \tag{1}$$

where ξ is a vector whose elements are polynomial functions of x and y and a is a parameter vector to be detected. Here, we have $\xi = (x^2, 2xy, y^2, 2x, 2y, 1)^\top$ for conics and $\xi = (x, y, 1)^\top$ for lines. If k samples are on the ith line a_i, we obtain $_kC_2$ independent equations to compute a_i. Thus we have systems of equations

$$\Xi_i a_i = 0, \quad i = 1, \ldots, m, \tag{2}$$

where $\Xi_i^\top = [\xi_{i(1)}, \ldots, \xi_{i(k)}]$. This selection of partial equations in eq. (2) is the key concept of the Hough transform [8].

In this paper, we demonstrate that the three-dimensional reconstruction of points and lines are all expressed in the same form as eq. (2), if we measure an image sequence, using a calibrated image system. Therefore, it is possible to reconstruct points, lines and planes in a space, which are usually linear features on a rigid object, using the same concept of the Hough transform.

Bober *et al.* [9] proposed an estimation method for the fundamental matrix, using the Hough transform. Torr and Zisserman proposed MLESAC [10], which is a generalization of RANSAC [11], for the estimation of image geometry. These methods deal with the image geometry of two views. Our method deals with the case of an image sequence. Furthermore, compared to these methods, our method has very simple implementation.

2   Geometry of Points and Lines

We assume that our imaging system is a pinhole camera. Using the pinhole camera model, in this section, we summarize the geometry between points and lines in space and their projections onto image planes.

2.1   Geometry of Points

Let $y = (X, Y, Z)^\top$ be a three-dimensional point in a world coordinate system and $x = (x, y)^\top$ be the projection of y onto an image plane. The relationship between y and x, where express a three-dimensional point and its image point, respectively, can be written as

$$\lambda \begin{bmatrix} x \\ 1 \end{bmatrix} = P \begin{bmatrix} y \\ 1 \end{bmatrix}, \tag{3}$$

where λ is an arbitrary nonzero scalar and P is a 3×4 matrix called the perspective projection matrix. Setting $\xi = (x, y, 1)^\top$ and $v = (X, Y, Z, 1)^\top$, we can compactly rewrite eq. (3) as

$$\lambda \xi = P v. \tag{4}$$

Eliminating scale factor λ in eq. (4), we obtain a pair of equations

$$(x\boldsymbol{p}_3 - \boldsymbol{p}_1)^\top \boldsymbol{v} = 0, \quad (y\boldsymbol{p}_3 - \boldsymbol{p}_2)^\top \boldsymbol{v} = 0, \tag{5}$$

where $\boldsymbol{P}^\top = [\, \boldsymbol{p}_1 \; \boldsymbol{p}_2 \; \boldsymbol{p}_2 \,]$.

If we have m points $\{\boldsymbol{y}_i\}_{i=1}^m$ in a space and observe m points from n cameras represented by $\{\boldsymbol{P}^j\}_{j=1}^n$, we obtain a collection of projections $\{\boldsymbol{x}_i^j\}_{i=1}^m{}_{j=1}^n$ on each image. Therefore, setting $\boldsymbol{v}_i = (\boldsymbol{y}_i, 1)^\top$, we have m systems of equations

$$(x_i^j \boldsymbol{p}_3^j - \boldsymbol{p}_1^j)^\top \boldsymbol{v}_i = 0, \quad (y_i^j \boldsymbol{p}_3^j - \boldsymbol{p}_2^j)^\top \boldsymbol{v}_i = 0, \tag{6}$$

where $\boldsymbol{x}_i^j = (x_i^j, y_i^j)^\top$ and $\boldsymbol{P}^{j\top} = [\boldsymbol{p}_1^j \; \boldsymbol{p}_2^j \; \boldsymbol{p}_3^j]$, for integers i and j such that $1 \leq i \leq m$ and $1 \leq j \leq n$. Note that point correspondences among image frames form sequences, $< \boldsymbol{x}_j^1, \boldsymbol{x}_j^2, \ldots, \boldsymbol{x}_j^n >$, $j = 1, \ldots, m$, in an image sequence.

2.2   Geometry of Lines

Setting $\boldsymbol{x}_1$ and $\boldsymbol{x}_2$ to be two points on an image plane, the planar line through these two points is defined by $\lambda \boldsymbol{\eta} = \boldsymbol{\xi}_1 \times \boldsymbol{\xi}_2$ up to a scale factor, where $\boldsymbol{\xi}_1^\top = (\boldsymbol{x}_1^\top, 1)$ and $\boldsymbol{\xi}_2^\top = (\boldsymbol{x}_2^\top, 1)$. Therefore, from eq. (4), we obtain the following relationship,

$$\lambda \boldsymbol{\eta} = \boldsymbol{P} \boldsymbol{v}_1 \times \boldsymbol{P} \boldsymbol{v}_2 = \begin{bmatrix} (\boldsymbol{p}_2 \wedge \boldsymbol{p}_3)^\top \\ (\boldsymbol{p}_3 \wedge \boldsymbol{p}_1)^\top \\ (\boldsymbol{p}_1 \wedge \boldsymbol{p}_2)^\top \end{bmatrix} \left[\boldsymbol{v}_1 \wedge \boldsymbol{v}_2 \right], \tag{7}$$

where $\wedge$ is the exterior product [12]. In eq. (7), a 6×1 vector $\lambda \boldsymbol{\rho} = \boldsymbol{v}_1 \wedge \boldsymbol{v}_2$ expresses a spatial line through the two points $\boldsymbol{y}_1$ and $\boldsymbol{y}_2$ in a space [13,12]. The coordinates of $\boldsymbol{\rho}$ are called the Plücker coordinates of a spatial line. The relationship between $\boldsymbol{\rho}$ and $\boldsymbol{\eta}$, where express a spatial line and its image line, respectively, can be written as

$$\lambda \boldsymbol{\eta} = \boldsymbol{P}_L \boldsymbol{\rho}, \tag{8}$$

where $\boldsymbol{P}_L$ is the 3×6 matrix defined by $\boldsymbol{P}_L^\top = [\boldsymbol{p}_2 \wedge \boldsymbol{p}_3 \; \boldsymbol{p}_3 \wedge \boldsymbol{p}_1 \; \boldsymbol{p}_1 \wedge \boldsymbol{p}_2]$. Eliminating scale factor λ in eq. (8), we obtain a pair of equations,

$$(a\,\boldsymbol{p}_1 \wedge \boldsymbol{p}_2 - c\,\boldsymbol{p}_2 \wedge \boldsymbol{p}_3)^\top \boldsymbol{\rho} = 0, \quad (b\,\boldsymbol{p}_1 \wedge \boldsymbol{p}_2 - c\,\boldsymbol{p}_3 \wedge \boldsymbol{p}_1)^\top \boldsymbol{\rho} = 0, \tag{9}$$

where $\boldsymbol{\eta} = (a, b, c)^\top$.

If we have m lines $\{\boldsymbol{\rho}_i\}_{i=1}^m$ in a space and observe m points from n cameras represented by $\{\boldsymbol{P}_L^j\}_{j=1}^n$, we obtain a collection of projections $\{\boldsymbol{\eta}_i^j\}_{i=1}^m{}_{j=1}^n$ on each image. Therefore, we have m systems of equations

$$(a_i^j \boldsymbol{p}_1^j \wedge \boldsymbol{p}_2^j - c_i^j \boldsymbol{p}_2^j \wedge \boldsymbol{p}_3^j)^\top \boldsymbol{\rho}_i = 0, \quad (b_i^j \boldsymbol{p}_1^j \wedge \boldsymbol{p}_2^j - c_i^j \boldsymbol{p}_3^j \wedge \boldsymbol{p}_1^j)^\top \boldsymbol{\rho}_i = 0 \tag{10}$$

where $\boldsymbol{\eta}_i^j = (a_i^j, b_i^j, c_i^j)^\top$ and $\boldsymbol{P}_L^{j\top} = [\boldsymbol{p}_2^j \wedge \boldsymbol{p}_3^j \; \boldsymbol{p}_3^j \wedge \boldsymbol{p}_1^j \; \boldsymbol{p}_1^j \wedge \boldsymbol{p}_2^j]$, where integers i and j such that $1 \leq i \leq m$ and $1 \leq j \leq n$. Note that line correspondences among image frames form sequences, $< \boldsymbol{\eta}_i^1, \boldsymbol{\eta}_i^2, \ldots, \boldsymbol{\eta}_i^n >$, $i = 1, \ldots, n$, in an image sequence.

3   Shape for Image Set

Equations (6) and (10) have the same form with eq. (2) which is the fundamental form for the Hough transform. Thus, if we predetermine the perspective
projection matries $\{\boldsymbol{P}^j\}_{j=1}^n$, the reconstruction of points and lines is the similar problem of the detection of lines and conics on an image set. This property
means that the Hough transform can achieve the reconstruction of points and
lines from an image sequence. In this section, we review the off-line algorithm
for three-dimensional reconstruction [1,2].

Equations (6) and (10) have $2n\times4$ and $2n\times6$ coefficient matrices, respectively.
Selecting four rows from eq. (6) and six rows from eq. (10), we can solve the
following systems of equations, for $k \in \{j_1, j_2\}$,

$$(x_i^k \boldsymbol{p}_3^k - \boldsymbol{p}_1^k)^\top \boldsymbol{v}_i = 0, \quad (y_i^k \boldsymbol{p}_3^k - \boldsymbol{p}_2^k)^\top \boldsymbol{v}_i = 0, \tag{11}$$

and for $k \in \{j_1, j_2, j_3\}$,

$$(a_i^k \boldsymbol{p}_1^k \wedge \boldsymbol{p}_2^k - c_i^k \boldsymbol{p}_2^k \wedge \boldsymbol{p}_3^k)^\top \boldsymbol{\rho}_i = 0, \quad (b_i^k \boldsymbol{p}_1^k \wedge \boldsymbol{p}_2^k - c_i^k \boldsymbol{p}_3^k \wedge \boldsymbol{p}_1^k)^\top \boldsymbol{\rho}_i = 0. \tag{12}$$

If we assume that points and lines on images are not occluded, there are $_nC_2$ and
$_nC_3$ combinations for the selection of rows. Therefore, we obtain $_nC_2$ and $_nC_3$
systems of equations for obtaining $\boldsymbol{v}_i$ and $\boldsymbol{\rho}_i$ for each i. On the other hand, if a
pair $(\boldsymbol{\nu}_i^{j_1}, \boldsymbol{\nu}_{i'}^{j_2})$ and a triple $(\boldsymbol{\eta}_i^{j_1}, \boldsymbol{\eta}_{i'}^{j_2}, \boldsymbol{\eta}_{i''}^{j_3})$ are not combinations of corresponding
features, respectively, the parameters which are computed from the false combi-
nations do not coincide with parameters for other pairs and triplets. Therefore,
if the number of images, n, is large, we obtain many correct pairs and triples for
randomly selected pairs and triplets.

It is well known that eqs. (11) and (12) have constraint equations irrelevant
to parameters of $\boldsymbol{v}_i$ and $\boldsymbol{\rho}_i$, which are called epipolar constraints and trifocal
constraints, respectively. These constraints provide conditions which allow the
existence of $\boldsymbol{v}_i$ and $\boldsymbol{\rho}_i$. Using these constraints, we can avoid meaningless voting
process. This property derives the following algorithm based on random sampling
and voting for the reconstruction of points and lines. This idea can be found in
references [14].

Algorithm 1
1. Set $k = 2$ for points (or $k = 3$ for lines).
2. Randomly select k samples from different images.
3. Check whether the k samples satisfy the constraint equations. If they are
   satisfied then go to **4**, otherwise return to **2**.
4. Solve the system of equations for the selected k samples.
5. Vote 1 to this solution in the accumulator space.
6. Repeat **2** to **5** for a predefined number of iterations.
7. Detect peaks in the accumulator space.

4   Shape for Image Sequence

In the previous section, we dealt with the algorithm, for which sample data are precollected from entire images. We call this algorithm "Shape from image set." However, if images are successively measured, the type of data is an image sequence. In this section, assuming that images are successively measured, we deal with the problem of "Shape from image sequence." One of the authors developed an on-line algorithm for the detection of piecewise-linear signals, using random sampling and voting [15]. Here, using the common concept with the previous papaer [15] for the feature extraction from time-varying data, we develop an algorithm for the reconstruction of three-dimensional shape from the image sequence. In the following, for simplicity, we will deal with the three-dimensional reconstruction of points. The arguement is easily extended to line reconstruction.

Let I_t be an image observed at time t. Since we deal with a collection of points for image I_t, image I_t can be considerd as a collection of points. Therefore, we set

$$I_t = \{x_1^t, \ldots, x_{m(t)}^t\}, \tag{13}$$

where $m(t)$ is the number of points in image I_t. For image sequence $\langle I_1, I_2, \ldots, I_t \rangle$, we assume that

$$s_i = \left\langle x_{i(t-k)}^{t-k}, x_{i(t-k+1)}^{t-k+1}, \ldots, x_{i(t)}^t \right\rangle \tag{14}$$

is a sequence of corresponding points for $0 < k < t$. Since projections of a point in a space determine a sequence of corresponding points, s_i is the collection of projections of a point, say y_i, in a space. In eq. (14), y_i is observed over the past k times.

If new image I_{t+1} is observed with holding a sequence of images $\langle I_1, I_2, \ldots, I_t \rangle$ and a sequence of corresponding point s_i in the memory, eq. (14) should be updated as

$$s_i \leftarrow s_i \circ \left\langle x_i^{t+1} \right\rangle, \tag{15}$$

where x_i^{t+1} is the projection of y_i to I_{t+1}. Here, $\circ$ is the concatenation operation.[1] However, since we do not know point correspondences among image frames, we cannot find x_i^{t+1} which is the projection of y_i from image I_{t+1}. The determination of point correspondences is an inverse problem. For infernce of point correspondences between a point in s_i and a point in I_{t+1}, we adopt random sampling and voting process.

Selecting a pair of points x_j^{t+1} and $x_i^{t'} \in s_i$, we compute three-dimensional point y_i according to the algorithm described in the previous section. After an

---

[1]   The concatenation operation $\circ$ is defined by,

$$\left\langle x_{i_1}, \ldots, x_{i_k} \right\rangle \circ \left\langle x_{i_{k+1}}, \ldots, x_{i_n} \right\rangle = \left\langle x_{i_1}, \ldots, x_{i_k}, x_{i_{k+1}}, \ldots, x_{i_n} \right\rangle. \tag{16}$$

appropriate number of iterations of random sampling and voting, if we have a peak at the position $\boldsymbol{y}_i$ in the accumulator space, we conclude that $\boldsymbol{x}_j^{t+1}$ corresponds to points in sequence s_i. However, if the height of the peak in the accumulator space is lower than a predefined threshold, we conclude that $\boldsymbol{x}_j^{t+1}$ has no corresponding points in s_i. Setting $A(\boldsymbol{y}_i)$ to be the height of the peak in the accumulator space for $\boldsymbol{y}_i$ and γ to be an predefined threshold, this property of sample points in the image sequence derives the following updating rule

$$s_i \leftarrow \begin{cases} s_i \circ \langle \boldsymbol{x}_i^{t+1} \rangle, & \text{if } A(\boldsymbol{y}_i) > \gamma, \\ s_i, & \text{otherwise.} \end{cases} \tag{17}$$

Using the updating rule in eq. (17), we develop Algorithm 3 for "Shape from image sequence".

Algorithm 2
1. Set t=1.
2. Randomly select a pair of sample points $(\boldsymbol{x}_j^{t+1}, \boldsymbol{x}_{j'}^{t'})$ such that $0 < t' \le t$.
3. Check whether the pair $(\boldsymbol{x}_j^{t+1}, \boldsymbol{x}_{j'}^{t'})$ satisfy the epipolar constraint. If they are satisfied then go to **4**, otherwise return to **2**.
4. Compute the three-dimensional point $\boldsymbol{y}_l$ from the pair $(\boldsymbol{x}_j^{t+1}, \boldsymbol{x}_{j'}^{t'})$.
5. Increment the accumulator $A(\boldsymbol{y})$ by one.
6. If $\gamma > A(\boldsymbol{y}_l)$, update $s_l \leftarrow s_l \circ \langle \boldsymbol{y}_j^{t+1} \rangle$ and remove $\boldsymbol{y}_j^{t+1}$ from I_{t+1}.
7. Repeat from **2** to **6** for a predefined number of iterations.
8. If $A(\boldsymbol{y}_l) > \gamma$, detect $\boldsymbol{y}_l$ and s_l.
9. Set t:=t+1 and go to **2**.

Algorithm 2 tracks corresponding points frame by frame, using multiple epipolar constraints for an image sequence. Since the epipolar constraint is a necessary condition for point correspondences, Algorithm 2 infers point correspondences, collecting multiple necessary conditions.

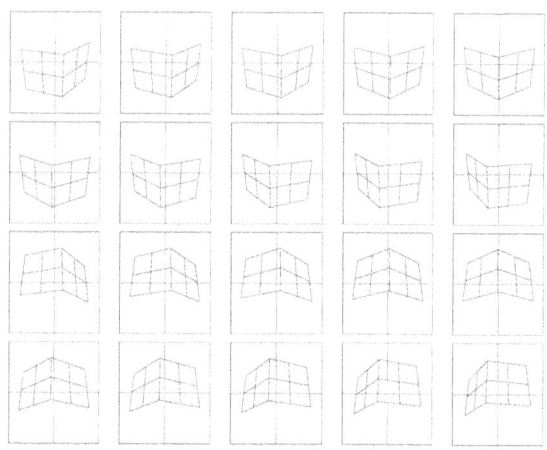

Fig. 1. A sequence of grid pattern images

5   Experiments

Using synthetic data, we evaluated the performance of our method. We reconstructed points and lines which form a grid pattern in three-dimensional space. We uesd images from 20 different views. Figure 1 shows a sequence of images taken from 20 views. The images are digitized as 256×256 pixels. In the experiments, we assume that the coordinates of points and lines on each image are predetected, using a corner detector and a line detector, respectively. For the detection of these features in each image, we adopt the conventional Hough transform.

We evaluated the error using the criterion

$$\Delta y = \sqrt{\frac{1}{m} \sum_{\alpha=1}^{m} ||\bar{\boldsymbol{y}}_{\alpha} - \boldsymbol{y}_{\alpha}||}, \qquad (18)$$

where $\bar{\boldsymbol{y}}_{\alpha}$ and $\boldsymbol{y}_{\alpha}$ are the original and estimated spatial points on the grid pattern, respectively. According to eq. (18), we obtained $\Delta y = 1.726041$.

The Plücker coordinates of a line in a space are not intuitively interpreted for geometric meaning. The line L in a space is represented by the point $\boldsymbol{r}$ on L which is closest to the coordinate origin and the unit vector $\boldsymbol{u}$ which is the orientation of L. For the evaluation of estimated lines in a space, we decomposed the Plücker coordinates $\boldsymbol{\rho}$, of L into $\boldsymbol{r}$ and $\boldsymbol{u}$. We evaluated the error using the criterions

$$\Delta r = \sqrt{\frac{1}{m} \sum_{\alpha=1}^{m} ||\bar{\boldsymbol{r}}_{\alpha} - \boldsymbol{r}_{\alpha}||}, \ \Delta u = \sqrt{\frac{1}{m} \sum_{\alpha=1}^{m} (\bar{\boldsymbol{u}}_{\alpha}^{\top} \boldsymbol{u}_{\alpha})}, \qquad (19)$$

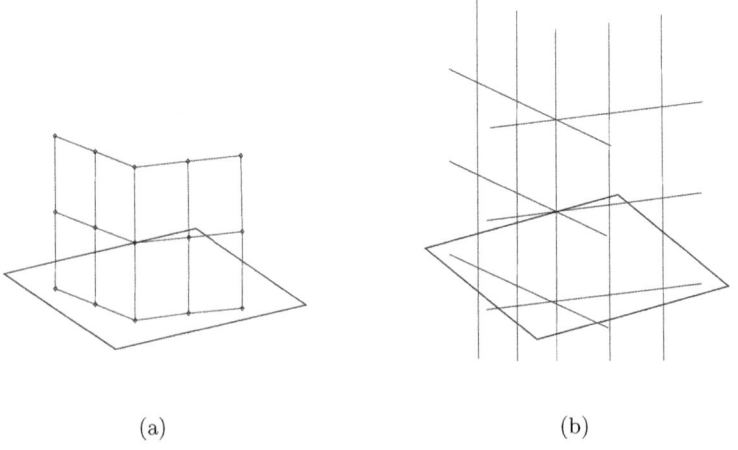

(a)   (b)

Fig. 2. (a) Reconstructed points. (b) Reconstructed lines

where $\bar{r}_\alpha$ and $\bar{u}_\alpha$ are the original parameters and r_α and u_α are the estimated parameters. According to eq. (19), we obtained $\Delta r = 1.844175$ and $\Delta u = 0.999895$.

6   Conclusions

In this paper, we first shown that the three-dimensional reconstruction of points and lines is expressed in a form which is suitable for the Hough transform if we measure an image sequence using a calibrated image system. Next, we proposed off-line algorithms for three-dimensinaol reconstruction which is based on the the Hough transform, using epipolar and trifocal constraints. Finally, for time-varying data, we proposed on-line algorithms for three-dimensinaol reconstruction which is based on the the Hough transform, using epipolar and trifocal constraints.

References

1. K. Kawamoto, A. Imiya: The detection of spatial points and lines by random sampling and voting procedure. (accepted for PRL). 193, 196
2. A. Imiya, K. Kawamoto: Performance analysis of shape recovery by random sampling and voting. *Performance Characterization in Computer Vision*, Kluwer Academic Publishers, Dordrecht, The Netherlands, (2000) 227–240. 193, 196
3. J. Illingworth, J. Kittler: A survey of the Hough transform. *CVGIP*, **44** (1988) 87–116. 193
4. V. F. Leavers: Which Hough Transform? *CVGIP-IU*, **58** (1993) 250–264. 193
5. H. Kälviäinen, P. Hirvonen, L. Xu, E. Oja: Probabilistic and non-probabilistic Hough transforms: overview and comparisons. *IVC*, **13** (1995) 239–252. 193
6. H. Kälviäinen: Motion detection by the RHT method: probability mechanisms and extensions. Research Report, **32**, Lappeenranta University of Technology, Lappeenranta, Finland, (1992). 194
7. A. Imiya, I. Fermin: Motion analysis by random sampling and voting process. *CVIU*, **73** (1999) 309–328. 194
8. A. Imiya, K. Kawamoto: A dynamics of the Hough transform and artificial neural networks. *Lecture Notes in Artificial Intelligence*, Springer, Berlin, **1715** (1999) 36–50. 194
9. M. Bober, N. Georgis, J. Kittler: On accurate and robust estimation of fundamental matrix. *CVIU*, **72** (1998) 39–53. 194
10. P. H. S. Torr, A. Zisserman: MLESAC: a new robust estimator with application to estimating image geometry. *CVIU*, **78** (2000) 138–156. 194
11. M. A. Fischler, O. Firschein: Parallel guessing: a strategy for high-speed computation. *PR*, **20** (1987) 257–263. 194
12. S. Carlsson: Multiple image invariance using the double algebra. *Lecture Notes in Computer Science*, Springer, Berlin, **825** (1994) 145–164. 195
13. O. D. Faugeras, B. Mourrain: On the geometry and algebra of the point and line correspondences between N images. Technical Report N2665, INRIA, (1995). 195
14. L. Xu, E. Oja: Randomized Hough transform: basic mechanisms, algorithms, and computational complexities. *CVGIP-IU*, **57** (1993) 131–154. 196
15. A. Imiya: Detection of piecewise-linear signals by the randomized Hough transform. *PRL*, **17** (1996) 771–776. 197

Binocular Stereo by Maximizing the Likelihood Ratio Relative to a Random Terrain

Georgy Gimel'farb

CITR, Department of Computer Science, Tamaki Campus, University of Auckland
Private Bag 92019, Auckland, New Zealand
g.gimelfarb@auckland.ac.nz

Abstract. A novel approach to computational binocular stereo based on the Neyman–Pearson criterion for discriminating between statistical hypotheses is proposed. An epipolar terrain profile is reconstructed by maximizing its likelihood ratio with respect to a purely random profile. A simple generative Markov-chain model of an image-driven profile that extends the model of a random profile is introduced. The extended model relates transition probabilities for binocularly and monocularly visible points along the profile to grey level differences between corresponding pixels in mutually adapted stereo images. This allows for regularizing the ill-posed stereo problem with respect to partial occlusions.

1 Introduction

Computational binocular stereo that reconstructs 3D terrains from stereo pairs of images is an ill-posed inverse photometric problem because a rich variety of different optical surfaces can produce the same stereo pair [11,12]. The ill-posedness is caused mainly by partial occlusions hindering stereo observation of some terrain points and by uniform or repetitive coloring of the surface. To partially regularize the problem, stereo images have to be matched with due regard to binocular and monocular visibility of terrain points.

Most of the known stereo matching algorithms (see, for instance, [1,2,8,9,13]) state and solve the stereo problem as a statistical problem of estimating a hidden Markov model of an epipolar terrain profile. The prior profile model is combined with the conditional model of stereo images, given the profile, to derive the posterior model and use it for measuring similarity between the stereo images for each possible profile. Then the reconstruction is conducted by maximizing the similarity between the images. In many cases the similarity is measured with no explicit account of possible partial occlusions.

Symmetric Dynamic Programming Stereo (SDPS) discussed in [3,6] follows the same scheme but allows for discriminating between the binocularly (BVP) and only monocularly visible points (MVP) along the profile during the reconstruction. All variants of an epipolar profile are represented by continuous paths in a specific graph of profile variants (GPV). Each GPV-node has three states specifying whether it represents the BVP yielding two corresponding pixels in a

R. Klette, S. Peleg, G. Sommer (Eds.): Robot Vision 2001, LNCS 1998, pp. 201–209, 2001.
© Springer-Verlag Berlin Heidelberg 2001

stereo pair or the MVP depicted only in the left or in the right stereo image. The allowable transitions between the successive GPV-nodes along the profile depend on their visibility states. In this case the similarity for the BVPs is obtained by comparing the corresponding pixels but some heuristic weights for the MVPs have to be involved to search for a profile yielding the best similarity between the images [4].

This paper proposes another approach that is based on the Neyman–Pearson criterion [10] and involves explicit Markov models of an epipolar profile: the reconstructed profile has to maximize the likelihood ratio with respect to a purely random one. The Markov-chain model of a random epipolar profile introduced in [4,5] is extended below by relating the transition probabilities for each GPV-node to grey values in the corresponding pixels of a stereo pair. Transitions to the GPV-nodes representing BVPs depend on grey level deviations between the mutually adapted stereo images, the adaptation tending to reduce relative photometric distortions of the binocularly visible corresponding parts of the images. Each BVP-transition specifies also the probabilities of transitions to the adjacent nodes representing MVPs so that all the transition probabilities in the GPV are related to the images. Thus the proposed model allows for a partial probabilistic regularization of terrain reconstruction such that the transition probabilities for purely random profiles constitute the regularizing parameters [7].

The paper is organized as follows. Section 2 considers the profile reconstruction based on maximizing the likelihood ratio. The extended probabilistic model of an epipolar profile is presented in Section 3. Experimental results and conclusions are given in Section 4.

2   Profile Reconstruction Using Log-Likelihood Ratio

Let x denote the x-coordinate of the GPV-node, p be the integer x-parallax, or disparity between the corresponding pixels with integer coordinates x_L and x_R in stereo images, and $s \in \{B, ML, MR\}$ be the visibility state indicating the BVP or MVP visible only in the left or right image, respectively. It holds for the symmetric stereo geometry [3,6] that $x = (x_L + x_R)/2$, $p = x_L - x_R$, $x_L = x + p/2$, and $x_R = x - p/2$.

Figure 1 shows a fragment of the GPV, each GPV-node (x, p, s) having seven admissible transitions to the next three nodes $(x + 0.5, p - 1, s')$, $(x + 0.5, p + 1, s')$, and $(x + 1, p, s')$. According to the explicit generative Markov model of the profile variants, every epipolar profile $\mathbf{p} = [(x_i, p_i) : i = 1, \ldots, n]$ with n GPV-nodes is generated as a Markov chain of $n - 1$ successive admissible transitions from each current GPV-node (x_i, p_i, s_i) to the next one $(x_{i+1}, p_{i+1}, s_{i+1})$; $i = 1, \ldots, n - 1$.

Let g_L and g_R be the left and right images of a stereo pair. Let $\Pr(\mathbf{p}|g_L, g_R)$ and $\Pr(\mathbf{p})$ specify the probability distributions of profiles in the GPV under two simple statistical hypotheses: an "image-driven" or a purely random profile, respectively. Thus the profile reconstruction can be based on the Neyman–Pearson criterion [10] of choosing the first hypothesis by comparing to a particular threshold Θ the likelihood ratio or, what is the same, the log-likelihood ratio $L(\mathbf{p}|g_\mathrm{L}, g_\mathrm{R})$ of an image-driven profile with respect to a random profile:

$$L(\mathbf{p}|g_\mathrm{L}, g_\mathrm{R}) = \log \Pr(\mathbf{p}|g_\mathrm{L}, g_\mathrm{R}) - \log \Pr(\mathbf{p}) \geq \Theta. \qquad (1)$$

The most adequate profile in the GPV can be chosen by maximizing the log-likelihood ratio in Equ. (1):

$$\mathbf{p}^* = \arg \max_{\mathbf{p}} L(\mathbf{p}|g_\mathrm{L}, g_\mathrm{R}). \qquad (2)$$

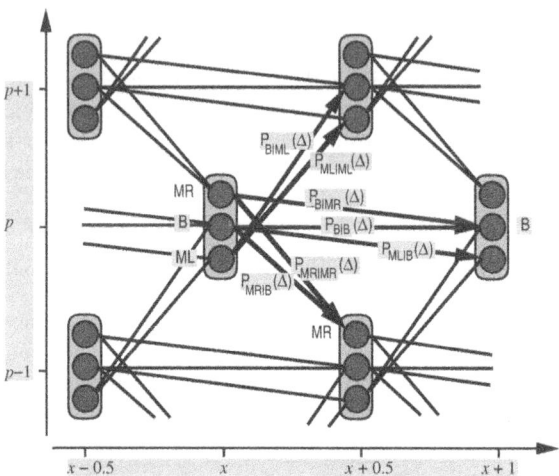

Fig. 1. Transition probabilities for the GPV

Because the similarity between stereo images is given in terms of the likelihood ratio, this approach allows for comparing profiles of different length and solving Equ. (2) with sequential decision rules to accelerate the reconstruction. In this latter case some profile variants can be rejected after comparing their initial parts of length $m < n$ if the likelihood ratio becomes lower than a particular threshold that depends generally on m.

The SDPS algorithm in [3,6] is easily modified to implement the reconstruction process of Equ. (2) if both the probability distributions $\Pr(\mathbf{p}|g_\mathrm{L}, g_\mathrm{R})$ and $\Pr(\mathbf{p})$ in Equ. (1) are represented by the products of transition probabilities specified by particular explicit Markov-chain models.

3 Markov-Chain Models of a Profile

A purely random profile is described in [4,5] as a stationary Markov chain of the GPV-nodes with the transition probabilities $P_{s'|s}$ that depend only on the visibility states. Both the monocular cases ML and MR are equivalent by symmetry and can be denoted as M:

$$
\begin{aligned}
P_{\mathrm{ML|ML}} &= P_{\mathrm{MR|MR}} \equiv P_{\mathrm{M|M}}; \\
P_{\mathrm{B|ML}} &= P_{\mathrm{B|MR}} \quad\equiv P_{\mathrm{B|M}} = 1 - P_{\mathrm{M|M}}; \\
P_{\mathrm{ML|B}} &= P_{\mathrm{MR|B}} \quad\equiv P_{\mathrm{M|B}} = 0.5\left(1 - P_{\mathrm{B|B}}\right).
\end{aligned}
\tag{3}
$$

Thus the stationary Markov chain producing purely random profiles is specified by the two transition probabilities $P_{\mathrm{B|B}}$ and $P_{\mathrm{M|M}}$.

For the image-driven Markov chain generating the actual terrain profiles, stereo images are assumed to be mutually adapted along each profile. The adaptation is performed within a given range $E = [2 - e_{\max}, e_{\max}]; \ 1 \le e_{\max} < 2$, of admissible ratios between grey level increments for the corresponding successive BVPs in both stereo images (see [3,6] for more detail). It has a goal of excluding or reducing relative photometric image distortions. Then the probability of transition from a current GPV-node (x, p, s) to the next node representing a BVP (x', p', B) can be related to the residual grey level difference $\Delta_{x',p'}$ between the corresponding points x'_{L} and x'_{R} in the adapted images for each profile under consideration.

The transition probabilities $\mathrm{Pr}_{s'|s}(\Delta_{x',p'}) \equiv \mathrm{Pr}_{s'|s}(\Delta)$ satisfy the obvious conditions of Equ. (3), the two last MVP-transitions being equalized from the same considerations of symmetry:

$$
\begin{aligned}
\mathrm{Pr}_{\mathrm{ML|ML}}(\Delta_{x+0.5,p+1}) &= 1 - \mathrm{Pr}_{\mathrm{B|ML}}(\Delta_{x+0.5,p+1}); \\
\mathrm{Pr}_{\mathrm{MR|MR}}(\Delta_{x+0.5,p-1}) &= 1 - \mathrm{Pr}_{\mathrm{B|MR}}(\Delta_{x+1,p}); \\
\mathrm{Pr}_{\mathrm{MR|B}}(\Delta_{x+0.5,p-1}) &= \mathrm{Pr}_{\mathrm{ML|B}}(\Delta_{x+1,p}) \\
&= 0.5 \cdot \left(1 - \mathrm{Pr}_{\mathrm{B|B}}(\Delta_{x+1,p})\right).
\end{aligned}
\tag{4}
$$

The log-likelihood ratio $l(x_i, p_i, s_i; x_{i-1}, p_{i-1}, s_{i-1}|g_{\mathrm{L}}, g_{\mathrm{R}})$ for the i-th transition along a profile $\mathbf{p}$ combines the transition probabilities of Equ. (4) depending on a given stereo pair $(g_{\mathrm{L}}, g_{\mathrm{R}})$ with the probabilities of Equ. (3) that specify a purely random profile:

$$
\begin{aligned}
l(x_i, p_i, s_i; x_{i-1}, p_{i-1}, s_{i-1}|g_{\mathrm{L}}, g_{\mathrm{R}}) = \\
\log \mathrm{Pr}_{s_i|s_{i-1}}(\Delta_{x_i,p_i}) - \log P_{s_i|s_{i-1}}.
\end{aligned}
\tag{5}
$$

Thus the log-likelihood ratio for the total profile $\mathbf{p}$ is an additive functional with respect to $\mathbf{p}$

$$
L(\mathbf{p}|g_{\mathrm{L}}, g_{\mathrm{R}}) = \sum_{i=1}^{n} l(x_i, p_i, s_i; x_{i-1}, p_{i-1}, s_{i-1}|g_{\mathrm{L}}, g_{\mathrm{R}})
\tag{6}
$$

that can be maximized by dynamic programming techniques.

Experiments in Section 4 show that the probabilities $P_{\mathrm{M|M}}$ and $P_{\mathrm{B|B}}$ of Equ. (3) can be considered in Equ. (2) as regularizing parameters that define smoothness and visual quality of the reconstructed terrain. The final reconstruction results depend also on the image adaptation range E and on the chosen probability function $\mathrm{Pr}_{s'|s}(\Delta)$ in Equ. (4).

4   Experimental Results and Concluding Remarks

Several digital x-parallax maps (DPM) consisting of the epipolar terrain profiles reconstructed from the artificial stereo pair "Corridor" and the natural stereo pair "Pentagon" in Fig. 2 are presented in Fig. 3. The "ground truth" with the total disparity range [0,10] in Fig. 2, (c), allows for checking the actual quality of reconstruction of the "Corridor" scene.

    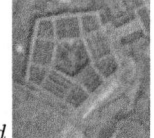

a   *b*   *c*   *d*   *e*

Fig. 2. Artificial stereo pair 256×256 "Corridor" (a, b) with the known "ground truth" in terms of ideal integer disparities shown in the range image (c) and the natural stereo pair 512×512 "Pentagon" (d, e)

These and other experiments (see, for instance, [7]) show that the reconstructed DPMs are in close agreement with visual perception within a sufficiently wide "triangular" domain of the regularizing parameter space. But it is worth noting that the orthoimages of these scenes, formed from the stereo pairs in line with the DPMs, are quite similar over almost the total parameter space. Thus even visually unacceptable solutions yield close similarity between the images as could be expected from the ill-posedness of the problem.

The reconstruction results are rather similar for the following three different probability functions with $\sigma = 5, \ldots, 15$:

$$\Pr_{\mathrm{B}|s}(\Delta) = 0.998 \exp\left(-\frac{\Delta^2}{\sigma^2}\right) + 0.001; \tag{7}$$

$$\Pr_{\mathrm{B}|s}(\Delta) = 0.998 \exp\left(-\frac{|\Delta|}{\sigma}\right) + 0.001; \tag{8}$$

$$\Pr_{\mathrm{B}|s}(\Delta) = 0.998 \exp\left(-\sqrt{\frac{|\Delta|}{\sigma}}\right) + 0.001, \tag{9}$$

and for the different adaptation ranges $E = [0.8, 1.2] \ldots [0.4, 1.6]$ although the reconstructed terrains become smoother for the larger values of σ and larger adaptation ranges. The "Corridor" and "Pentagon" scenes in Fig. 3 are reconstructed using the same image adaptation range $E = [0.8, 1.2]$ but different transition probabilities of Equ. (8) and Equ. (7), respectively, with $\sigma = 10$.

In principle, the functions $\Pr_{\mathrm{B}|s}(\Delta)$ can be empirically estimated using training samples of the epipolar stereo pairs with known terrain models. But the

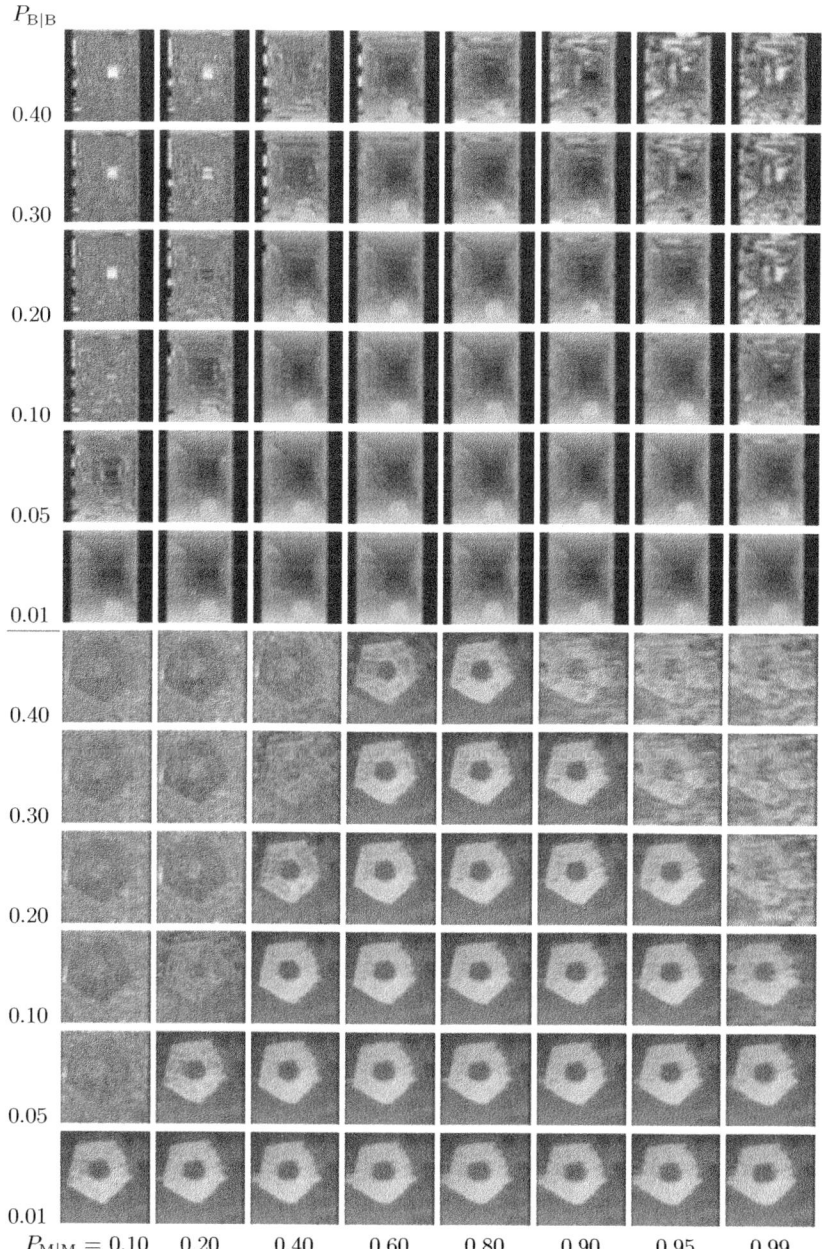

Fig. 3. Range images of the reconstructed scenes "Corridor" and "Pentagon"

experiments show that reconstruction results depend much more on the regularizing parameters.

Table 1. Mean absolute differences (m.a.d.) and standard deviations (st.d.) of the reconstructed "Corridor" from the ground truth in Fig. 2, (c)

		Transition probabilities of Equ. (8) with $\sigma = 5$							
	$P_{\text{M}|\text{M}}$	0.10	0.20	0.40	0.60	0.80	0.90	0.95	0.99
$P_{\text{B}|\text{B}} = 0.40$	m.a.d.	2.99	2.93	2.07	1.26	1.03	1.41	1.91	1.97
	st.d.	1.96	2.05	2.12	1.74	1.35	1.49	1.64	1.61
0.30	m.a.d.	2.97	2.89	1.50	0.92	0.86	0.97	1.48	1.95
	st.d.	1.99	2.08	1.94	1.47	1.29	1.28	1.51	1.61
0.20	m.a.d.	2.96	2.54	0.97	0.68	0.65	0.76	0.91	1.88
	st.d.	2.01	2.13	1.60	1.25	1.10	1.14	1.22	1.61
0.10	m.a.d.	2.87	1.27	0.56	0.46	0.44	0.48	0.56	1.08
	st.d.	2.08	1.84	1.06	0.85	0.44	0.48	0.56	1.08
0.05	m.a.d.	1.88	0.65	0.43	0.38	0.38	0.37	0.40	0.70
	st.d.	2.09	1.19	0.79	0.67	0.65	0.62	0.66	1.02
0.01	m.a.d.	0.48	0.37	0.35	0.35	0.34	0.34	0.34	0.37
	st.d.	0.90	0.62	0.54	0.55	0.53	0.54	0.53	0.55

		Transition probabilities of Equ. (8) with $\sigma = 10$							
	$P_{\text{M}|\text{M}}$	0.10	0.20	0.40	0.60	0.80	0.90	0.95	0.99
$P_{\text{B}|\text{B}} = 0.40$	m.a.d.	3.02	2.98	1.79	0.80	0.74	1.42	2.08	2.22
	st.d.	2.05	2.09	2.04	1.39	1.08	1.51	1.65	1.71
0.30	m.a.d.	3.02	2.87	0.94	0.48	0.54	0.75	1.52	2.20
	st.d.	2.05	2.12	1.58	0.91	0.87	1.06	1.53	1.71
0.20	m.a.d.	3.01	2.42	0.51	0.36	0.39	0.49	0.71	2.16
	st.d.	2.08	2.12	1.00	0.63	0.66	0.76	1.01	1.72
0.10	m.a.d.	2.83	0.75	0.37	0.36	0.35	0.36	0.41	0.98
	st.d.	2.12	1.40	0.62	0.55	0.57	0.57	0.62	1.23
0.05	m.a.d.	1.50	0.39	0.35	0.35	0.35	0.34	0.35	0.51
	st.d.	1.90	0.66	0.54	0.54	0.54	0.53	0.55	0.78
0.01	m.a.d.	0.35	0.35	0.34	0.34	0.34	0.33	0.34	0.35
	st.d.	0.54	0.53	0.54	0.54	0.53	0.53	0.53	0.53

Table 1 shows the mean absolute differences and standard deviations between the reconstructed digital x-parallax models of the "Corridor" scene and the ground truth of Fig. 2,c for the transition probabilities of Equ. (8) with $\sigma = 5$ and $\sigma = 10$. The latter reconstruction results are shown in Fig. 3. This scene has the dominant uniform or almost uniform coloring, and the best results with the mean absolute error 0.34 and standard deviation 0.53 – 0.54 in the total disparity range $[0, 10]$ are obtained for $P_{\text{B}|\text{B}} = 0.01$ and $P_{\text{M}|\text{M}} = 0.80 \ldots 0.95$ ($\sigma = 5$) or $0.40 \ldots 0.95$ ($\sigma = 10$). The "Pentagon" scene with more textured coloring gives the apparently best visual results for larger values of $P_{B|B}$, e.g., $P_{B|B} = 0.10\ldots0.20$ and $P_{M|M} = 0.80$ in Fig. 3, and the like results are even better for $\sigma = 5$ as shown in [7].

These experiments suggest that the proposed approach offers advantages over conventional reconstruction techniques based on stereo matching that allows for partial occlusions. Also, it shows promise of taking account of image coloring uniformity if this latter could be properly related to the regularizing parameters.

Acknowledgements

This work was supported in part by the University of Auckland Research Committee research grant XXXX/9343/3414108.

References

1. H. H. Baker: Surfaces from mono and stereo images. *Photogrammetria* **39** (1984) 217–237.   201
2. W. Förstner: Image matching. In: R. M. Haralick and L. G. Shapiro: *Computer and Robot Vision.* Vol. 2, chapter 16. Reading, Addison-Wesley (1993) 289–378. 201
3. G. L. Gimel'farb: Intensity-based computer binocular stereo vision: signal models and algorithms. *Int. J. of Imaging Systems and Technology* **3** (1991) 189–200.   201, 202, 203, 204
4. G. L. Gimel'farb: Regularization of low-level binocular stereo vision considering surface smoothness and dissimilarity of superimposed stereo images. In: C. Arcelli, L. P. Cordella, G. Sanniti di Baja (eds.): *Aspects of Visual Form Processing.* Singapore, World Scientific (1994) 231–240.   202, 203
5. G. L. Gimel'farb: Symmetric bi- and trinocular stereo: tradeoffs between theoretical foundations and heuristics. *Computing Supplement* **11** (1996) 53–72.   202, 203
6. G. Gimel'farb: Stereo terrain reconstruction by dynamic programming. In: B. Jaehne, H. Haussecker, P. Geisser (eds.): *Handbook of Computer Vision and Applications* **2**: *Signal Processing and Pattern Recognition.* San Diego, Academic Press (1999) 505–530.   201, 202, 203, 204
7. G. Gimel'farb, H. Li: Probabilistic regularization in symmetric dynamic programming stereo. In: *Proc. of the Image and Vision Computing New Zealand'2000 Conf.*, November 2000, Hamilton, New Zealand. (2000) [to appear].   202, 205, 208
8. M. J. Hannah: Digital stereo image matching techniques. *Int. Archives on Photogrammetry and Remote Sensing* **27** (1988) 280–293.   201
9. T. Kanade, M. Okutomi: A stereo matching algorithm with an adaptive window: theory and experiment. *IEEE Trans. on Pattern Analysis and Machine Intelligence* **16** (1994) 920–932.   201
10. M. G. Kendall, A. Stuart: *The Advanced Theory of Statistics.* **2**: *Inference and Relationship.* London, Charles Griffin (1967).   202
11. V. R. Kyreitov: *Inverse Problems of Photometry.* Novosibirsk, Computing Center of the Siberian Branch of the Academy of Sciences of the USSR (1983) [In Russian]. 201
12. T. Poggio, V. Torre, C. Koch: Computational vision and regularization theory. *Nature* **317** (1985) 317–319.   201

13. G.-Q. Wei, W. Brauer, , G. Hirzinger: Intensity- and gradient-based stereo matching using hierarchical Gaussian basis functions. *IEEE Trans. on Pattern Analysis and Machine Intelligence* **20** (1998) 1143–1160. 201

Stereo Reconstruction from Polycentric Panoramas

Fay Huang, Shou Kang Wei, and Reinhard Klette

CITR, Computer Science Department, The University of Auckland
Tamaki Campus, Auckland, New Zealand
{fay,shoukang,rklette}@citr.auckland.ac.nz

Abstract. The paper defines polycentric panoramas as a generalized model of panoramic images which covers a wide range of previously introduced models such as single-center panoramas, multi-perspective panoramic images, or concentric panoramic images. This paper presents geometric fundamentals towards the stereo reconstruction of scenes or objects based on captured pairs of polycentric panoramas. The paper discusses the image acquisition model, epipolar geometry and basics of correspondence analysis. The derived general epipolar curve theorem holds for pairs of polycentric panoramas. This is briefly illustrated by adjusting the general formula of the theorem to specific panoramic image pairs. Experimental results (for two different types of polycentric panoramas) towards stereo reconstruction also demonstrate the practical relevance of the derived geometric fundamentals.

Keywords: Panoramic images, stereo reconstruction, epipolar geometry.

1 Introduction

A 360 degrees full-view panoramic image can be acquired in different ways [17]. Image stitching techniques introduce errors into the resulting panorama [1]. A better ways is using a slit camera [12]. A slit camera is characterized geometrically by a single focal point and a 1D linear image slit. To acquire a polycentric panoramic image, a slit camera rotates with respect to a fixed 3D axis (e.g. the rotation axis of a turntable) and captures one slit image in every constant angular interval. Each slit image contributes to one column of a polycentric panoramic image. To be more precise, we assume a rotation of a slit camera with respect to a fixed rotation axis and taking slit images consecutively at equidistant angles α_t, for $1 \leq t \leq W_{\mathcal{P}}$. Let the distance between the slit camera's focal point and the rotation axis be $R + \varepsilon_t$ for any of these angles α_t. We assume that these focal points are in a plane (exactly) orthogonal to the rotation axis, and this paper only considers cases with $\varepsilon_t = 0$, for all t, i.e. accurate positionings of the camera's focal point on a circle of radius R. We call this the *base circle* of this panoramic image. The optical axes of the slit camera are always assumed to be in the plane of the base circle. Let ω_t be the angle between the normal vector of the

R. Klette, S. Peleg, G. Sommer (Eds.): Robot Vision 2001, LNCS 1998, pp. 209–218, 2001.
© Springer-Verlag Berlin Heidelberg 2001

base circle at the associated focal point and the optical axis of the slit camera. We assume that all angles ω_t are constant for one panoramic image. Figure 1 depicts the assumed acquisition situation. Thus a *panoramic image* is characterized by radius R of the base circle, angle ω, an effective focal length f (which always remains constant for all $W_\mathcal{P}$ positions of the slit camera for capturing one panoramic image), and the number $W_\mathcal{P}$ of image columns. Panoramic images are captured on a straight cylinder of radius R, where the resolution $H_\mathcal{P}$ of the line camera specifies the number $H_\mathcal{P}$ of image rows. Altogether, a panoramic image can be considered as being a planar $H_\mathcal{P} \times W_\mathcal{P}$ rectangular array.

A panoramic image acquired with a slit camera having its focal point exactly on the rotation axis, i.e. $R = 0$, is referred to as a *single-center panoramic image*. The base circle degenerates to a point. Studies about stereo reconstructions based on given pairs of single-center panoramic images have been carried out in [7,5]. In this case, the epipolar lines can be described by sine curves, and all of them pass through both epipoles. Curve parameterization and image resampling may be carried out before stereo matching takes place.

A panoramic image with a radius $R > 0$ is referred to as a *multi-perspective panoramic image*. This type of panoramic images is receiving much attention recently for applications of 3D scene visualizations and reconstructions, for instance, [3,10,13,15,16,11].

In particular, a set of panoramic images all acquired with respect to the same base circle, a constant focal length f, but different values ω, are referred to as a set of *concentric panoramic images*. H-Y. Shum and R. Szeliski [16] have shown that epipolar geometry consists of horizontal lines if two concentric panoramic images are *symmetric* with respect to the normal vector of the base circle at the associated focal point (i.e. angles ω and $-\omega$).

The authors of this paper are not aware of further studies of epipolar geometry of panoramic images. We define a set of *polycentric panoramic images* as a collection of panoramic images acquired with respect to different (but parallel) rotation axes, where the associated value R for each image can be either greater than or equal to zero, and where angles ω and the focal length f of these images may differ, but where it is assumed that all base circles are coplanar.

Since single-center, multi-perspective, or concentric panoramic images can be regarded as special cases of these polycentric panoramic images, the epipolar curve equation derived for a pair of polycentric panoramas provides a unified approach for computing epipolar curves in those more specific types of panoramic images. In this paper, we elaborate the derivation of the general epipolar curve equation and provide examples for two typical panoramic image cases visualizing a synthetic 3D scene.

Material about stereo matching in panoramic image pairs can be found in [4,2,9]. Among them, only S.B. Kang and R. Szeliski [4] attempt to find corresponding points along epipolar curves. They attribute their negative result to the inaccurate estimation of epipolar geometry and the difficulty of correct matching in case of *singular situations* when the projection ray of the second camera passes through the focal point of the first camera. J. Gluckman et al. [2] use a

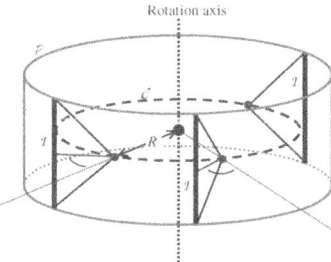

Fig. 1. The image acquisition model of a panoramic image. To acquire a panoramic image $\mathcal{P}$ a slit camera rotates with respect to a fixed 3D axis and captures one slit image $\mathcal{I}$ in every constant angular interval. Each slit image contributes to one column of $\mathcal{P}$. The orientation of the slit camera is defined by the angle ω between the normal vector of the dashed base circle $\mathcal{C}$ at the associated focal point and the optical axis of the slit camera

co-axis panoramic pair[1] such that the epipolar lines are straight and emitting radially from the epipole to the boundary of the image. Matching is done based on the correlation method with a rectangular support-region. M. Ollis et al. [9] studied various configurations of catadioptric panoramic image-acquisition models. Nevertheless, all of them are of co-axis type; hence, the epipolar geometry is basically the same as in the work reported by J. Gluckman et al. They propose a skewed-rectangular shape of a support-region, which is reported as giving a better matching result.

Unlike the co-axis situations there is no one-to-one correspondence between epipolar curves in a pair of polycentric panoramic images. We derive an equation of the epipolar curve in one image corresponding to a given point in the other image. This search space restriction can then be used to detect corresponding points. We illustrate that by using an adaptive support-region matching scheme as described in [14,6]. As a result, the accuracy of point matches seems to be acceptable, see Fig. 3 and Fig. 4.

The paper is organized as follows. The acquisition model of polycentric panoramic images has been defined in Section 1. The derivation of the epipolar curve equation through various geometric transformations is elaborated in Section 2. The 3D reconstruction formula is given in Section 3. Some experimental results of stereo reconstructions for polycentric panoramas are provided in Section 4. Future work and conclusions are drawn in Section 5.

2   Epipolar Curve Equation

Consider a pair of polycentric panoramic images, a *source image* $\mathcal{P}$ and a *destination image* $\mathcal{P}'$. If an image point $\mathbf{p}$ of $\mathcal{P}$ is given, then a 3D projection ray,

---

[1] Two panoramic images are acquired using the same rotation axes.

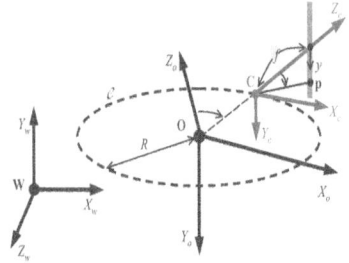

Fig. 2. The geometrical relationship among a slit camera coordinate system with the origin **C**, an associated turning-rig coordinate system with the origin **O**, and the world coordinate system with the origin **W**

denoted as ℓ_c, emitting from a focal point **C** through the point **p** with respect to the slit camera coordinate system can be described by $\mathbf{P} + \lambda\mathbf{D}$, where **P** is a 3-vector, $\lambda \in \Re$ is any scalar, and **D** is a unit directional vector, we have

$$\mathbf{D} = (0, \sin\phi, \cos\phi)^{\mathrm{T}},$$

where ϕ is the angular coordinate for the point **p** on the slit image (see Figure 2).

We define a 3D turning-rig[2] coordinate system, as shown in Figure 2, for the source panoramic image. The origin, denoted as **O**, coincides with the center of the base circle $\mathcal{C}$. The projection ray ℓ_c is first transformed from the slit camera coordinate system to the turning-rig coordinate system. The resulting ray is denoted as ℓ_o. The transformation formula is as follows:

$$
\begin{aligned}
\ell_o &= \mathbf{R}_{oc}^{-1}\mathbf{P} + \mathbf{T}_{oc} + \lambda\mathbf{R}_{oc}^{-1}\mathbf{D}, \\
&= \mathbf{T}_{oc} + \lambda\mathbf{R}_{oc}^{-1}\mathbf{D}, \\
&= \begin{pmatrix} R\sin\theta \\ 0 \\ R\cos\theta \end{pmatrix} + \lambda \begin{bmatrix} \cos(\theta+\omega) & 0 & -\sin(\theta+\omega) \\ 0 & 1 & 0 \\ \sin(\theta+\omega) & 0 & \cos(\theta+\omega) \end{bmatrix}^{-1} \begin{bmatrix} 0 \\ \sin\phi \\ \cos\phi \end{bmatrix},
\end{aligned} \tag{1}
$$

where the 3×3 rotation matrix $\mathbf{R}_{oc}$ and the 3×1 translation vector $\mathbf{T}_{oc}$ specify the orientation and the location of a slit camera coordinate system with respect to the turning-rig coordinate system (see Figure 2).

The projection ray ℓ_o is then transformed to the turning-rig coordinate system of the destination panoramic image $\mathcal{P}'$ through the world coordinate system, denoted as $\ell_{o'}$,

$$
\begin{aligned}
\ell_{o'} &= \mathbf{R}_{wo'}(\mathbf{R}_{wo}^{-1}\mathbf{T}_{oc} + \mathbf{T}_{wo} - \mathbf{T}_{wo'}) + \lambda\mathbf{R}_{wo'}\mathbf{R}_{wo}^{-1}\mathbf{R}_{oc}^{-1}\mathbf{D} \\
&= \mathbf{R}_{wo'}\left[\mathbf{R}_{wo}^{-1}\begin{bmatrix} R\sin\theta \\ 0 \\ R\cos\theta \end{bmatrix} + \mathbf{T}_{wo} - \mathbf{T}_{wo'}\right] + \lambda\mathbf{R}_{wo'}\mathbf{R}_{wo}^{-1}\begin{bmatrix} \sin(\theta+\omega)\cos\phi \\ \sin\phi \\ \cos(\theta+\omega)\cos\phi \end{bmatrix},
\end{aligned} \tag{2}
$$

---

[2] For example, a turntable or a turning head on a tripod.

where the 3×3 rotation matrix $\mathbf{R}_{wo}$ ($\mathbf{R}_{wo'}$) and the 3×1 translation vector $\mathbf{T}_{wo}$ ($\mathbf{T}_{wo'}$) specify the orientation and the location of the turning-rig coordinate system of the source panoramic image $\mathcal{P}$ (the destination panoramic image $\mathcal{P}'$) with respect to the world coordinate system.

The epipolar curve equation is an equation in terms of x' and y' which are the image coordinates of the destination panoramic image $\mathcal{P}'$. Every point (x', y') is the projection of some 3D point on the ray $\ell_{o'}$. In other words, every (x', y') is possibly the corresponding point of the given image point $\mathbf{p}$ (i.e. with the image coordinates (x, y)) of $\mathcal{P}$. Let $\mathcal{I}'$ denote the slit image which contributed to the column x' of $\mathcal{P}'$ and let $\mathbf{C}'$ denote the associated slit camera's focal point. For each column x', the corresponding y' value can be found by the following two steps. First find the intersection point, denoted as $\mathbf{Q}'$, of the ray $\ell_{o'}$ and the plane $\wp'$ passing through $\mathbf{C}'$ and $\mathcal{I}'$. Second, obtain the value of y' by projecting point $\mathbf{Q}'$ to the slit image $\mathcal{I}'$.

The position of the focal point $\mathbf{C}'$ with respect to the turning-rig coordinate system of $\mathcal{P}'$ can be described by $(R' \sin \theta', 0, R' \cos \theta')$, where R' is the radius of the circle $\mathcal{C}'$ and $\theta' = (2\pi x')/(W'_{\mathcal{P}})$, where $W'_{\mathcal{P}}$ is the width of the destination panoramic image. A unit vector perpendicular to the plane $\wp'$ is $(-\cos(\theta' + \omega'), 0, \sin(\theta' + \omega'))$, where ω' is the angle between the normal vector of $\mathcal{C}'$ at $\mathbf{C}'$ and plane $\wp'$. Therefore, the equation of plane $\wp'$ is

$$- \cos(\theta' + \omega') X + \sin(\theta' + \omega') Z = R' \sin \omega' , \tag{3}$$

where the variables X and Z are with respect to the turning-rig coordinate system of the destination panoramic image $\mathcal{P}'$.

We substitute the x- and z-components of the projection ray $\ell_{o'}$ in Equ. 2 into the plane equation Equ. 3, and solve the value of λ. The intersection point $\mathbf{Q}'$ can then be calculated from Equ. 2. We denote the obtained coordinates of $\mathbf{Q}'$ as $(X_{o'}, Y_{o'}, Z_{o'})$. We have

$$\begin{bmatrix} X_{c'} \\ Y_{c'} \\ Z_{c'} \end{bmatrix} = \begin{bmatrix} X_{o'} \cos(\theta' + \omega') - Z_{o'} \sin(\theta' + \omega') + R' \sin \omega' \\ Y_{o'} \\ X_{o'} \sin(\theta' + \omega') + Z_{o'} \cos(\theta' + \omega') - R' \cos \omega' \end{bmatrix},$$

which transforms the point $\mathbf{Q}'$ to the slit camera coordinate system associated to the slit image. For each image column x', now we project the 3D point $(X_{c'}, Y_{c'}, Z_{c'})$ on the associated slit image $\mathcal{I}'$. This allows us to formulate the following general epipolar curve theorem for polycentric panoramic images:

Theorem 1. *The corresponding value of y' for each x' can be obtained by*

$$y' = \frac{f' Y_{o'}}{X_{o'} \sin(\frac{2\pi x'}{W'_{\mathcal{P}}} + \omega') + Z_{o'} \cos(\frac{2\pi x'}{W'_{\mathcal{P}}} + \omega') - R' \cos \omega'} , \tag{4}$$

where f' is the effective focal length of the slit camera acquiring $\mathbf{p}'$.

Note that the corresponding point $\mathbf{p}'$ with coordinates (x', y') is only valid if the value of the denominator of y' is greater than zero.

3   Reconstruction Formula

We consider a pair of corresponding points (x, y) and (x', y') in the source panoramic image $\mathcal{P}$ and in the destination panoramic image $\mathcal{P}'$ respectively. We denote the intersecting point of the projection rays defined by these two image points by $\mathbf{Q}$. We are interested to calculate the coordinates of $\mathbf{Q}$ with respect to the turning-rig coordinate system of the source panoramic image. The equation of the projection ray defined by (x, y) is given in Equ. 1. The point $\mathbf{Q}$ can be described by Equ. 1 with a particular value of λ, say λ_Q. The value of λ_Q can be interpreted as being the length of the line segment $\overline{\mathbf{CQ}}$, where $\mathbf{C}$ is the focal point of the slit camera associated to the image column x. Previously, we have calculated a λ value by substituting the x- and z-components of the projection ray $\ell_{o'}$ in Equ. 2 into the plane equation Equ. 3, which actually gives the desired λ_Q. Thus, we may summarize it in the following reconstruction theorem:

Theorem 2. *The point $\mathbf{Q}$ in Euclidean space, identified by such a pair of corresponding points, is*

$$\mathbf{Q} = \begin{pmatrix} R\sin(\frac{2\pi x}{W_{\mathcal{P}}}) \\ 0 \\ R\cos(\frac{2\pi x}{W_{\mathcal{P}}}) \end{pmatrix} + k \begin{bmatrix} \sin(\frac{2\pi x}{W_{\mathcal{P}}} + \omega)\cos(\tan^{-1}(\frac{y}{f})) \\ \sin(\tan^{-1}(\frac{y}{f})) \\ \cos(\frac{2\pi x}{W_{\mathcal{P}}} + \omega)\cos(\tan^{-1}(\frac{y}{f})) \end{bmatrix}$$

with respect to the source panorama's turning-rig coordinate system, where

$$k = \frac{R'\sin\omega' + \cos(\frac{2\pi x'}{W'_{\mathcal{P}}} + \omega')\mathbf{r}_1^T(\mathbf{T}_{oc} - \mathbf{T}_{oo'}) - \sin(\frac{2\pi x'}{W'_{\mathcal{P}}} + \omega')\mathbf{r}_3^T(\mathbf{T}_{oc} - \mathbf{T}_{oo'})}{\sin(\frac{2\pi x'}{W'_{\mathcal{P}}} + \omega')\mathbf{r}_3^T(\mathbf{R}_{oc}^{-1}\mathbf{D}) - \cos(\frac{2\pi x'}{W'_{\mathcal{P}}} + \omega')\mathbf{r}_1^T(\mathbf{R}_{oc}^{-1}\mathbf{D})}.$$

In this theorem, the 3×3 rotation matrix $\mathbf{R}_{oo'}$ and the 3×1 translation vector $\mathbf{T}_{oo'}$ specify the orientation and the location of the destination turning-rig coordinate system of origin $\mathbf{O}'$ with respect to the source turning-rig coordinate system of origin $\mathbf{O}$ (i.e. $\mathbf{R}_{oo'} = \mathbf{R}_{wo'}\mathbf{R}_{wo}^{-1}$ and $\mathbf{T}_{oo'} = \mathbf{T}_{wo'}\mathbf{T}_{wo}^{-1}$), where $\mathbf{r}_1^T$ and $\mathbf{r}_3^T$ are the first and third row vectors of the matrix $\mathbf{R}_{oo'}$, respectively.

4   Experimental Results

We present two stereo reconstruction results of two typical types of polycentric panoramic image pairs. One is for $R > 0$, which is referred to as a multi-perspective panoramic pair, and the other is for $R = 0$, which is referred to as a single-center panoramic pair. All of the panoramic images for our experiments are captured in a 3D synthetic scene: a squared room containing different objects such as sphere, box, knot etc. with mapped real images.

Consider two multi-perspective panoramic images, a source panoramic image $\mathcal{P}$ and a destination panoramic image $\mathcal{P}'$, as shown on the top and in the middle of Fig. 3, respectively. The middle panoramic image $\mathcal{P}'$ was acquired at the same

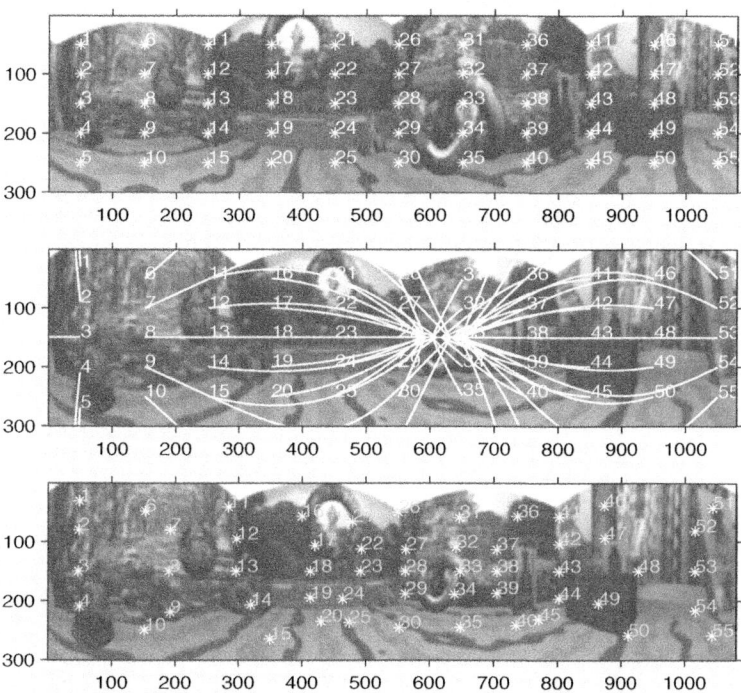

Fig. 3. An example of a polycentric panoramic image pair, a multi-perspective panoramic pair: The top (source) panoramic image shows 55 test points in labeled '⁎' and enumerated positions, the middle (destination) panoramic image shows the corresponding epipolar curves and the bottom panoramic image shows the corresponding points found in the destination panoramic image for those 55 test points in the source panoramic image

height[3] with $2m$ to the east and $1m$ to the north of the top panoramic image $\mathcal{P}$. The effective focal lengths of the slit cameras used for acquiring these two panoramic images are both equal to 35.704 mm. The radiuses of the base circles, where slit camera's focal points lie on, are both equal to 40 mm. The orientations of the slit cameras with respect to each rotation axes are both equal to $45°$. Each slit camera takes 1080 slit images for one panoramic image.

The orientations and the positions of the turning-rig coordinate systems of $\mathcal{P}$ and $\mathcal{P}'$ with respect to the world coordinate system can be defined as: $\mathbf{R}_{wo} = \mathbf{R}_{wo'} = \mathbf{I}_{3\times3}$ and $\mathbf{T}_{wo} = (0, 0, 0)^{\mathrm{T}}$ and $\mathbf{T}_{wo'} = (t_x, 0, t_z)^{\mathrm{T}}$, respectively. Given

---

[3] The y-components of their associated rotation centers' coordinates with respect to the world coordinate system are equal.

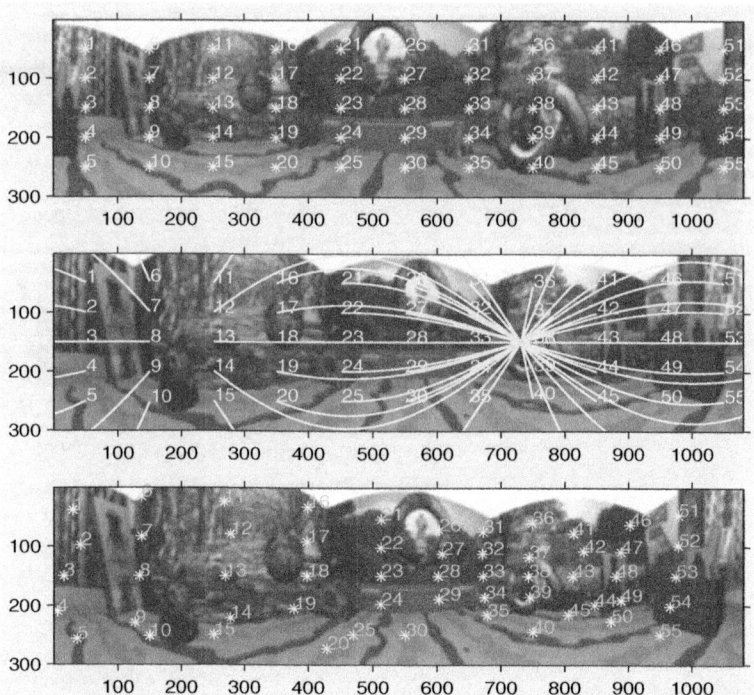

Fig. 4. An example of a polycentric panoramic image pair, a single-center panoramic pair: The top (source) panoramic image shows 55 test points in labeled '∗' and enumerated positions, the middle (destination) panoramic image shows the corresponding epipolar curves and the bottom panoramic image shows the corresponding points found in the destination panoramic image for those 55 test points in the source panoramic image

an image point (x, y) on $\mathcal{P}$, Equ. 4 of the epipolar curve on $\mathcal{P}'$ is

$$
y' = y \frac{f'}{f} \left(\frac{R' \sin \omega' - R \sin(\frac{2\pi x'}{W_\mathcal{P}'} - \frac{2\pi x}{W_\mathcal{P}} + \omega') - t_x \cos(\frac{2\pi x'}{W_\mathcal{P}'} + \omega') + t_z \sin(\frac{2\pi x'}{W_\mathcal{P}'} + \omega')}{-R \sin \omega - R' \sin(\frac{2\pi x'}{W_\mathcal{P}'} - \frac{2\pi x}{W_\mathcal{P}} - \omega) - t_x \cos(\frac{2\pi x}{W_\mathcal{P}} + \omega) + t_z \sin(\frac{2\pi x}{W_\mathcal{P}} + \omega)} \right) .
$$

In Fig. 3, the destination panoramic image in the middle shows the corresponding epipolar curves of those 55 test points in the source panoramic image. We observed that the epipolar curves associated to the image points on the same column of the source image converge to the same point. Thus, there are multiple epipolar curve converging points (i.e. multiple epipoles). All the converging points are clustered together in a small line segment. We also observed that those epipolar curves are different in length and are crossing each other. An important observation of the epipolar curves in this type of the panoramic images is that there is no one-to-one correspondence between epipolar curves. For example, an image point $\mathbf{p}$ in the source panoramic image defines an epipolar curve $\mathcal{L}'$ in the destination panoramic image. Choose any two image points $\mathbf{p}_1'$ and $\mathbf{p}_2'$ on the epipolar curve $\mathcal{L}'$. The corresponding epipolar curves for $\mathbf{p}_1'$ and $\mathbf{p}_2'$ in the source panoramic image only intersect at point $\mathbf{p}$. Based on these observations, the adaptive support-region matching scheme [14,6] and correlation cost function are used. The overall results of the matching are reasonable good under such a serious image distortion. In Fig. 3, the bottom image shows the corresponding points found in the destination panoramic image for those 55 test points selected in the source panoramic image.

As a second example consider two single-center panoramic images, a source panoramic image $\mathcal{P}$ and a destination panoramic image $\mathcal{P}'$, as shown on the top and in the middle of Fig. 4, respectively. The image acquisition settings of these two are the same as the ones for acquiring multi-perspective panoramas except that any of the focal points of the used two slit cameras is incident with the respective rotation axis. Given an image point (x, y) on $\mathcal{P}$, the equation of the epipolar curve in $\mathcal{P}'$ is

$$y' = y \left(\frac{f'}{f}\right) \left(\frac{-t_x \cos(\frac{2\pi x'}{W_{\mathcal{P}}'}) + t_z \sin(\frac{2\pi x'}{W_{\mathcal{P}}'})}{-t_x \cos(\frac{2\pi x}{W_{\mathcal{P}}}) + t_z \sin(\frac{2\pi x}{W_{\mathcal{P}}})}\right) = y \cdot k \left(t_z \sin\left(\frac{2\pi x'}{W_{\mathcal{P}}'}\right) - t_x \cos\left(\frac{2\pi x'}{W_{\mathcal{P}}'}\right)\right),$$

where

$$k = \left(\frac{f'}{f \left(t_z \sin(\frac{2\pi x}{W_{\mathcal{P}}}) - t_x \cos(\frac{2\pi x}{W_{\mathcal{P}}})\right)}\right)$$

is a scalar.

From Fig. 4 we observed that all the epipolar curves converge to a single point (i.e. the epipole). The epipolar curves are different in length, but they do not cross each other in this case. An important feature of the epipolar curves in the single-center panorama pair, which is not true in case of multi-perspective panoramic pairs, is that corresponding points of image points which lie on the same epipolar curve in the destination image, also lie on the same epipolar curve in the source image. Thus, in this special case of single-center panoramic images there is a one-to-one correspondence between epipolar curves.

5   Future Work and Conclusion

In this paper, we defined a basic acquisition model of polycentric panoramas. The general epipolar curve equation is derived and can easily be modified with respect to any more specific image acquisition case (e.g. single-center panoramas). In this study, we assume that the motion (i.e. the 3D spatial relationship) between a panoramic pair is given, but practically it needs to be calculated in advance by calibration of the panoramic image acquisition process. For a set of uncalibrated panoramic images, it is interesting to know how many corresponding points are necessary to calibrate the desired parameters. There is no one-to-one correspondence between epipolar curves of a pair of polycentric panoramic images, except for a few very special cases such as single-center panoramic images, due to the nature of multiple optical centers. The adaptive support-region matching scheme is used to obtain reasonable results. However, to improve the matching quality, the shape of support-region can be adapted non-uniformly along epipolar curves and an algorithm also incorporating global optimization should be developed.

References

1. Ch.-Y. Chen, R. Klette: Image stitching - comparisons and new techniques. In: Proceed. *CAIP*, (September 1999) 615–622. 209
2. J. Gluckman, S. K. Nayar, K. J. Thorek: Real-time panoramic stereo. *DARPA98*, (1998) 299–303. 210
3. H. Ishiguro, M. Yamamoto, S. Tsuji: Omni-directional stereo. *IEEE-PAMI*, **14** (1992) 257–262. 210
4. S. B. Kang, R. Szeliski: 3-d scene data recovery using omnidirectional multibaseline stereo. IEEE Conf. on *Computer Vision and Pattern Recognition*, (1996) 364–370. 210
5. S. B. Kang, R. Szeliski: 3-d scene data recovery using omnidirectional multibaseline stereo. *IJCV*, **25** (1997) 167–183. 210
6. R. Klette, K. Schlüns, A. Koschan: *Computer Vision - Three-Dimensional Data from Images*. Springer, Singapore, (1998). 211, 217
7. L. McMillan, G. Bishop: Plenoptic modeling: an image-based rendering system. *Proc. SIGGRAPH'95*, (1995) 39–46. 210
8. T. Matsuyama, T. Wada: Cooperative distributed vision - dynamic integration of visual perception, action, and communication. *2nd Internat. Workshop in Cooperative Distributed Vision*, Kyoto, (1998) 1–40.
9. M. Ollis, H. Herman, S. Singh: Analysis and design of panoramic stereo vision using equi-angular pixel cameras. Technical Report CMU-RI-TR-99-04, Carnegie Mellon University, Pittsburgh, USA, (1999). 210, 211
10. S. Peleg, M. Ben-Ezra: Stereo panorama with a single camera. *CVPR99*, (1999) I:395–401. 210
11. S. Peleg, Y. Pritch, M. Ben-Ezra: Cameras for stereo panoramic imaging. Technical report, Computer Vision Lab, Institute of Computer Science of The Hebrew University, (2000). 210
12. R. Sandau et al.: Design principles of the lh systems ads40 airborne digital sensor. *Internat. Archives of Photogrammetry and Remote Sensing*, XXXIII, Part B1, Commission I (2000) 258–265. 209
13. H.-Y. Shum, L.-W. He: Rendering with concentric mosaics. *Proc. SIGGRAPH'98*, Los Angeles, (August 1999) 299–306. 210
14. Y. Shirai: *Three-Dimensional Computer Vision*. Springer, Berlin, (1986). 211, 217
15. H. Shum, A. Kalai, S. Seitz: Omnivergent stereo. *ICCV99*, Korfu/Greece, (1999) 22–29. 210
16. H. Shum, R. Szeliski: Stereo reconstruction from multiperspective panoramas. *ICCV99*, Korfu/Greece, (1999) 14–21. 210
17. S. K. Wei, F. Huang, R. Klette: Classification and characterization of image acquisition for 3d scene visualization and reconstruction applications. Technical Report 59, CITR, Auckland University, New Zealand, (June 2000). 209

Two Modules of a Vision–Based Robotic System: Attention and Accumulation of Object Representations

Norbert Krüger, Daniel Wendorff, and Gerald Sommer

Lehrstuhl für Kognitive Systeme, Institut für Informatik,
Christian–Albrechts–Universität zu Kiel
Preusserstrasse 1-9, 24105 Kiel, Germany
{nkr,dw,gs}@ks.informatik.uni-kiel.de

Abstract. In this paper, two modules of a behavior based robotic–vision system are described: An attention mechanism, and an accumulation algorithm to extract stable object representations within a perception–action cycle.

1 Introduction

The aim of our research is the design and implementation of an active vision system coupled with a robot arm (see Fig. 1a) which is able to recognize and grasp objects with autonomously learned representations. The system shall gain robot control over new objects (i.e., grasp a new object in a scene) by an instinctive and rudimentary behavior pattern and use the control over the object to accumulate a representation of the object and finally apply these representations to robustly track, grasp and recognize the object in a complex scene.

In this paper, two modules of such a system are described: A visual and (potentially) haptic attention mechanism, and an accumulation algorithm to extract stable object representations. In the first module (described in section 2) the system directs its attention to new objects and manipulates the active components (i.e., cameras and grasper) such that a situation is achieved in which grasping becomes easier: grasper and object appear in the center of a zoomed stereo image pair (see Fig. 1h). In this situation grasping of the object can be performed using only relative positions between grasper and object. The high resolution allows to accurately extract 3D–Information about the relative position and orientation of grasper and object by stereo. Note that our attention mechanism is planned not to be only vision–based. We are currently redeveloping a haptic sensor [16] which allows to explore an object haptically. Therefore, our attention mechanism potentially focuses visual *and* haptic attention to the new object. The attention mechanism is to a wide degree predetermined but also contains adaptable components: The grasper is permanently tracked by the system. The information of motor commands and tracking results allow a self–calibration during the perception–action cycle [9,17].

R. Klette, S. Peleg, G. Sommer (Eds.): Robot Vision 2001, LNCS 1998, pp. 219–226, 2001.
© Springer-Verlag Berlin Heidelberg 2001

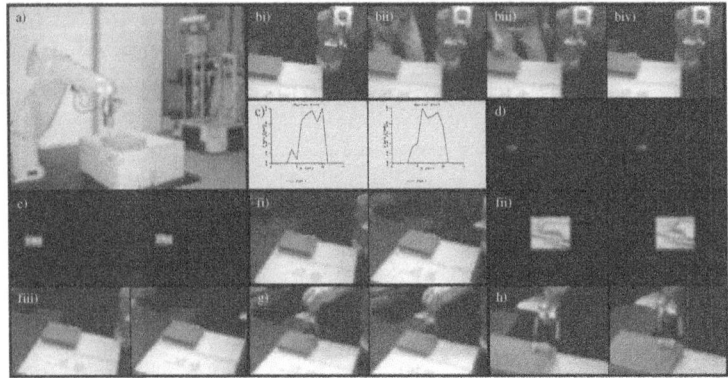

Fig. 1. a) Active binocular head with robot arm. bi-biv) Images of a person entering the scene, putting an object into the scene and leaving the scene. c) Graph indicating a dynamic period by the magnitude of differences between images. d-h) Stereo images: d) Difference image before and after the dynamic period e) Similarities of a Gabor jet extracted from the center of gravity in the left image to the jets extracted from other pixel positions of the difference area. Maxima are defined as corresponding points. fi) Fixation of the new object. fii) Similarities of the Gabor jets for fine tuning of fixation. fiii) Fixation after a second camera action. g) Movement of the robot arm to a position near the object. h) Zoom

The second module (described in section 3) uses control over the object to extract a stable representation. We account for the vagueness of semantic information extracted from single images by assigning confidences to this information and accumulating this information over an image sequence of a controlled moving object. Although the information extracted from single images contains errors (see the representations on the left hand side of Fig. 3) a more stable representation can be achieved by combining information from different images (see right hand side of Fig. 3). Because the object can change its position and orientation — and this change might be wanted because another view of the object gives new information which might not be extractable from former ones — we face the correspondence problem: Correspondences between entities describing the object in different images (or 3D interpretations extracted from stereo images) are not known. However, the parameters of motion are known since the robot manipulates the object and the transformations of entities can be compensated for each frame of the sequence. Knowing the correspondences, an algorithm can be applied to update and improve the object representation iteratively within a perception–action–cycle.

One important aspect of the design of a complex behavior based vision system is the interaction of modules developed by different people within one software package to derive complex competences from the combination of more primitive competences. We are currently developing a C++–library (KiViGraP, **Ki**eler **Vi**sion and **Gras**ping **P**roject) in which this interaction is going to occur (for details see [13]).

2 Attention Mechanism Based on Visual–Robotic Perception–Action Cycles

Our basic behavior aim at a tactile contact with a new object can be divided into a number of more simple competences (described below). The behavior pattern can be understood to a wide degree as a reflex action: The system shall "aim" to get in contact to new objects to explore them visually and haptically. Going even further, it "aims" to grasp the object using a rudimentary representation to learn a more sophisticated and efficient representation (see section 3). During robot actions a permanent tracking of the grasper allows to permanently recalibrate the system.

The module described in this section is going to initiate a situation in which grasping and tactile exploration is facilitated. Since for the accumulation scheme (section 3) it is essential that the system has physical control over the object, the module described in this section can be understood as part of a bootstrapping process, that (once the system's experience has been grown) can be substituted by or transformed into a more goal–oriented behavior pattern. However, the bridge between attention and grasping has not yet been built and is part of current research.

In the following we describe some submodules used to achieve tactile contact. The modules described here are not understood to be performed in a sequential process but as competences which interact with each other (e.g., tracking and self–calibration) and which can be applied depending on the actual system's goal. It is likely that at the very beginning of the bootstrapping process the structure and relations of the competencies are more predetermined than after a period of adaptation.

- **Detection of a new object and detection of a suitable time interval for robot action:** A new object is detected by the difference in each of the two stereo images before and after a dynamic period, i.e., a period in which people or other objects enter the scene (see Fig. 1bi–iv). For reasons of grasping success and maintaining safety for people interacting with the robot, it is necessary not to intervene in a dynamic situation. The system searches for a new object when a dynamic period occurred — a person puts a new object into the scene — followed by a stable period — the person leaves the scene (see Fig. 1bi–iv). Figure 1c shows a graph indicating the dynamic in a scene. During a period in which the graph shows high values the robot is not allowed to intervene. The behavior pattern, responsible for robot and people safety can be understood as a permanent (self)protection expert which restricts all other robot processes.

  In case that the person puts a new object into the scene, the object is detected by the difference in the images before and after the dynamic period (see Fig. 1d). Since simple differences of grey–level images are unstable due to little movements of the camera or variation of illumination, we also compute the difference of the magnitude of Gabor wavelet responses. For efficient computation we make use of the separability of quaternionic Gabor wavelets [5].

– **Fixation, approaching and zooming:** In case of detection of a change in the images before and after the dynamic period we fixate the new object. The internal camera parameters of our binocular camera-head are calibrated at an initialization stage. Then the system recalibrates itself after a movement by computing the new projection parameters from the motion commands given to the camera head. This recalibration is relatively stable even after a number of movements.

The two areas which represent differences in the image (or more precisely their center of gravity) give us two corresponding points for which we can compute a 3D–position with our calibrated system. Knowing its 3D–position we could easily fixate the object. However, since the correspondence of two objects is defined by the center of gravity of areas (which might not be very precise), the system may additionally use information about similarities within a small area around our difference areas. We compare image patches (with a method similar to [12] based on Gabor wavelets and jets) to find more precise correspondences in the two stereo images (see Fig. 1e). The system can achieve a higher robustness by iteratively computing the distance of the object and the image center after fixation. Note that these distances also can be used as a measure for the performance of the system, i.e., can also be used in a more global feedback loop to optimize the system.

Finally, the robot arm is moved to a position near the computed 3D–position of the object (see Fig. 1g) and the system can perform a zoom to get a higher resolution of both, the object and the grasper (see Fig. 1h). Object and grasper appear magnified and their relative distance can be used for grasper manipulation with high accuracy. It is expected that this relative distance can be extracted with higher accuracy than absolute distances from stereo images.

– **Tracking and self–calibration:** The system is equipped with a permanent grasper–tracking mechanism which is also based on the jet–representation in [12]. The 2D–tracking results and the motion parameters given to the robot can be compared to recalibrate the system by a simple update rule. It seems to be important that calibration does not only occur at the beginning of a process (often with an artificial calibration pattern) but is performed permanently during the normal perception–action cycle. Therefore, we have to face the tracking of the grasper in our quite uncontrolled environment. This is known as a very hard matching task which we are able to solve even with our rudimentary object representation by allowing only 'sure' matches to be used for self–calibration (for details see [18]). Here again, the system's ability to measure the success of performing competences is of significant importance.

We would like to finish this section with the remark that, although in its current state the behavior pattern is to a huge degree predetermined, we do intend to achieve a more robust, more flexible and more–goal oriented behavior pattern in a complex system through learning. Self–calibration by grasper tracking already supports an even better estimate of internal and external parameters and therefore a more robust behavior. Furthermore, the module can measure its success of fixation (Fig. 1fi-iii) and is intended to detect the success of tactile contact and grasping. Therefore, this information can be used as feedback for a more global learning which may allow to achieve direct contact and successful grasping more frequently by optimizing free parameters of the system. Finally, after achieving robot control more complex object representations can be learned (see section 3) and the original reflex behavior can be transformed into a more goal–oriented behavior, e.g., an object is only grasped when it hasn't been learned so far.

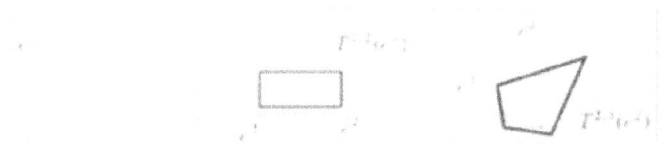

Fig. 2. The accumulation scheme. The entity e^1 (here represented as a square) is transformed to $T^{1,2}(e^1)$. Note that without this transformation it is nearly impossible to find a correspondence between the entities e^1 and e^2 because the entities show significant differences in appearance and position. Here a correspondence between $T^{1,2}(e^1)$ and e^2 is found because a similar square can be found close to $T^{1,2}(e^1)$ and both entities are merged to the entity $\hat{e}^2$. The confidence assigned to $\hat{e}^2$ is set to a higher value than the confidence assigned to e^1 indicated by the width of the lines of the square. The same procedure is then applied for the next frame for which again a correspondence has been found. By this scheme information can be accumulated to achieve robust representations

3   Accumulation of Inaccurate Information to a Robust Object Representation

After grasping the object, an accumulation scheme can be applied to extract a representation of the object (see Figs. 2 and 3). Feature extraction faces the problem that semantic information extracted by artificial systems from a single image or stereo images even under optimal conditions is necessarily imperfect. For instance, although there exist a large amount of edge detectors none of them is comparable to human performance. Moreover, we see it as an important problem to extract object representations in *real* situations and not in artificially adapted conditions (such as homogeneous background, controlled pose etc.), i.e.,

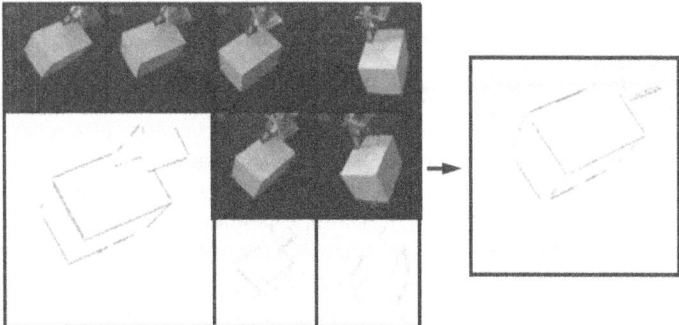

Fig. 3. left) top: left and right image of an object. bottom: the projected 3D representation extracted from the stereo images. **middle)** Two pairs of stereo images (top: left camera image, middle: right camera image) and the the projected 3D representation (bottom). **right)** Projected 3D Representation accumulated over a set of stereo images. The system's confidence for the presence of line segments is represented as grey value (Dark values represent high confidences)

we intend to fulfill the requirements *situatedness* formulated by Brooks [3]. One important reason for the extremely good performance of humans on these tasks in even very difficult situations is that the human visual system applies *constraints* to interpret a certain scene or situation [7,10]. An important constraint is the utilization of the coherence of objects during a rigid body motion which allows to accumulate information over time. Furthermore, in an active vision–based robot system we are able, instead of only passively perceiving a certain situation, to support learning by our own actions. This corresponds to *embodiment* as another requirement formulated by Brooks [3].

Our accumulation algorithm can be defined independently of the entities used to represent objects. The algorithm also is independent of the concrete equivalence relation or transformation used to define correspondences. It only requires an object representation by certain entities for which a metric is defined and to which certain transformations or equivalence relations (such as rigid body motion) can be applied. This accumulation algorithm is an extension of an algorithm introduced in [11,14] which has only dealt with 2D representation and translational motion.

Let $e \in E$ be an entity used to describe objects (for instance a 2D–line segment, a structure tensor [8] extracted from an image, 3D–line segments extracted from a stereo image pair or any other kind of object descriptor) and $d(e, e')$ be a distance measure on the space of entities E. Furthermore, let T be a transformation or equivalence relation, for instance a rigid body motion or the projective map corresponding to a rigid body motion. If e^i is an entity extracted from frame i of a sequence of events then $T^{i,i+1}(e^i)$ is the transformation $T^{i,i+1}$ from the i–th to the $(i+1)$–th frame applied to e^i.

Let e^{i+1} be an entity extracted from the $(i+1)$–th frame of the sequence. We say that e^i and e^{i+1} are likely to correspond to each other if $d(T(e^i), e^{i+1})$ is small. Often it might not be possible to find an exact correspondence with $d(T(e^i), e^{i+1}) = 0$. For example, if we want to compare local image patches in two images knowing the exact projective transformation corresponding to the rigid body motion of an object from the first to the second frame, the corresponding image patches can not be expected to be exactly equal because of factors such as noise during the image acquisition, changing illumination, non–Lambertian surfaces or discretization errors, i.e., the features are quasi–invariant. The problem may even become more severe when we extract more complex entities such as 3D or 2D line segments or 3D–surface patches. Therefore it is advantageous to formalize a confidence of correspondence by using a metric.

The accumulation of information can now simply be achieved by the following update rule: If there exists an entity e^{i+1} in the $(i+1)$–th frame for which $d(T(e^i), e^{i+1})$ is small (i.e., a correspondence is likely), then merge $T(e^i)$ and e^{i+1} by some kind of average operator, $\hat{e}^{i+1} = \mathrm{merge}(T(e^i), e^{i+1})$, and set the confidence for $\hat{e}^{i+1}$ to a higher value than the confidence assigned to e^i. If there exists no entity e^{i+1} in the $(i+1)$–th frame for which $d(T(e^i), e^{i+1})$ is small, the confidence for entity e^i to be part of the object is decreased. In Fig. 2 a schematic representation of the algorithm is shown for two iterations.

The accumulation scheme could also be interpreted as an iterative clustering scheme with an in build equivalence relation to compensate the motion of the object. It is also related to, so called 'dynamic neural nets' [6,4], in which cells appear or vanish according to some kind of confidence measure.

Figure 3 shows the application of this scheme to representations consisting of 3D line–segments extracted from stereo images. For these entities the change of the transformation (i.e., $T^{i,i+1}(e)$) and a metric can be computed explicitly (for details see [1]). Up to now, only one aspect of an object can be accumulated because correspondences are needed which are not granted when occlusion does occur. That means, that when the robot rotates the object by a larger degree, it is likely that new edges occur in the stereo images and other edges disappear. In the current state we ensure that the same aspect is presented to the system by only allowing movements within a small subspace of the space of rigid body motions. To define such a subspace of possible rigid–body motions we make explicitly use of the metric defined on the space of unit–quaternions corresponding to rotations in Euclidean space [2].

4   Outlook

We have introduced two basic competences of an object recognition and manipulation system. In both modules perception and action are tightly intertwined within perception–action cycles [9,17].

Important components of such a system are still missing, such as performing grasping of the object after the attention mechanism. However, for such a grasp the attention mechanism gives a good starting point, because we have only to operate with relative positions and since we gained high resolution of the important aspects of the scene by active control of the camera. A further important problem is the application of our extracted representations to recognition and grasping tasks. In [15] we could successfully apply one of our accumulated representations to the tracking problem.

References

1. M. Ackermann: Akkumulieren von Objektrepräsentationen im Wahrnehmungs–Handlungs Zyklus. Christian–Albrechts Universität zu Kiel, Institut für Informatik und Praktische Mathematik (Diplomarbeit), (2000). 225
2. W. Blaschke: *Kinematik und Quaternionen.* VEB Deutscher Verlag der Wissenschaften, (1960). 225
3. R. A. Brooks: Intelligence without reason. *Internat. Joint Conf. on Artificial Intelligence,* (1991) 569–595. 224
4. J. Bruske, G. Sommer: Dynamic cell structure learns perfectly topology preserving map. *Neural Computation,* **7** (1985) 845–865. 225
5. T. Bülow, G. Sommer: Quaternionic gabor filters for local structure classification. In: A. K. Jain, S. Venkatesh, and B. C. Lovell, Eds., *14th Internat. Conf. on Pattern Recognition,* ICPR'98, vol. 1, Brisbane, Australia, August 16-20, IEEE Computer Society, (1998) 808–810. 222
6. B. Fritzke: Growing cell structures – a self-organizing network for unsupervised and supervised learning. *Neural Networks,* **7** (1994) 1441–1460. 225
7. S. Geman, E. Bienenstock, R. Doursat: Neural networks and the bias/variance dilemma. *Neural Computation,* **4** (1995) 1–58. 224
8. B. Jähne (Ed.): *Digitale Bildverarbeitung.* Springer, (1997). 224
9. J.J Koenderink: Wechsler's vision: An essay review of computational vision by Harry Wechsler. *Ecological Psychology,* **4** (1992) 121—128. 219, 225
10. N. Krüger: *Visual Learning with a priori constraints.* PhD Thesis, Shaker Verlag, Germany, (1998). 224
11. N. Krüger, G. Peters: Orassyll: Object recognition with autonomously learned and sparse symbolic representations based on metrically organized local line detectors. *Computer Vision and Image Understanding,* **77** (2000). 224
12. M. Lades, J. C. Vorbrüggen, J. Buhmann, J. Lange, C. von der Malsburg, R. P. Würtz, W. Konen: Distortion invariant object recognition in the dynamik link architecture. *IEEE Trans. on Computers,* **42** (1993) 300–311. 222
13. KiViGraP (Homepage of the Kieler Vision and Grasping Project). http://www.ks.informatik.uni-kiel.de/~kivi/kivi.html. 221
14. M. Pötzsch, N. Krüger, C. von der Malsburg: A procedure for automatic analysis of images and image sequences based on two–dimensional shape primitives. *U. S. Patent Application,* (1999). 224
15. B. Rosenhahn, N. Krüger, T. Rabsch, G. Sommer: Automatic tracking with a novel pose estimation algorithm. In: Proceed. *Robot Vision 2001,* Auckland, (2001). 226
16. P. Schmidt: Entwicklung und Aufbau von taktiler Sensorik für eine Roboterhand. Institut für Neuroinformatik Bochum (Internal Report), (2000). 219
17. G. Sommer: Algebraic aspects of designing behaviour based systems. In: G. Sommer and J. J. Koenderink, Eds., *Algebraic Frames for the Perception and Action Cycle,* Springer, (1997) 1–28. 219, 225
18. D. Wendorff: Diplomarbeit (in progress). 222

Compatibilities for the Perception-Action Cycle

Josef Pauli and Gerald Sommer

Institut für Informatik und Praktische Mathematik
Christian-Albrechts-Universität zu Kiel
Preußerstraße 1–9, D-24105 Kiel, Germany
jpa@ks.informatik.uni-kiel.de

Abstract. We apply an eye-on-hand Robot Vision system for treating the following three tasks: (a) Tracking objects for obstacle avoidance; (b) Arranging certain viewing conditions; (c) Acquiring an object recognition function. The novelty is the use of so-called compatibilities between motion features and view sequence features. Under real image formation, compatibilities are more general and appropriate than exact invariants. We demonstrate the usefulness for constraining the search for corresponding features, for parameterizing correlation matching procedures, and for fine-tuning approximations of appearance manifolds.

1 Introduction

During the late eighties Computer Vision scientists realized that the human intelligence underlying the perception of the environment is not only based on views but also on accompanying actions. Since then, cameras have been mounted on agile devices in order to enable active viewing and study vision in combination with actions. Although this new paradigm of Robot Vision (or Active Vision) produced exciting solutions for problems which are too difficult for static vision, the potential usefulness is far from being fully realized [1].

Our work demonstrates the usefulness of controlled camera movements for three exemplary applications, i.e. tracking objects for obstacle avoidance, arranging certain viewing conditions, acquiring an object recognition function. In this context the theoretical concept of *invariance* is relaxed into the practical concept of *compatibility*. Regarding this, the first attempt has been undertaken by Binford and Levitt [3], who introduced *quasi-invariance* under transformations of geometric features. Our compatibility concept considers more general transformations, maybe with different types of features prior and after the mapping, and considers robot actions as the source of the transformations, and thus integrates real-world actions and perception.

We focus on compatibilities between 3D motion features and 2D view sequence features. Based on visual demonstration, statistical measurements are taken to evaluate the deviation from the exact invariance and thus specify the compatibility, which can be used in subsequent online applications. We study compatibilities for typical sub-tasks of the mentioned applications, i.e. constraining the search for corresponding features (section 2), parameterizing correlation

R. Klette, S. Peleg, G. Sommer (Eds.): Robot Vision 2001, LNCS 1998, pp. 227–236, 2001.

Fig. 1. (a) and (b) Two consecutive images with gray value corners: (c) Image with motion vectors at the gray value corners

matching procedures (section 3), and fine-tuning approximations of appearance manifolds (section 4). For the applications we used a 6-DOF robot arm (Stäubli-RX90) and a monochrome video camera mounted on the back of the robot hand. Within a working space of a cube with sidelength $500mm$ the camera can be arranged in any position and orientation.

2   Constraining the Search for Corresponding Features

We would like to acquire depth features from a collection of objects, e.g. bottles and cans in a refrigerator. For this purpose the camera will be translated continually in front of the objects. Gray value corners can be extracted (e.g. with SUSAN [7]) and must be tracked along the image sequence. Based on correspondences, shape-from-motion strategies can be applied to obtain the relevant information. For example, Fig. 1 shows two consecutive images (left and middle) with gray value corners extracted by SUSAN, and the right image depicts motion vectors at these points. We are interested to restrict the search for corresponding corners, i.e. determine an individual disparity range for each corner.

In an experimentation (offline) phase we put a calibration pattern onto the ground plane. It depicts a regular distributed set of black dots. Both at nearest and farthest distance to the ground (i.e. the top and bottom borders of the viewing space), the camera makes a certain step of movement, respectively. Motion vectors for the calibration dots are determined in the images, resulting in two vector fields V^1 and V^2. Figure 2 shows images of the calibration pattern prior and after lateral camera translation (at top border of viewing space). The flow of dots from left to right results in vector field V^1 (not depicted). Figure 3 just shows the lengths of motion vectors of V^1, which are not constant due to large image distortions (caused by a lense with small focal length, $4mm$). For the specified camera movement, the two vector fields impose *expectations on motion vectors* which can be used later on during the online phase. Let us assume an image point p_i which originates from an arbitrary 3D point within the viewing space, and assume a step of camera movement as specified according to the calibration phase. For the image point the *angles of the motion vector* taken from V^1 or V^2 are approximately the same. Furthermore, the *length of the motion vector* must be in the interval of the relevant lengths given in V^1 and V^2. Consequently, a point p_i in the first image and a point q_j in the second image is a candidate pair for correspondence, only if the following constraints hold:

$$\Phi(V^1(p_i)) \approx \Phi(q_j - p_i) \approx \Phi(V^2(p_i)) \tag{1}$$

$$L(V^1(p_i)) \geq \|q_j - p_i\| \geq L(V^2(p_i)) \tag{2}$$

Symbol Φ denotes the angle and L the length of a vector. Just these carefully selected candidate pairs are taken for applying normalized cross correlation in order to determine the most appropriate one, as shown in Fig. 1 (right image). The compatibility is represented by the two equations (1) and (2).

3   Parameterizing Correlation Matching Procedures

The robot hand including the hand-mounted camera should be arranged in a certain relation to the object. This can be regarded as a sub-task of a grasping process or a sub-task leading to optimal viewing of an object. A servoing mechanism will be applied which does the arrangement step by step and is based on continual visual feedback and correlation matching in the series of images. In section 2 we treated exemplary the case of camera translation, and now we consider compatibilities for the case of camera rotation. If a camera is rotating around the optical axis, which is normal to the object surface, then *log-polar transformation (LPT)* can be applied to the gray value images [4]. The motivation is that the transformed object pattern is shifting instead of rotating, which makes the correlation matching more efficient during the servoing process. Figure 4 shows two images of an integrated circuit (IC) object under rotation by a turning angle of 90°. These are two examples from a collection of 24 images taken under angle offset of 15°, respectively. Figure 5 shows the horizontal translation of the log-polar transformed pattern of the rotating object.

However, in a view sequence perfect invariance only holds for a flat 2D object without any side faces, and a simulated pinhole camera is assumed whose optical axis must be kept normal to the object surface. In realistic applications,

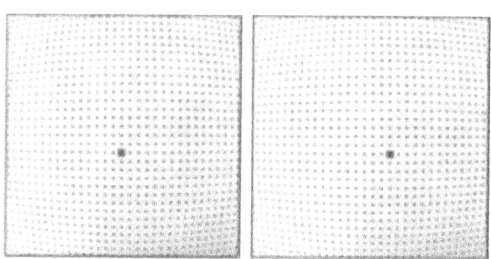

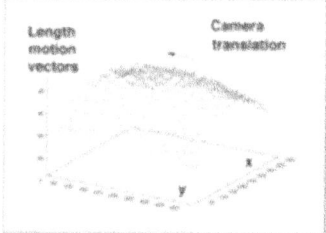

Fig. 3. Lengths of mot. vectors for lateral camera translation

Fig. 2. Calibration pattern prior/after lateral camera translation, flow of dots from left to right

Fig. 4. Integrated circuit object under rotation by turning angle 90°   **Fig. 5.** Horizontal translation of LPT pattern

resampling error occur certainly, the objects are of three-dimensional shape presumably, the camera objectives may cause unexpected distortions, and possibly the optical axis is not exact normal to the object surface (misalignment). Because of these realistic imponderables, certain variations of the LPT patterns will occur. We are interested in determining the real deviations from invariance in order to obtain tolerance parameters for correlation matching.

By demonstrating sample objects and performing typical camera rotations relative to the objects, one can make measurements of real deviations from invariance, i.e. *actual variations of the LPT patterns*. Despite of these variations, it is expected that the manifold of LPT patterns is much more compact and easier to describe than the original manifold of appearance patterns. For example, presumably, a *single multi-dimensional Gaussian*, specified by a center vector and a certain covariance matrix, may approximate the variation.

For illustration, we perform a simple experiment which is based on histograms of edge orientations. Specifically, orientations of gray value edges are considered in order to demonstrate the influence of LPT to a rotating 3D object, i.e. measuring the deviation from pure pattern translation in the log-polar transformed image. The image library consists of 24 images, as mentioned above. The histograms should be computed from the relevant area of the LPT image containing the object pattern, respectively. To simplify this sub-task a nearly homogeneous background has been used such that it is easy to extract the gray value structure of the IC object. We compute for the extracted LPT patterns a histogram of gradient angles of the gray value edges, respectively.

Figure 6 (left) shows a histogram determined from an arbitrary image in the library. The mean histogram is computed from the LPT patterns of the whole set of 24 images, shown in Fig. 6 (middle). Next, we compute for each histogram the deviation vector from the mean histogram. From the whole set of deviations once again a histogram is computed, which is shown in Fig. 6 (right).

This latter histogram can be approximated as a Gaussian with the maximum value at 0 and the Gaussian turning point approximately at the value $\sigma = 5$. Under ideal (simulated) conditions the Gaussian would be an impulse function with extent 0. However, the real value of σ is a measure for the deviation from perfect invariance. It can be used to parameterize approaches of pattern matching, e.g. specifying thresholds for the coefficient of normalized cross correlation in order to obtain reasonable matching hypotheses.

4 Fine-Tuning Manifold Approximations for Recognition

For the recognition of a scene object in an image we need to have an appropriate recognition function. This function can hardly be implemented manually and instead should be learned automatically in the task-relevant environment. Based on a robot-controlled process of taking sample views we can incorporate action-related information for improving the generalization in the learning mechanism.

Appearance-Based Object Recognition

A holistic learning approach can be applied which is based on 2D appearance patterns of the relevant objects or response patterns resulting from specific filter operations. The main interest is to represent or approximate the pattern manifolds such that an optimal compromise between *efficiency, invariance and discriminability* of object recognition is achieved. It is essential to keep these manifolds as simple as possible, because the complexity is correlated to the time needed for object recognition. In section 3 we restricted camera poses and movements and thereby reduced the manifold complexity by LPT. However, in this section we accept general viewing poses. Apart from the efficiency criterion the recognition function must respond with constant high values for any appearance of the object (invariance criterion), and must be able to discriminate between target and other objects (discriminability criterion).

The most popular approach of manifold approximation is based on principal component analysis (PCA) for a collection of views [5]. This is done for each object leading to an object-specific Eigenspace, respectively. An unknown view can be recognized by computing proximity values to the training samples in the Eigenspaces, and determining the most relevant manifold. An improvement of

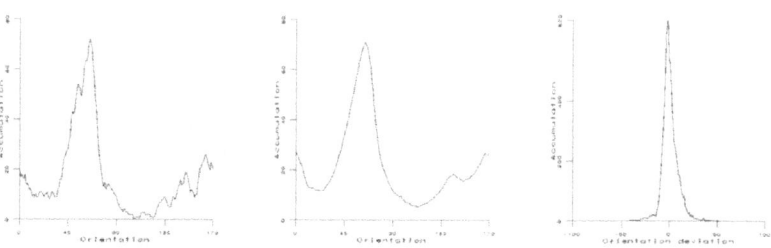

Fig. 6. (a) Histogram of edge orientations under LPT for one image; (b) Mean histogram for several images; (c) Accumulation of orientation deviations

this one-nearest-neighbor approach is obtained by applying a clustering approach for the purpose of generalization. Closely located training samples are clustered and the clusters approximated as a multi-dimensional Gaussian, respectively. However, clustering procedures such as ISODATA search for neighboring elements according to simple metric, and do not consider any inherent topology between training samples. For example, if training samples are acquired by consecutively rotating the camera around the object, then we know in advance that the pattern variation can be approximated as a one-dimensional course in the space of patterns. Consequently, the clustering procedure should generate segments of this course by taking the succession of training views into account. By *imposing a topology* onto the collection of sample views, which is obtained from the process of image taking, we can cluster more adequately.

Role of Temporal Context in Object Recognition

In addition to the one-dimensional topology we also take advantage of the *temporal continuity* of gray values between the views in the image sequence.[1] For an object under rotation the temporal continuity can be observed exemplary in a series of histograms of orientations of gray value edges. Figure 7 shows four gray value images (a,b,c,d) of a transceiver box which has been rotated slightly in four discrete steps of $5°$. Figure 8 depicts the overlay of four histograms of edge orientations for these four images (but suppressing the gray values of the background). The histogram curves move to the right continually under slight object rotation.[2] These sequential correlations between consecutive images hold for small changes in the relation of object and camera. They are considered for fine-tuning the manifold approximation.

Incorporating Temporal Context for Manifold Approximation

Let us assume that the clustering is already performed under the constraint of a one-dimensional topology. This leads to a representative view for each cluster, respectively, which will be taken as *seed views* for manifold approximation. A *sequence of Gaussian basis functions* is used for approximating the one-dimensional course in the space of patterns. Each seed view is the basis for specifying the center of a multi-dimensional Gaussian with the dimension equal to the number of pixels. Each Gaussian is almost hyper-spherical except for one direction whose Gaussian extent is stretched. The exceptional direction at the current seed view is determined on the basis of the difference vector between the previous and the next seed view. For illustrating the principle, we take two-dimensional points which represent the seed views. Figure 9 shows a series of three seed views, i.e. previous, current and next seed view (X_{i-1}^s, X_i^s and X_{i+1}^s). At the current seed view the construction of an elongated Gaussian is depicted. Actually, an ellipse is shown which represents the contour related to a certain Gaussian altitude.

---

[1] The importance of temporal context in object recognition is well-known [2].

[2] The variation of the accumulation values is due to changing lighting conditions or due to the appearing or disappearing of object faces.

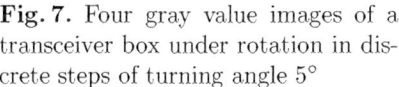

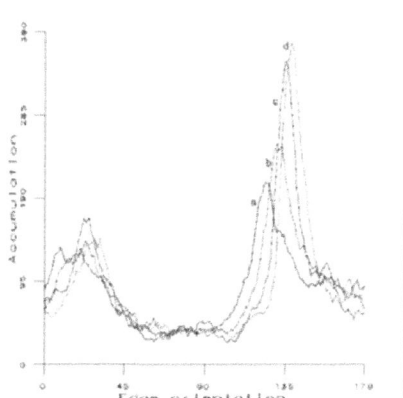

Fig. 7. Four gray value images of a transceiver box under rotation in discrete steps of turning angle 5°

Fig. 8. Overlay of four histograms of edge orientations computed for the four images in Figure 7

The Gaussian extent along this exceptional direction must be defined such that the significant variations between successive seed views are considered. For orthogonal directions the Gaussian extents are only responsible for taking random imponderables into account such as lighting variations. Consequently, the Gaussian extent along the exceptional direction must be set larger than the extent along the orthogonal directions. It is reasonable to determine the exceptional Gaussian extent dependent on the euclidean distance measurement between the previous and the next seed view. We avoid mathematical details because they are simple. However, it is worth to mention a similarity of this approach of manifold approximation with the so-called "oriented particle system" for surface modeling, introduced by Szeliski and Tonnesen [8].

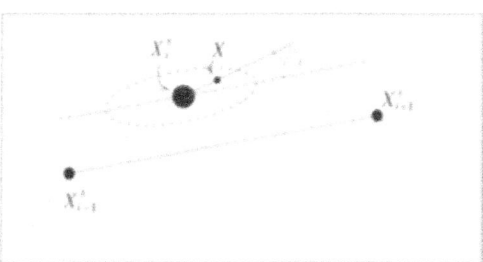

Fig. 9. Constructing hyper-ellipsoidal basis functions for time-series of seed vectors

Applying the Fine-Tuned Manifold for Object Recognition

Although our approach is very simple, both efficiency and robustness of the recognition function increases significantly. The usefulness of constructing elongated Gaussians is illustrated for recognizing the transceiver box in Fig. 7. For learning the recognition function the object is rotated in steps of $10°$ leading to 36 training images. All of them are used as seed images (for simplicity). The computation of gradient magnitudes followed by a thresholding procedure yields a set of gray value edges for each seed image. From each thresholded seed image a histogram of edge orientations is computed. A Gaussian basis function (GBF) network is installed by defining elongated GBFs according to the mentioned approach. Histograms of the seed images are used as the Gaussian center vectors and the Gaussians are modified based on previous and next seed histograms (and applying a user-defined stretching factor). In the GBF network the combination factors for the Gaussians are determined by the pseudo inverse technique.

For assessing the network of elongated Gaussians, we also construct a network of spherical Gaussians and compare the recognition results computed by the two GBF networks. The testing views are taken from the transceiver box but different from the training images. The testing data are subdivided in two categories. The first category consists of histograms of edge orientations arising from images with a certain *angle offset* relative to the training images. Temporal continuity of object rotation is considered purely. For these situations the relevant recognition function has been trained particularly. The second category consists of histograms of edge orientations arising from images with *angle offset and are scaled*, additionally. The recognition function composed of elongated Gaussians should recognize histograms of the first category robustly, and should discriminate clearly the histograms of the second category. The recognition function composed of spherical Gaussians should not be able to discriminate between both categories due to an increased generalization effect, i.e. accepting not only the angle offsets but also scaling effects. The desired results are shown in the diagrams of Fig. 10. By applying the recognition function of spherical Gaussians to all testing histograms, we can hardly discriminate between the two categories (left). Instead, by applying the recognition function of elongated Gaussians, we can define a threshold for discriminating between both categories (right).

5   Summary and Discussion

For an eye-on-hand system we presented three typical applications, i.e. tracking objects for obstacle avoidance, arranging certain viewing conditions, and acquiring an object recognition function. The concrete tasks have been to constrain the search for corresponding features, to parameterize correlation matching, and to fine-tune appearance manifolds. For solving the first task, specific steps of motion are performed during an experimentation phase in order to acquire constraints for motion vectors. By restricting the kind of motion to these specific ones during the online phase we can exploit the acquired constraints in the search for correspondences. For solving the second task, a specific course of motion is

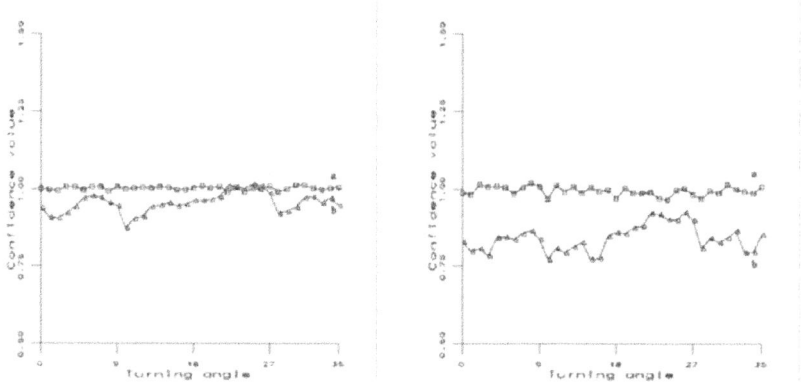

Fig. 10. Confidence values of recognizing an object based on histograms of edge orientations. For testing, the object has been rotated by an offset angle relative to the training images (result in curve a), or the object has been rotated and the image has been scaled additionally relative to the training images (result in curve b). (Left) Curves show the courses under the use of spherical Gaussians, both categories of testing data can hardly be distinguished; (Right) Curves show the courses under the use of elongated Gaussians, both categories of testing data can be distinguished clearly

performed during an experimentation phase in order to determine real deviations from a theoretical invariance, i.e. the variation of an LPT pattern under camera rotation. The distribution of deviations has been approximated by a Gaussian. The Gaussian extent can be used to determine a threshold in procedures which make use of correlation matching. For solving the third task, the camera is moved step by step for acquiring appearance patterns from an object. The pattern manifold is fine-tuned by making use of the known one-dimensional topology and the temporal continuity of the gray-values.

All three examples have in common that specific movements of the camera lead to certain changes in the images. From an abstract point of view, these are compatibilities between 3D motion features and 2D view sequence features. They are approximated during an experimentation phase based on statistical evaluations. Also, the examples show the usefulness of repeatable actions, i.e. the pre-specified actions in the experimentation phase must be repeatable in the application phase. The usefulness is due to the applicability of action-based information for supporting image processing in the application phase.

Apart form compatibilities for the perception-action cycle, which have been treated exemplary in this work, we also studied other compatibilities for the purpose of boundary extraction (published in [6]). The advantage is to reduce the amount of object-specific knowledge and instead make extensive use of constraints which are inherent in the three-dimensional nature of objects and in the process of image formation. For high-level Robot Vision applications a further category of compatibility is of interest. It is the compatibility between a deliberate plan (e.g. a strategy for solving a task) and the concrete servoing process (which is based on visual feedback). Generally, a compromise is needed between plan fulfillment and plan adjustment with the latter being triggered by requirements in the observed reality. Our approach of considering such compatibilities is based on dynamic potential fields (a publication will be prepared soon).

References

1. Y. Aloimonos, C. Fermüller: Analyzing action representations. Workshop on *Algebraic Frames for the Perception-Action Cycle*, LNCS **1888** (2000) 1–21. 227
2. S. Becker: Implicit learning in 3D object recognition – The importance of temporal context. *Neural Computation* **11** (1999) 347–374. 232
3. T. Binford, T. Levitt: Quasi-invariants – Theory and exploitation. *Image Understanding Workshop* (1993) 819–829. 227
4. M. Bolduc, M. Levine: A review of biologically motivated space-variant data reduction models for robot vision. *Comp. Vis. and Image Understanding* **69** (1998) 170–184. 229
5. H. Murase, S. Nayar: Visual learning and recognition of 3D objects from appearance. *Internat. J. of Computer Vision* **14** (1995) 5–24. 231
6. J. Pauli: Compatibilities for boundary extraction, Symp. der Deutschen Arbeitsgem. für Mustererkennung, *Informatik aktuell*, Springer-Verlag (2000) 468–475. 235
7. S. Smith, J. Brady: SUSAN – A new approach to low level image processing. *Internat. J. of Computer Vision* **23** (1997) 45–78. 228
8. R. Szeliski, D. Tonnesen: Surface modeling with oriented particle systems. *ACM SIGGRAPH Computer Graphics Annual Conf.* (1992) 185–194. 233

Trifocal Tensors with Grassmann-Cayley Algebra

Hongbo Li

Academy of Mathematics and System Sciences, Chinese Academy of Sciences
Beijing 100080, P. R. China
hli@mmrc.iss.ac.cn

Abstract. In this paper we study trifocal tensors with Grassmann-Cayley algebra. We propose a new method to derive relations among epipoles, fundamental tensors and trifocal tensors of three pinhole cameras. By this method we can find some new constraints satisfied by trifocal tensors.

1   Introduction

The trifocal tensor appears to be one of the most interesting theoretical constructions for three view vision. It was discovered by Shashua (1994) and Hartley (1994) using epipolar geometry. A systematic way of deriving its necessary and sufficient conditions was proposed by Faugeras and Mourrain (1995a, b). In the same publications three groups of degree-six constraints of the trifocal tensor were also discovered, and were later used by Faugeras and Papadopoulo (1997, 1998) in projective estimation.

The major mathematical tool used by Faugeras and Mourrain (1995a, b) is the so-called *Grassmann-Cayley algebra* of $\mathcal{R}^4$, which is a 16-dimensional graded vector space equipped with two multilinear, associative and grade-dependent anti-commutative products: the *wedge product* "$\wedge$" and the *meet product* "$\vee$". Elements in the algebra are called multivectors. When $\mathcal{R}^4$ is taken as a 3-dimensional projective space, the wedge product of two (or three) vectors represents the line (or plane) passing through them. The meet product of two (or three) planes is a vector, which represents the line (or point) of intersection of the planes. The meet product of a line and a plane is a vector representing their point of intersection.

In this paper, we also use the Grassmann-Cayley algebra to study trifocal tensors. The difference is that we rely on different expansions of the meet product to derive constraints satisfied by the trifocal tensor, instead of the Cramer's rule used by Faugeras and Mourrain (1995a, b). This approach is very efficient. With only seven meet products we can derive all the constraints on fundamental tensors and trifocal tensors of three cameras mentioned in (Faugeras and Mourrain, 1995a, b; Faugeras and Papadopoulo, 1997, 1998), and make considerable generalization by finding new constraints that have not appeared in the literature.

In section 2 we discuss various coordinate systems. In section 3 we use brackets to express epipoles, fundamental tensors and trifocal tensors. In section 4

R. Klette, S. Peleg, G. Sommer (Eds.): Robot Vision 2001, LNCS 1998, pp. 237–244, 2001.

we propose a method to derive relations among epipoles, fundamental tensors and trifocal tensors, which is based on different expansions of meet products into brackets. In section 5 we present some new constraints on trifocal tensors derived with the method.

2   Coordinate Systems

Let $\mathbf{C}$ be a vector in $\mathcal{R}^4$ representing the optical center of a pinhole camera, $\mathbf{p}_1, \mathbf{p}_2, \mathbf{p}_3$ be vectors in $\mathcal{R}^4$ representing points or directions in the image plane. Assume that

$$\det(\mathbf{C}, \mathbf{p}_1, \mathbf{p}_2, \mathbf{p}_3) = 1. \tag{1}$$

Then $\{\mathbf{C}, \mathbf{p}_1, \mathbf{p}_2, \mathbf{p}_3\}$ is called a *projective camera coordinate system*. The space $\mathbf{C} \wedge \mathcal{R}^4$ is 3-dimensional, called the *projective space of images*. It is spanned by $\{\mathbf{C} \wedge \mathbf{p}_1, \mathbf{C} \wedge \mathbf{p}_2, \mathbf{C} \wedge \mathbf{p}_3\}$, the latter being called a *projective image coordinate system*.

The *bracket* $[\mathbf{x}_1\mathbf{x}_2\mathbf{x}_3\mathbf{x}_4]$ of four vectors $\mathbf{x}_1, \mathbf{x}_2, \mathbf{x}_3, \mathbf{x}_4$ refers to $\det(\mathbf{x}_1, \mathbf{x}_2, \mathbf{x}_3, \mathbf{x}_4)$. It is also a linear function of the wedge product $X = \mathbf{x}_1 \wedge \mathbf{x}_2 \wedge \mathbf{x}_3 \wedge \mathbf{x}_4$, called the *dual* of X and also denoted by $[X]$.

The dual operator can be extended linearly to the whole Grassmann-Cayley algebra. The following linear function $[\mathbf{x}_1\wedge\mathbf{x}_2\wedge\mathbf{x}_3]$ is called the *dual* of $\mathbf{x}_1\wedge\mathbf{x}_2\wedge\mathbf{x}_3$:

$$\mathbf{x} \mapsto [\mathbf{x}_1\mathbf{x}_2\mathbf{x}_3\mathbf{x}], \quad \text{for } \mathbf{x} in \mathcal{R}^4. \tag{2}$$

The *dual* $[\mathbf{x}_1\mathbf{x}_2]$ of $\mathbf{x}_1 \wedge \mathbf{x}_2$, and the *dual* $[\mathbf{x}_1]$ of $\mathbf{x}_1$, can be defined similarly. The importance of dual operator lies in that, for any multivectors x, y,

$$[x \wedge y] = [x] \vee [y], \quad [x \vee y] = [x] \wedge [y], \tag{3}$$

i.e., the wedge product and the meet product are dual to each other.

For a coordinate system $\{\mathbf{x}_1, \mathbf{x}_2, \mathbf{x}_3, \mathbf{x}_4\}$ of $\mathcal{R}^4$, the linear function defined by

$$\mathbf{x}_4^* = \frac{[\mathbf{x}_1 \wedge \mathbf{x}_2 \wedge \mathbf{x}_3]}{[\mathbf{x}_1\mathbf{x}_2\mathbf{x}_3\mathbf{x}_4]} \tag{4}$$

satisfies $\mathbf{x}_4^*(\mathbf{x}_4) = 1$, $\mathbf{x}_4^*(\mathbf{x}_1) = \mathbf{x}_4^*(\mathbf{x}_2) = \mathbf{x}_4^*(\mathbf{x}_3) = 0$, so $\mathbf{x}_4^*$ is just the *reciprocal vector* of $\mathbf{x}_4$ with respect to the basis $\{\mathbf{x}_1, \mathbf{x}_2, \mathbf{x}_3, \mathbf{x}_4\}$. The reciprocal vectors of $\mathbf{x}_1, \mathbf{x}_2, \mathbf{x}_3$ can be obtained similarly. $\{\mathbf{x}_1^*, \mathbf{x}_2^*, \mathbf{x}_3^*, \mathbf{x}_4^*\}$ is the *reciprocal coordinate system*, or *reciprocal basis*, of $\{\mathbf{x}_1, \mathbf{x}_2, \mathbf{x}_3, \mathbf{x}_4\}$.

In this paper, all computations are carried out in the reciprocal coordinate systems of projective camera coordinate systems and projective image coordinate systems.

3   Epipoles, Fundamental Tensors and Trifocal Tensors by Brackets

For two cameras with optical centers $\mathbf{C}, \mathbf{C}'$ respectively, the image of point $\mathbf{C}'$ in camera $\mathbf{C}$ is $E^{\mathbf{C}\mathbf{C}'} = \mathbf{C} \wedge \mathbf{C}'$, called the *epipole* of $\mathbf{C}'$ in camera $\mathbf{C}$. Similarly, the *epipole* of $\mathbf{C}$ in camera $\mathbf{C}'$ is $E^{\mathbf{C}'\mathbf{C}} = \mathbf{C}' \wedge \mathbf{C}$.

In the projective image coordinate system $\{\mathbf{C} \wedge \mathbf{p}_1, \mathbf{C} \wedge \mathbf{p}_2, \mathbf{C} \wedge \mathbf{p}_3\}$, $E^{\mathbf{CC}'}$ is the following vector:

$$\mathbf{E}^{\mathbf{CC}'} = \begin{pmatrix} \mathbf{E}_1^{\mathbf{CC}'} \\ \mathbf{E}_2^{\mathbf{CC}'} \\ \mathbf{E}_3^{\mathbf{CC}'} \end{pmatrix} = \begin{pmatrix} [\mathbf{C}\mathbf{p}_2\mathbf{p}_3\mathbf{C}'] \\ [\mathbf{C}\mathbf{p}_3\mathbf{p}_1\mathbf{C}'] \\ [\mathbf{C}\mathbf{p}_1\mathbf{p}_2\mathbf{C}'] \end{pmatrix}. \tag{5}$$

Let $\{\mathbf{C}^*, \mathbf{1}, \mathbf{2}, \mathbf{3}\}$ be the reciprocal basis of $\{\mathbf{C}, \mathbf{p}_1, \mathbf{p}_2, \mathbf{p}_3\}$. Let $\{\mathbf{C}', \mathbf{p}_1', \mathbf{p}_2', \mathbf{p}_3'\}$ be a projective coordinate system of camera $\mathbf{C}'$, and let $\{\mathbf{C}'^*, \mathbf{1}', \mathbf{2}', \mathbf{3}'\}$ be the corresponding reciprocal basis. Using (3), we obtain

$$\mathbf{E}^{\mathbf{CC}'} = \begin{pmatrix} [\mathbf{11}'\mathbf{2}'\mathbf{3}'] \\ [\mathbf{21}'\mathbf{2}'\mathbf{3}'] \\ [\mathbf{31}'\mathbf{2}'\mathbf{3}'] \end{pmatrix}. \tag{6}$$

Any image x in camera $\mathbf{C}$ can be represented by $\mathbf{C} \wedge \mathbf{x}$, where $\mathbf{x} in \mathcal{R}^4$. Similarly, Any image x' in camera $\mathbf{C}'$ can be represented by $\mathbf{C}' \wedge \mathbf{x}'$. The *fundamental tensor* $F^{\mathbf{CC}'}$ of the two cameras is the following bilinear symmetric function:

$$F^{\mathbf{CC}'}(\mathbf{C} \wedge \mathbf{x}, \mathbf{C}' \wedge \mathbf{x}') = [\mathbf{C}\mathbf{x}\mathbf{C}'\mathbf{x}']. \tag{7}$$

In the projective coordinate systems $\{\mathbf{C} \wedge \mathbf{p}_1, \mathbf{C} \wedge \mathbf{p}_2, \mathbf{C} \wedge \mathbf{p}_3\}$, $\{\mathbf{C}' \wedge \mathbf{p}_1', \mathbf{C}' \wedge \mathbf{p}_2', \mathbf{C}' \wedge \mathbf{p}_3'\}$ of the two image planes, the fundamental tensor takes the following matrix form, called the *fundamental matrix*:

$$\mathbf{F}^{\mathbf{CC}'} = \begin{pmatrix} \mathbf{F}_{11}^{\mathbf{CC}'} & \mathbf{F}_{12}^{\mathbf{CC}'} & \mathbf{F}_{13}^{\mathbf{CC}'} \\ \mathbf{F}_{21}^{\mathbf{CC}'} & \mathbf{F}_{22}^{\mathbf{CC}'} & \mathbf{F}_{23}^{\mathbf{CC}'} \\ \mathbf{F}_{31}^{\mathbf{CC}'} & \mathbf{F}_{32}^{\mathbf{CC}'} & \mathbf{F}_{33}^{\mathbf{CC}'} \end{pmatrix} = \begin{pmatrix} [\mathbf{C}\mathbf{p}_1\mathbf{C}'\mathbf{p}_1'] & [\mathbf{C}\mathbf{p}_1\mathbf{C}'\mathbf{p}_2'] & [\mathbf{C}\mathbf{p}_1\mathbf{C}'\mathbf{p}_3'] \\ [\mathbf{C}\mathbf{p}_2\mathbf{C}'\mathbf{p}_1'] & [\mathbf{C}\mathbf{p}_2\mathbf{C}'\mathbf{p}_2'] & [\mathbf{C}\mathbf{p}_2\mathbf{C}'\mathbf{p}_3'] \\ [\mathbf{C}\mathbf{p}_3\mathbf{C}'\mathbf{p}_1'] & [\mathbf{C}\mathbf{p}_3\mathbf{C}'\mathbf{p}_2'] & [\mathbf{C}\mathbf{p}_3\mathbf{C}'\mathbf{p}_3'] \end{pmatrix}. \tag{8}$$

In the corresponding reciprocal bases $\{\mathbf{2} \wedge \mathbf{3}, \mathbf{3} \wedge \mathbf{1}, \mathbf{1} \wedge \mathbf{2}\}$, $\{\mathbf{2}' \wedge \mathbf{3}', \mathbf{3}' \wedge \mathbf{1}', \mathbf{1}' \wedge \mathbf{2}'\}$, the fundamental matrix is in the following form:

$$\mathbf{F}^{\mathbf{CC}'} = \begin{pmatrix} [\mathbf{23}2'\mathbf{3}'] & [\mathbf{23}3'\mathbf{1}'] & [\mathbf{23}1'\mathbf{2}'] \\ [\mathbf{31}2'\mathbf{3}'] & [\mathbf{31}3'\mathbf{1}'] & [\mathbf{31}1'\mathbf{2}'] \\ [\mathbf{12}2'\mathbf{3}'] & [\mathbf{12}3'\mathbf{1}'] & [\mathbf{12}1'\mathbf{2}'] \end{pmatrix}. \tag{9}$$

For three cameras with optical centers $\mathbf{C}, \mathbf{C}', \mathbf{C}''$ respectively, the *trifocal tensor* of camera $\mathbf{C}$ with respect to cameras $\mathbf{C}', \mathbf{C}''$ is the following trilinear function, which is antisymmetric with respect to its last two arguments:

$$T(\mathbf{C} \wedge \mathbf{x}, \mathbf{C}' \wedge \mathbf{x}' \wedge \mathbf{y}', \mathbf{C}'' \wedge \mathbf{x}'' \wedge \mathbf{y}'') = (\mathbf{C} \wedge \mathbf{x}) \vee (\mathbf{C}' \wedge \mathbf{x}' \wedge \mathbf{y}') \vee (\mathbf{C}'' \wedge \mathbf{x}'' \wedge \mathbf{y}''). \tag{10}$$

Two other trifocal tensors can be defined by interchanging $\mathbf{C}$ with $\mathbf{C}', \mathbf{C}''$:

$$T'(\mathbf{C}' \wedge \mathbf{x}', \mathbf{C} \wedge \mathbf{x} \wedge \mathbf{y}, \mathbf{C}'' \wedge \mathbf{x}'' \wedge \mathbf{y}'') = (\mathbf{C}' \wedge \mathbf{x}') \vee (\mathbf{C} \wedge \mathbf{x} \wedge \mathbf{y}) \vee (\mathbf{C}'' \wedge \mathbf{x}'' \wedge \mathbf{y}''), \tag{11}$$

$$T''(\mathbf{C}'' \wedge \mathbf{x}'', \mathbf{C} \wedge \mathbf{x} \wedge \mathbf{y}, \mathbf{C}' \wedge \mathbf{x}' \wedge \mathbf{y}') = (\mathbf{C}'' \wedge \mathbf{x}'') \vee (\mathbf{C} \wedge \mathbf{x} \wedge \mathbf{y}) \vee (\mathbf{C}' \wedge \mathbf{x}' \wedge \mathbf{y}'). \tag{12}$$

In the projective coordinate systems $\{\mathbf{C} \wedge \mathbf{p}_1, \mathbf{C} \wedge \mathbf{p}_2, \mathbf{C} \wedge \mathbf{p}_3\}$, $\{\mathbf{C}' \wedge \mathbf{p}_1', \mathbf{C}' \wedge \mathbf{p}_2', \mathbf{C}' \wedge \mathbf{p}_3'\}$, $\{\mathbf{C}'' \wedge \mathbf{p}_1'', \mathbf{C}' \wedge \mathbf{p}_2'', \mathbf{C}' \wedge \mathbf{p}_3''\}$ of the three image planes, the trifocal tensor T has the following components:

$$\mathbf{T}_{ijk} = (\mathbf{C} \wedge \mathbf{p}_i) \vee (\mathbf{C}' \wedge \check{\mathbf{p}}_j') \vee (\mathbf{C}'' \wedge \check{\mathbf{p}}_k''), \tag{13}$$

where $1 \leq i, j, k \leq 3$, $\check{\mathbf{p}}_1 = \mathbf{p}_2 \wedge \mathbf{p}_3$, $\check{\mathbf{p}}_2 = \mathbf{p}_3 \wedge \mathbf{p}_1$, $\check{\mathbf{p}}_3 = \mathbf{p}_1 \wedge \mathbf{p}_2$. Using the reciprocal bases, we get

$$(\mathbf{T}_{1jk} \quad \mathbf{T}_{2jk} \quad \mathbf{T}_{3jk}) = ([\mathbf{23j'k''}] \quad [\mathbf{31j'k''}] \quad [\mathbf{12j'k''}]). \tag{14}$$

4   Deriving Relations on Epipoles, Fundamental Tensors and Trifocal Tensors

Consider the following vectors in $\mathcal{R}^4$:

$$\{1, 2, 3, 1', 2', 3', 1'', 2'', 3''\}. \tag{15}$$

According to (6), (9) and (14), in the reciprocal bases of the projective coordinate systems of the three image planes, any epipole, fundamental tensor or trifocal tensor of the three cameras has it components in the form of brackets of vectors in (15). Conversely, any nonzero bracket of vectors in (15) equals a component of one of the epipoles, fundamental tensors and trifocal tensors up to the sign.

$$\begin{aligned}
\mathbf{E}^{\mathbf{CC}'} &\leftrightarrow \{\mathbf{ij'k'l'}\}, & \mathbf{E}^{\mathbf{C'C}} &\leftrightarrow \{\mathbf{i'jkl}\}, & \mathbf{E}^{\mathbf{CC}''} &\leftrightarrow \{\mathbf{ij''k''l''}\}, \\
\mathbf{E}^{\mathbf{C''C}} &\leftrightarrow \{\mathbf{i''jkl}\}, & \mathbf{E}^{\mathbf{C'C}''} &\leftrightarrow \{\mathbf{i'j''k''l''}\}, & \mathbf{E}^{\mathbf{C''C}'} &\leftrightarrow \{\mathbf{i''j'k'l'}\}, \\
\mathbf{F}^{\mathbf{CC}'} &\leftrightarrow \{\mathbf{ijk'l'}\}, & \mathbf{F}^{\mathbf{CC}''} &\leftrightarrow \{\mathbf{ijk''l''}\}, & \mathbf{F}^{\mathbf{C'C}''} &\leftrightarrow \{\mathbf{i'j'k''l''}\}, \\
\mathbf{T} &\leftrightarrow \{\mathbf{ijk'l''}\}, & \mathbf{T}' &\leftrightarrow \{\mathbf{i'j'kl''}\}, & \mathbf{T}'' &\leftrightarrow \{\mathbf{i''j''kl'}\}.
\end{aligned} \tag{16}$$

The only constraints on vectors in (15) are that $\{1, 2, 3\}$, $\{1', 2', 3'\}$, $\{1'', 2'', 3''\}$ are triplets of linear independent vectors. Thus, all relations among the epipoles, epipolar tensors and trifocal tensors can be proved using the so-called *Grassmann-Plücker relations* (see Li and Wu, 1999). Although it is easy to *verify* geometric relations with the Grassmann-Plücker relations, it is difficult to *derive* geometric relations.

In this paper we follow a *Grassmann-Cayley algebra approach* (see Li and Sommer, 2000) to finding geometric relations on epipoles, epipolar tensors and trifocal tensors, and in particular, to finding constraints of the trifocal tensors. We consider the set of *meet products of all possible wedge products* of vectors in (15). Since the meet product is associative and grade-dependent anti-commutative, for the same meet product there are different ways to *expand* it. The equalities of these expansions lead to equalities of brackets. Those equalities which are *independent of indices* of vectors in (15) can be changed into equalities on epipoles, epipolar tensors and trifocal tensors.

The following is an example on deriving a relation among the fundamental tensor $F^{\mathbf{CC}''}$ and the epipoles $E^{\mathbf{CC}'}, E^{\mathbf{C''C}}, E^{\mathbf{C''C}'}$. The wedge product $\mathbf{C}' \wedge$

$\mathbf{C} \wedge \mathbf{C}''$ represents the projective plane passing through the three optical centers. Using (3), we get

$$[\mathbf{C}' \wedge \mathbf{C} \wedge \mathbf{C}''] = (\mathbf{1} \wedge \mathbf{2} \wedge \mathbf{3}) \vee (\mathbf{1}' \wedge \mathbf{2}' \wedge \mathbf{3}') \vee (\mathbf{1}'' \wedge \mathbf{2}'' \wedge \mathbf{3}''). \qquad (17)$$

Expanding the meet product differently, we get the conclusion that for any $1 leq i leq 3$ and any even permutation j_-, j, j_+ of $1, 2, 3$,

$$lambda \sum limits_{j=1}^{3} \mathbf{E}_i^{\mathbf{CC}'} \mathbf{F}_{ij}^{\mathbf{CC}''} = \mathbf{E}_{j_+}^{\mathbf{C}''\mathbf{C}} \mathbf{E}_{j_-}^{\mathbf{C}''\mathbf{C}'} - \mathbf{E}_{j_-}^{\mathbf{C}''\mathbf{C}} \mathbf{E}_{j_+}^{\mathbf{C}''\mathbf{C}'}, \qquad (18)$$

where $lambda$ is a nonzero scale independent of the indices. In matrix form, (18) can be written as

$$lambda (\mathbf{F}^{\mathbf{CC}''})^T \mathbf{E}^{\mathbf{CC}'} = \mathbf{E}^{\mathbf{C}''\mathbf{C}} \times \mathbf{E}^{\mathbf{C}''\mathbf{C}'}. \qquad (19)$$

The geometric meaning is that the epipolar line of $\mathbf{E}^{\mathbf{CC}'}$ in camera $\mathbf{C}''$ passes through epipoles $\mathbf{E}^{\mathbf{C}''\mathbf{C}}$, $\mathbf{E}^{\mathbf{C}''\mathbf{C}'}$.

5   Some New Constraints on Trifocal Tensors

(1) Degree-four constraints.

Let $k_1 \neq k_2$, and let $1 leq i, j_1, j_2 leq 3$. Then

$$t_{ij_1k_1} t_{ij_2k_2} = t_{ij_1k_2} t_{ij_2k_1}, \qquad (20)$$

where

$$t_{ijk} = T_{ij_+k_+} T_{ij_-k_-} - T_{ij_+k_-} T_{ij_-k_+}, \qquad (21)$$

j_-, j, j_+ and k_-, k, k_+ being even permutations of $1, 2, 3$.

(2) The first group of degree-six constraints.

2.a. Let $i_1 \neq i_2$, $i_3 \neq i_4$, and let $1 leq j_1, j_2, j_3, j_4 leq 3$. Then

$$u''_{i_1 i_2 j_1 j_2} u''_{i_3 i_4 j_3 j_4} = u''_{i_1 i_2 j_1 j_4} u''_{i_3 i_4 j_3 j_2}, \qquad (22)$$

where

$$u''_{i_1 i_2 j_1 j_2} = t_{i_1 j_1 1} T_{i_2 j_2 1} + t_{i_1 j_1 2} T_{i_2 j_2 2} + t_{i_1 j_1 3} T_{i_2 j_2 3}. \qquad (23)$$

2.b. Let i_-, i, i_+ be an even permutation of $1, 2, 3$, and let $1 leq j_1, j_2, j_3, j_4 leq 4$. Then

$$u''_{ii_+ j_1 j_2} u''_{ii_- j_3 j_4} = u''_{ii_- j_1 j_2} u''_{ii_+ j_3 j_4}. \qquad (24)$$

Remark. (22) includes as a special case the first group of degree-six constraints of Faugeras and Mourrain (1995a, b): for any $1 leq k_1, k_2, l_1, l_2 leq 3$, if $k_1 \neq l_1, k_2 \neq l_2$, then

$$\begin{aligned} &|\mathbf{T}_{k_1 k_2 \cdot}\ \mathbf{T}_{k_1 l_2 \cdot}\ \mathbf{T}_{l_1 l_2 \cdot}|\,|\mathbf{T}_{k_1 k_2 \cdot}\ \mathbf{T}_{l_1 k_2 \cdot}\ \mathbf{T}_{l_1 l_2 \cdot}| \\ &= |\mathbf{T}_{k_1 k_2 \cdot}\ \mathbf{T}_{l_1 k_2 \cdot}\ \mathbf{T}_{k_1 l_2 \cdot}|\,|\mathbf{T}_{k_1 k_2 \cdot}\ \mathbf{T}_{k_1 l_2 \cdot}\ \mathbf{T}_{l_1 l_2 \cdot}|, \end{aligned} \qquad (25)$$

where $\mathbf{T}_{k_1 k_2.} = (\mathbf{T}_{k_1 k_2 1}, \mathbf{T}_{k_1 k_2 2}, \mathbf{T}_{k_1 k_2 3})^T$. The reason is as follows.

Let $i_1 \neq i_2$, and let j, j_1, j_2 be a permutation of $1, 2, 3$ with permutation sign $\epsilon(j, j_1, j_2)$. Define

$$v''_{i_1 i_2 j_1 j_2} = \epsilon(j, j_1, j_2) u''_{i_1 i_2 j j_2}. \tag{26}$$

It can be proved that

$$
\begin{aligned}
|\mathbf{T}_{k_1 k_2.} \ \mathbf{T}_{k_1 l_2.} \ \mathbf{T}_{l_1 l_2.}| &= v''_{k_1 l_1 k_2 l_2}, \\
|\mathbf{T}_{k_1 k_2.} \ \mathbf{T}_{l_1 k_2.} \ \mathbf{T}_{l_1 l_2.}| &= v''_{l_1 k_1 l_2 k_2}, \\
|\mathbf{T}_{k_1 k_2.} \ \mathbf{T}_{l_1 k_2.} \ \mathbf{T}_{k_1 l_2.}| &= v''_{k_1 l_1 l_2 k_2}, \\
|\mathbf{T}_{l_1 k_2.} \ \mathbf{T}_{k_1 l_2.} \ \mathbf{T}_{l_1 l_2.}| &= v''_{l_1 k_1 k_2 l_2}.
\end{aligned}
\tag{27}
$$

So (25) can be written as

$$v''_{k_1 l_1 k_2 l_2} v''_{l_1 k_1 l_2 k_2} = v''_{k_1 l_1 l_2 k_2} v''_{l_1 k_1 k_2 l_2}, \tag{28}$$

or using (26) and assuming $k = \{1, 2, 3\} - \{k_2, l_2\}$,

$$u''_{k_1 l_1 k l} u''_{l_1 k_1 k k_2} = u''_{k_1 l_1 k k_2} u''_{l_1 k_1 k l_2}, \tag{29}$$

which is a special case of (22) under the correspondences

$$k_1 \leftrightarrow i_1 = i_4, \ l_1 \leftrightarrow i_2 = i_3, \ k \leftrightarrow j_1 = j_3, \ l_2 \leftrightarrow j_2 \neq j_1, \ k_2 \leftrightarrow j_4 \neq j_3. \tag{30}$$

(3) The second group of degree-six constraints.

3.a. *Let $i_1 \neq i_2$, $i_3 \neq i_4$, and let $1 \leq k_1, k_2, k_3, k_4 \leq 3$. Then*

$$u'_{i_1 i_2 k_1 k_2} u'_{i_3 i_4 k_3 k_4} = u'_{i_1 i_2 k_1 k_4} u'_{i_3 i_4 k_3 k_2}, \tag{31}$$

where

$$u'_{i_1 i_2 k_1 k_2} = t_{i_1 1 k_1} T_{i_2 1 k_2} + t_{i_1 2 k_1} T_{i_2 2 k_2} + t_{i_1 3 k_1} T_{i_2 3 k_2}. \tag{32}$$

3.b. *Let i_-, i, i_+ be an even permutation of $1, 2, 3$, and let $1 \leq k_1, k_2, k_3, k_4 \leq 4$. Then*

$$u'_{i i_+ k_1 k_2} u'_{i i_- k_3 k_4} = u'_{i i_- k_1 k_2} u'_{i i_+ k_3 k_4}. \tag{33}$$

Remark. (31) includes as a special case the second group of degree-six constraints of Faugeras and Mourrain (1995a, b): for any $1 \leq k_1, k_2, l_1, l_2 \leq 3$, if $k_1 \neq l_1$, $k_2 \neq l_2$, then

$$
\begin{aligned}
&|\mathbf{T}_{k_1.k_2} \ \mathbf{T}_{k_1.l_2} \ \mathbf{T}_{l_1.l_2}| \, |\mathbf{T}_{k_1.k_2} \ \mathbf{T}_{l_1.k_2} \ \mathbf{T}_{l_1.l_2}| \\
&= |\mathbf{T}_{l_1.k_2} \ \mathbf{T}_{k_1.l_2} \ \mathbf{T}_{l_1.l_2}| \, |\mathbf{T}_{k_1.k_2} \ \mathbf{T}_{l_1.k_2} \ \mathbf{T}_{k_1.l_2}|,
\end{aligned}
\tag{34}
$$

where $\mathbf{T}_{k_1.k_2} = (\mathbf{T}_{k_1 1 k_2}, \mathbf{T}_{k_1 2 k_2}, \mathbf{T}_{k_1 3 k_2})^T$. The reason is as follows.

Let $i_1 \neq i_2$, and let k, k_1, k_2 be a permutation of $1, 2, 3$. Define

$$v'_{i_1 i_2 k_1 k_2} = \epsilon(k, k_1, k_2) u'_{i_1 i_2 k k_2}. \tag{35}$$

It can be proved that

$$
\begin{aligned}
|\mathbf{T}_{k_1.k_2}\ \mathbf{T}_{k_1.l_2}\ \mathbf{T}_{l_1.l_2}| &= v'_{k_1 l_1 k_2 l_2}, \\
|\mathbf{T}_{k_1.k_2}\ \mathbf{T}_{l_1.k_2}\ \mathbf{T}_{l_1.l_2}| &= v'_{l_1 k_1 l_2 k_2}, \\
|\mathbf{T}_{k_1.k_2}\ \mathbf{T}_{l_1.k_2}\ \mathbf{T}_{k_1.l_2}| &= v'_{k_1 l_1 l_2 k_2}, \\
|\mathbf{T}_{l_1.k_2}\ \mathbf{T}_{k_1.l_2}\ \mathbf{T}_{l_1.l_2}| &= v'_{l_1 k_1 k_2 l_2}.
\end{aligned}
\tag{36}
$$

So (34) can be written as

$$
v'_{k_1 l_1 k_2 l_2} v'_{l_1 k_1 l_2 k_2} = v'_{k_1 l_1 l_2 k_2} v'_{l_1 k_1 k_2 l_2},
\tag{37}
$$

or using (35) and assuming $k = \{1,2,3\} - \{k_2, l_2\}$,

$$
u'_{k_1 l_1 k l_2} u'_{l_1 k_1 k k_2} = u'_{k_1 l_1 k k_2} u'_{l_1 k_1 k l_2},
\tag{38}
$$

which is a special case of (31) under the correspondences

$$
k_1 \leftrightarrow i_1 = i_4,\ l_1 \leftrightarrow i_2 = i_3,\ k \leftrightarrow k_1 = k_3,\ l_2 \leftrightarrow k_2 \neq k_1,\ k_2 \leftrightarrow k_4 \neq k_3. \tag{39}
$$

(4) The third group of degree-six constraints.

4.a. *Let $j_1 \neq j_2, k_1 \neq k_2, k_3 \neq k_4$, then*

$$
u_{j_1 j_2 k_1 k_2} u_{j_2 j_1 k_3 k_4} = u_{j_1 j_2 k_3 k_4} u_{j_2 j_1 k_1 k_2},
\tag{40}
$$

where

$$
u_{j_1 j_2 k_1 k_2} = t'_{1 j_1 k_1} T_{1 j_2 k_2} + t'_{2 j_1 k_1} T_{2 j_2 k_2} + t'_{3 j_1 k_1} T_{3 j_2 k_2},
\tag{41}
$$

$$
t'_{ijk} = T_{i_+ j k_+} T_{i_- j k_-} - T_{i_+ j k_-} T_{i_- j k_+},
\tag{42}
$$

i_-, i, i_+ and k_-, k, k_+ being even permutations of $1, 2, 3$.
4.b. *Let $j_1 \neq j_2, j_3 \neq j_4$, and let k_-, k, k_+ be an even permutation of $1, 2, 3$. Then*

$$
u_{j_1 j_2 k k_+} u_{j_3 j_4 k k_-} = u_{j_1 j_2 k k_-} u_{j_3 j_4 k k_+}.
\tag{43}
$$

Remark. (40), (43) both include as a special case the third group of degree-six constraints of Faugeras and Mourrain (1995a, b): for any $1 \leq k_1, k_2, l_1, l_2 \leq 3$, if $k_1 \neq l_1, k_2 \neq l_2$, then

$$
\begin{aligned}
&|\mathbf{T}_{.k_1 k_2}\ \mathbf{T}_{.k_1 l_2}\ \mathbf{T}_{.l_1 l_2}||\mathbf{T}_{.k_1 k_2}\ \mathbf{T}_{.l_1 k_2}\ \mathbf{T}_{.l_1 l_2}| \\
&= |\mathbf{T}_{.k_1 k_2}\ \mathbf{T}_{.l_1 k_2}\ \mathbf{T}_{.k_1 l_2}||\mathbf{T}_{.l_1 k_2}\ \mathbf{T}_{.k_1 l_2}\ \mathbf{T}_{.l_1 l_2}|,
\end{aligned}
\tag{44}
$$

where $\mathbf{T}_{.k_1 k_2} = (\mathbf{T}_{1 k_1 k_2}, \mathbf{T}_{2 k_1 k_2}, \mathbf{T}_{3 k_1 k_2})^T$. The reason is as follows.

Let $j_1 \neq j_2$, and let k, k_1, k_2 be a permutation of $1, 2, 3$. Define

$$
v_{j_1 j_2 k_1 k_2} = -\epsilon(k, k_1, k_2) u_{j_1 j_2 k k_2}.
\tag{45}
$$

It can be proved that

$$
\begin{aligned}
|\mathbf{T}_{.k_1 k_2}\ \mathbf{T}_{.k_1 l_2}\ \mathbf{T}_{.l_1 l_2}| &= v_{k_1 l_1 k_2 l_2}, \\
|\mathbf{T}_{.k_1 k_2}\ \mathbf{T}_{.l_1 k_2}\ \mathbf{T}_{.l_1 l_2}| &= v_{l_1 k_1 l_2 k_2}, \\
|\mathbf{T}_{.k_1 k_2}\ \mathbf{T}_{.l_1 k_2}\ \mathbf{T}_{.k_1 l_2}| &= v_{k_1 l_1 l_2 k_2}, \\
|\mathbf{T}_{.l_1 k_2}\ \mathbf{T}_{.k_1 l_2}\ \mathbf{T}_{.l_1 l_2}| &= v_{l_1 k_1 k_2 l_2}.
\end{aligned}
\tag{46}
$$

So (44) can be written as

$$v_{k_1 l_1 k_2 l_2} v_{l_1 k_1 l_2 k_2} = v_{k_1 l_1 l_2 k_2} v_{l_1 k_1 k_2 l_2}, \qquad (47)$$

or using (45) and assuming $k = \{1, 2, 3\} - \{k_2, l_2\}$,

$$u_{k_1 l_1 k l_2} u_{l_1 k_1 k k_2} = u_{k_1 l_1 k k_2} u_{l_1 k_1 k l_2}, \qquad (48)$$

which is a special case of (40) under the correspondences

$$k_1 \leftrightarrow j_1, \ l_1 \leftrightarrow j_2, \ k \leftrightarrow k_1 = k_3, \ l_2 \leftrightarrow k_2, \ k_2 \leftrightarrow k_4, \qquad (49)$$

and also a special case of (43) under the correspondences

$$k_1 \leftrightarrow j_1 = j_4, \ l_1 \leftrightarrow j_2 = j_3, \ k \leftrightarrow k, \ l_2 \leftrightarrow k_+, \ k_2 \leftrightarrow k_-. \qquad (50)$$

References

1. E. Bayro-Corrochano, J. Lasenby, G. Sommer: Geometric Algebra: a framework for computing point and line correspondences and projective structure using n-uncalibrated cameras. In: *Proc. of ICPR'96*, Vienna, Vol. I, (1996) 334–338.
2. O. Faugeras, B. Mourrain: On the geometry and algebra of the point and line correspondences between n images. In: *Proc. ICCV'95*, (1995) 951–956.
3. O. Faugeras, B. Mourrain: On the geometry and algebra of the point and line correspondences between n images. Technical report **2665**, INRIA (1995).
4. O. Faugeras, T. Papadopoulo: Grassmann-Cayley algebra for modelling systems of cameras and the algebraic equations of the manifold of trifocal tensors. Technical report **3225**, INRIA (1997).
5. O. Faugeras, T. Papadopoulo: A nonlinear method for estimating the projective geometry of three views. In: *Proc. ICCV'98* (1998).
6. R. Hartley: Lines and points in three views – an integrated approach. In: *Proc. of ARPA Image Understanding Workshop*, Defense Advanced Research Projects Agency, Morgan Kaufmann Publ. Inc. (1994).
7. D. Hestenes, R. Ziegler: Projective geometry with Clifford algebra. *Acta Appl. Math.* **23** (1991) 25–63.
8. H. Li and Y. Wu (2000): Outer product factorization in Clifford algebra. *Proc. of ATCM'99*, Guangzhou, pp. 255–264.
9. H. Li and G. Sommer (2000): Coordinate-free projective geometry for computer vision. In *Geometric Computing with Clifford Algebra*, G. Sommer (*Ed.*), Springer Heidelberg, pp. 415–454.
10. A. Shashua: Trilinearity in visual recognition by alignment. In: LNCS **800**, J.-O. Eklundh (*Ed.*), Springer, (1994) 479–484.

Camera Calibration Using Rectangular Textures

Jacky Baltes

Center for Imaging Technology and Robotics, University of Auckland
Auckland, New Zealand
j.baltes@auckland.ac.nz

Abstract. This paper describes a practical method for the camera calibration given a single image of a regular texture. This paper uses the calibration of images of skyscrapers as an example. The paper introduces two algorithms for the assignment of real world coordinates to feature points. The first algorithm selects five closely connected feature points and determines the orientation of the rectangular pattern. The second algorithm iteratively sorts the feature points and assigns real world coordinates to them. Lastly, the Tsai camera calibration algorithm is used to compute the camera parameters.

1 Introduction

This paper describes an application of our camera calibration method, which was initially developed for RoboCup.

RoboCup is an international competition of fully autonomous robots playing soccer [4]. The first competition was organized in 1997, and it has rapidly increased in popularity.

Apart from the obvious challenges in robotics, control theory, path planning, artificial intelligence, and machine learning, RoboCup also presents an interesting domain for real-time computer vision. In the small league, robots are identified using a global vision system. To achieve adequate control, a vision system must track ten robots, a ball and compute their position, orientation, velocity with a cycle time of less than 20ms. More details about the All Botz videoserver can be found in [1].

This paper shows how the camera calibration method which was originally developed for the RoboCup domain can be used to compute the calibrate images of any regular textures or co-planar feature points using a single image.

Section 2 is an introduction to camera calibration in general and the challenges of calibration using regular textures. The extraction of calibration points is shown in section 3. The Tsai camera calibration used as back end to compute the final calibration parameters is described in 4. The paper concludes with section 5.

2 Camera Calibration

Accurate camera calibration is an essential ingredient in any computer vision system. Therefore, it has been a very active and productive research area. A

R. Klette, S. Peleg, G. Sommer (Eds.): Robot Vision 2001, LNCS 1998, pp. 245–251, 2001.

number of different camera calibration methods have been developed [5]. Most of these methods are based on the pin-hole camera model.

The input to camera calibration methods is a set of image features and their associated real-world attributes. For example, the well known Tsai calibration uses a set of image points with known real world coordinates as input. Haralick proposes a calibration method that use the image coordinates of parallel lines[3].

Most camera calibration methods use calibration objects, for example cubes with color patches. These calibration objects allow accurate control over their feature points and allow therefore accurate calibration.

Calibration of pictures of natural scenes requires a different approach because the calibration objects are too small. For example, assume that we need to calibrate the geometry of the images shown in Fig. 2. It is clearly impractical to build a calibration cube of this dimension.

As can be seen though, both images contain regular feature points that can be used for calibration. For example, if the distance between floors and between windows is known, then the position of the centers of the windows can be calculated. This is the basic method in our approach. However, the problem with real world scenes is that the features are hard to extract from the image. Furthermore, there are many feature points and to assign them manually would be time consuming and error prone. Lastly, not all feature points can be extracted from the image and some of them will be missing. The following section discusses a system to automatically deal with these problems.

Skyscraper (Singapore)   Wall Street (New York)

Fig. 1. Sample Calibration Pictures

3   Extraction of Calibration Points

This section describes the algorithm for sorting the feature points and assigning real world coordinates to them. Our system uses a semi-automatic approach. First, a manual preprocessing step is used to segment the image (Subsection 3.1). Secondly, an automatic routine is used to compute the centers of the feature points, to sort them, to correct for missing features, and to assign real world coordinates to the points.

3.1   Image Preprocessing

The first step in the camera calibration routine is a manual preprocessing step to clean up the image, to remove unwanted artifacts, and to select suitable parameter settings for the color segmentation routines. Figure 3.1 shows the output of the pre-processing step for the Singapore skyscraper (see Fig. 2).

As can be seen, some of the features were not distinct enough to be recognized and are missing in the preprocessed image.

3.2   Sorting of Feature Points

This section describes the heart of the calibration routine — the algorithm to assign real world coordinates to the feature points that were extracted in the

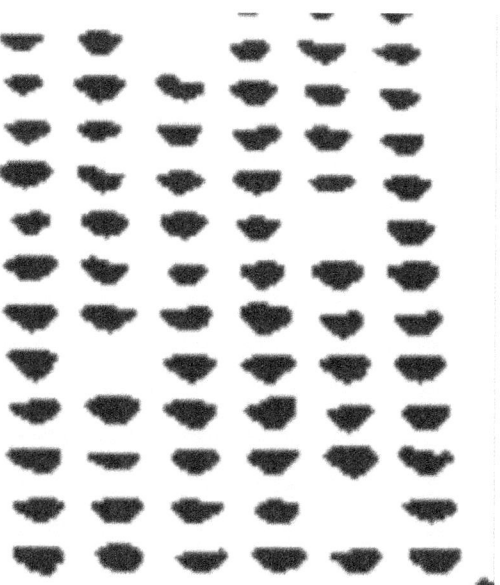

Fig. 2. Output of the Preprocessing Step

pre-processing step. The algorithm consists of two main parts: (a) algorithm 1 computes an initial guess of the translation matrix. (b) algorithm 2 is an iterative algorithm which assigns real world coordinates to the feature points and updates the guess of the transformation matrix.

Algorithm 1 Algorithm to Assign Real World Coordinates

1: $S = \text{extractCentres}(Image)$
2: $P = \text{selectSeedPoints}(S)$
3: $M_{XY} = \text{computeTransformationMatrix}(S, \text{realWorldCoords}(S))$
4: $M_{YX} = \text{computeTransformationMatrix}(S, \text{swapCoord}(\text{realWorldCoords}(S)))$
5: **if** $\text{error}(M_{XY}) < \text{error}(M_{YX})$ **then**
6:    $M_{Initial} = M_{XY}$
7: **else**
8:    $M_{Initial} = M_{YX}$
9:    $\text{swapCoordinates}(S)$
10: **end if**
11:   $C_S = \text{assignCentres}(S, M_{Initial})$

In line 1 of algorithm 1, the centers of the features are extracted. Then in line 2, a number of seed points are selected. These seed points are used to compute the initial transformation matrix. Five seed points are selected to form a cross in the two-dimensional plane. To guarantee a good estimate of the initial transformation matrix, the five centers that form a cross closest to the center of the image are selected. The routine `computeTransformationMatrix` called in line 3 uses the Boyer-Moore pseudo inverse method to compute a least means square (LMS) approximation for the 3×4 transformation matrix for the perspective projection. Since our system assumes that the distances between squares are not identical, the algorithm checks both possible orientations for the width and the height of the features and selects the best match (Line 4).

Algorithm 2 describes the method for the assignment of real world coordinates to the feature points. This algorithm iteratively selects the unassigned feature points that are closest to an already assigned point. In line 8, the variables b_x and b_y contain the closest number of blocks given the current estimate M of the transformation matrix. This estimate of the number of blocks allows the algorithm to compensate for missing feature points. Once the real world coordinates have been assigned to the new point, a new transformation matrix is computed (line 11). This process is repeated until all points have been assigned.

The output of the system is shown in Fig. 4 shows the output of our system. All feature points are correctly aligned and the system correctly corrects for the missing features. The initial five seed points are shown in the gray shaded region.

Algorithm 2 Algorithm to Assign Real World Coordinates($S, M_{Initial}$)

1: $M = M_{Initial}$
2: $N =$
3: **while** $S \neq$ **do**
4:     **for all** $s \in S$ **do**
5:         $s_{North}, s_{East}, s_{South}, s_{West} = \text{findNearestNeighbors}(s, M)$
6:     **end for**
7:     **for all** $n \in \{s_{North}, s_{East}, s_{South}, s_{West}\}$ **do**
8:         $b_x = dist((s - n)/width, M), b_y = dist((s - n)/height, M)$
9:         $n_x, n_y = s_x + bx * width, s_y = by * height$
10:        $N = N + n, S = S - n$
11:        $M = \text{computeTransformationMatrix}(N, \text{realWorldCoords}(S))$
12:    **end for**
13: **end while**
14: return N

---

4 Tsai Camera Calibration

After the computation of the matching points, we use a public domain implementation of Tsai's camera calibration to compute the extrinsic and intrinsic parameters of the camera model.

The Tsai calibration method uses a four step process to compute the parameters of a pin hole camera with radial lens distortion.

Firstly, the position (X_T, Y_T, Z_T) and the orientation (R_X, R_Y, R_Z) of the camera with respect to the world coordinate system is computed. This involves solving a simple system of linear equations. This step translates the 3D World coordinates into 3D camera coordinates and computes the six extrinsic parameters of the camera model.

In Step 2, the perspective distortion of a pin hole camera is compensated for. This step is a non-linear approximation and computes the focal length f of the camera. The output of this step are the ideal undistorted image coordinates.

Thirdly, the radial lens distortion parameters (κ_1, κ_2) are computed. These parameters compensate for the pin cushion effect of video cameras, that is straight lines along the edges of the camera are rounded. The output of step 3 are the distorted image coordinates.

Lastly, the image coordinates are discretized into the real image coordinates by taking the number of pixels in each row and column of an image into consideration.

The last three steps compute five intrinsic parameters of the camera model (focal length, lens distortion, scale factor for the rows, and the origin in the image plane).

The Tsai method is a very efficient, accurate, and versatile camera calibration method and is therefore very popular in computer vision. Nevertheless, one of the shortcomings of Tsai's method is that it is not able to compute the uncertainty factor S_X from only co-planar calibration points. In practice, this does

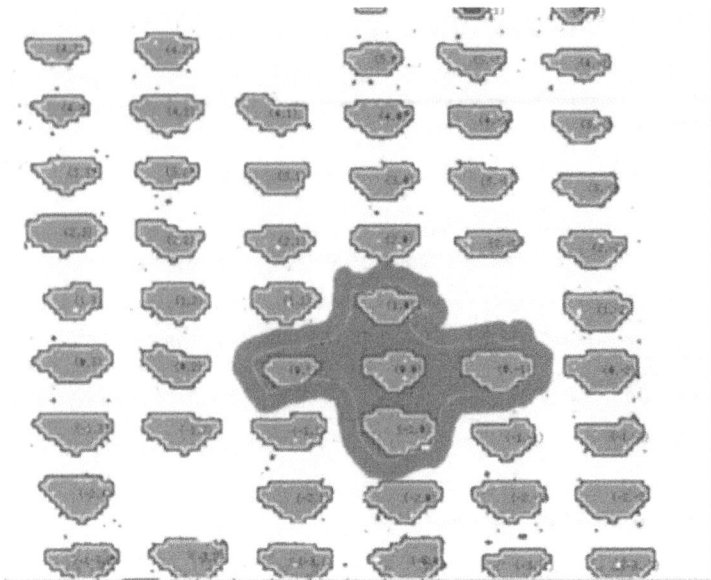

Fig. 3. Assignment of Real World Coordinates

not present a big problem, since most natural 3D textures consist of several planar textures. The calibration points that are extracted using our method can be concatenated into a large set of 3D calibration points.

Furthermore, during the sorting of the feature vectors, our extraction algorithms use a simple pin hole camera model to compute the perspective projection parameters. Therefore, our method is independent of the exact calibration method used.

5   Conclusion

This paper presents a practical system for the accurate calibration of real world scenes that have a regular texture. Our system is currently limited to rectangular textures, but this limitation is due to the current implementation. In theory, any regular texture is suitable.

The Tsai calibration does not provide an efficient method for calculation of the uncertainty factor S_x given co-planar calibration points. We are investigating other suitable methods as the one described in [2].

References

1. J. Baltes: Practical camera and color calibration for large scale rooms. In: *Proceed. of the IJCAI Workshop on RoboCup*, Stockholm, Sweden, (July 1999).   245
2. J. Batista, H. Araujo, A. T. Almeida: Monoplanar camera calibration: Iterative multistep approach. In: *BMVC93*, (1993).   250

3. R. Haralick: Determining camera parameters from the perspective projection of a rectangle. *Pattern Recognition*, **22** (1989). 246

4. H. Kitano (Ed.): *RoboCup-97: Robot Soccer World Cup I*. Springer Verlag, (1998). 245

5. R. Y. Tsai: An efficient and accurate camera calibration technique for 3d machine vision. In: *Proceed. of IEEE Conf. on Computer Vision and Pattern Recognition*, Miami Beach, FL, (1986) 364–374. 246

Optical Flow in Log-mapped Image Plane
A New Approach

Mohammed Yeasin[*]

Dept. of Computer Science and Engg.,
The Pennsylvania State University, University Park, PA-16802
yeasin@cse.psu.edu

Abstract. In this article we propose a novel approach to compute the optical flow directly on log-mapped images. We propose the use of a generalized dynamic image model (GDIM) based method for computing the optical flow as opposed to the brightness constancy model (BCM) based method. We introduce a new notion of "variable window" and use the space-variant form of gradient operator while computing the spatio-temporal gradient in log-mapped images for a better accuracy and to ensure that the local neighborhood is preserved. We emphasize that the proposed method must be numerically accurate, provides a consistent interpretation and is capable of computing the peripheral motion. Experimental results on both the synthetic and real images have been presented to show the efficacy of the proposed method.

1 Introduction

Computation of optical flow in log-mapped images is a challenging issue due to complex neighborhood connectivity and the lack of shift invariant processing. There are several implementations to compute optical flow in the log-polar plane, i.e. [1,2,3,4], but these implementations have a number of limitations:

1. Traditionally, optical flow on log-polar images has been computed based on the BCM using Cartesian domain gradient operator. The computed optical flow based on the BCM method fails to provide an un-ambiguous interpretation of optical flow and is numerically inaccurate.
2. The use of the standard Cartesian domain derivative operator to compute the spatio-temporal gradient on log-mapped image is incorrect as the spatial neighborhood is broken by the logarithmic mapping.
3. Existing implementations of computing optical flow in log-mapped images considers only the foveal part of the image and will produce erroneous estimate if one has to estimate motion using peripheral part of images.

To overcome the above mentioned problems, we present a novel approach to compute optical flow on log-mapped images. We propose the use of a generalized dynamic image model (GDIM) based method as opposed to the traditional

---

[*] This work was partially supported by NSF ITR grant IIS-0081935 and NSF CAREER grant IIS-97-33644.

R. Klette, S. Peleg, G. Sommer (Eds.): Robot Vision 2001, LNCS 1998, pp. 252–260, 2001.

BCM model based method and employ the space-variant form gradient operator (see [5] for details) to compute the optical flow directly on log-mapped image plane. We introduce a new notion called the *variable window* while computing the spatio-temporal gradient on a log-mapped image to ensure the local neighborhood structure is preserved. We show that the inclusion of above mentioned modifications significantly enhance the accuracy of optical flow computation directly on log-mapped images.

The rest of the article is organized as follows. In Sec. 2 we briefly review the literature related to the computation of optical flow. In the subsequent (Sec. 3) we propose a novel method to compute optical flow on a log-polar image. Following this we present the computed optical flow using the proposed method on a log-mapped images in Sec. 4. Finally, Sec. 5 concludes the article.

2   Computation of Optical Flow

Optical flow has commonly been defined as the apparent motion of image brightness patterns in an image sequence, but the common definition of optical flow as an image displacement field does not provide a correct interpretation. In the most recent effort to avoid this problem a *revised definition* of optical flow has been given in [6]. The revised definition of optical flow permits us to relax the brightness constancy model (BCM). To compute the optical flow, the so-called generalized dynamic image model (GDIM) has been proposed [6]:

$$I(\mathbf{x} + \delta\mathbf{x}) = M(\mathbf{x})I(\mathbf{x}) + C(\mathbf{x}) . \tag{1}$$

The radiometric transformation from $I(\mathbf{x})$ to $I(\mathbf{x} + \delta\mathbf{x})$ is explicitly defined in terms of the multiplier and offset fields $M(\mathbf{x})$ and $C(\mathbf{x})$, respectively. The geometric transformation is implicit in terms of the correspondence between points $\mathbf{x}$ and $\mathbf{x} + \delta\mathbf{x}$. If we write M and C in terms of variations from one and zero, respectively, $M(\mathbf{x}) = 1 + \delta m(\mathbf{x})$ and $C(\mathbf{x}) = 0 + \delta c(\mathbf{x})$, we can express GDIM explicitly in terms of the scene brightness variation field

$$I(\mathbf{x} + \delta\mathbf{x}) - I(\mathbf{x}) = \delta I(\mathbf{x}) = \delta m(\mathbf{x})I(\mathbf{x}) + \delta c(\mathbf{x}) . \tag{2}$$

Despite of a wide variety of approaches to compute optical flow, the algorithms can be classified into three main categories: gradient-based methods [10], matching techniques [11], and frequency-based approaches [12]. But a recent review [13] on the performance analysis of different kinds of algorithm suggests that the overall performance of the gradient-based techniques are superior. Hence, in this article we adopt gradient-based approach to compute the optical flow using the proposed GDIM-based method.

3   Proposed Method

In this article, we compute the optical flow on log-mapped images, in particular, we use the $log(z + a)$ mapping model proposed by [14]. To perform the mapping, the input image is divided into two half-planes along the vertical mid-line. The mapping for the two hemi-fields can be concisely given by the equation

$$\Omega = \log(z + \kappa a) - \log(a) , \tag{3}$$

where $z = x + iy$ is the retinal position and $\Omega = \xi + i\eta$ is the corresponding cortical point, while $\kappa = \text{sgn } x = \pm 1$ indicates left or right hemisphere. The combined mapping is conformal within each half plane. In a strict mathematical sense, the properties of scale and rotation invariance are not present in the mapping. However, if $|z| \gg a$, then $\log(z + a) \cong \log(z)$, and therefore, these properties hold. Also, since the $\log(z + a)$ template has a slice missing in the middle, circles concentric with and rays through the foveal center do not map to straight lines.

We use the revised definition of optical flow as opposed to the traditional definition of optical flow. This discrimination is very important as the mapping of the gray-value image $I(x, y)$ to the gray-value image $I(\xi, \eta)$ in the log-mapped image is by no means trivial. Using the coordinate transformation we can exactly map the Cartesian motion field on the log-mapped plane, however, the deformation of the gray-value image causes new apparent shifts of the gray-value function or eliminates the existing one. It is important to note that, by using the GDIM-based method we are not trying to model the nonuniform sampling by a multiplier and offset field, rather we adopt a systematic approach of computing optical flow which provides consistent interpretation. The problem of nonuniform sampling is partially tackled using the notion *'variable window'* and *'space-variant form of derivative operator'*.

In [1], the optical flow on log-polar plane was computed by assuming that the optical flow constraint based on the BCM has the same functional form:

$$v_\xi I_\xi + v_\eta I_\eta + I_t = 0, \tag{4}$$

where the symbols I_ξ, I_η and I_t are spatio-temporal derivatives of the image. Note that the brightness constancy assumption is severely violated in real world scenes. We address this problem by adopting the revised definition of optical flow, based on GDIM, which allows the variation of intensity in successive frames. By applying first order Taylor series approximation on the left hand side of Equ. 1 and using the similar argument used in [1] the GDIM based optical flow constraint equation on the log-mapped image can be written as

$$I_t + v_\xi I_\xi + v_\eta I_\eta - (I\delta m + \delta c) = 0, \tag{5}$$

where I_p stands for partial derivative with respect to p. By writing, $I_t = \delta g + \delta I$, where $\delta g = -(I_\xi v_\xi + I_\eta v_\eta)$, one can quantify explicitly the geometric and radiometric components of variation in an image sequence. The ratio $\delta g/\delta I$ can be used a measure of relative strength of geometric and radiometric cues.

Application of Equ. 5 for the computation of optical flow has been discussed in earlier work. One way to solve this is by imposing smoothness through minimization of energy functionals [8]. Alternatively, smoothness can be imposed based on the assumption that flow fields are constant within small regions around each point [6]. Requiring the flow field to be constant within a small region around each point and using least-square optimization we obtain the equations

$$
\sum_W
\begin{bmatrix}
I_\xi^2 & I_\xi I_\eta & -I_\xi I & -I_\xi \\
I_\xi I_\eta & I_\eta^2 & -I_\eta I & -I_\eta \\
-I_\xi I & I I_\eta & I^2 & I \\
-I_\xi & I_\eta & I & 1
\end{bmatrix}
\begin{bmatrix}
v_\xi \\
v_\eta \\
\delta m \\
\delta c
\end{bmatrix}
=
\sum_W
\begin{bmatrix}
-I_\xi I_t \\
-I_\eta I_t \\
I I_t \\
I_t
\end{bmatrix},
\tag{6}
$$

where W is a neighborhood region. Note that in a log-mapped image this neighborhood region is complicated and variable due to the nonlinear properties of the logarithmic mapping. We address this issue by using a new notion called a *variable window* (see Fig. 2(a)). By solving Equs. (6) we can compute the optical flow directly on log-mapped images.

4   Experimental Results

The log-mapping is conformal, i.e., it preserves local angles. In order to retain this property after discretization, we chose identical discretization steps in radial and angular directions. We have experimented with both synthetic and real image sequences. For real images we consider image sequences of an indoor laboratory, an outdoor and an underwater scene to show the efficacy of the proposed algorithm. Synthetically generated examples include the computed image motion using both the BCM and GDIM-based method to demonstrate the effect of neglecting the radiometric variations in an image sequence.

4.1   Synthetic Image Sequences

We use synthetic images in a first experiment to demonstrate a number of issues using artificial ground truth. The first image is that of a textured 256×256 face image (see Fig. 1(a)). Using a known motion (0.4 pixel horizontal motion in the Cartesian space which corresponds to $0 - 30$ pixel image motion in the log-mapped image) and a radiometric transformation field (a Gaussian distribution of radiometric transformation field (δm) in the range between $0.8 - 1.0$ and $\delta c = 0$), we compute the second image. The third image was derived from the first image using the above radiometric transformation only. Two sequences using frame $1 - 2$ and $1 - 3$ are considered. Figure 1(b) shows a sample transformed log-mapped image (derived from Fig. 1(a)).

We compute optical flow directly on log-mapped images by solving Equs. (6), using the weighted Least-square technique. The least square problems are solved using pseudo-inverse without any thresholding. Image gradients were calculated using the space-variant form of the gradient operator for a better numerical accuracy in optical flow computation. A new notion of variable window (for example, by transforming a 5×5 window using logarithmic mapping) has been introduced to preserve the local-neighborhood structure on a log-mapped image

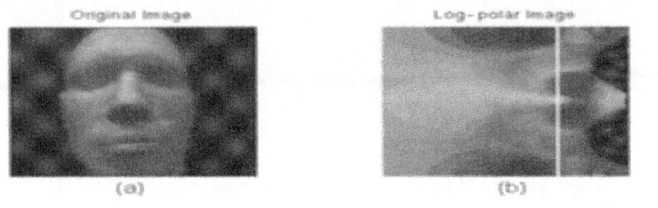

Fig. 1. Sample images: (a) sample synthetic image; (b) log-polar transformed image

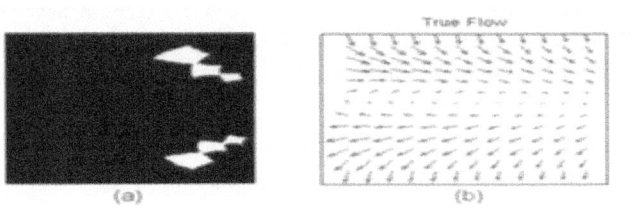

Fig. 2. Simulated optical flow: (a) Variable window: size of the window varies across the image, and (b) True image motion used to generate synthetic image sequence

(see Fig. 2(a)). In our experiment we use only the peripheral part of the image i.e., the portion of the log-mapped image right to the white vertical line for the computation of optical flow (see Fig. 1(b)). The idea of using the periphery stems from biological motivation and also to increase the computational efficiency.

To analyze the quantitative performance we compared the error statistics for both the BCM and GDIM methods. The error measurements used here are the root mean square (RMS) error, the average relative error (given in percentage), and the angular error (given in degrees). The average relative error in some sense gives the accuracy of the magnitude part while the angular error provides information related to phase of the flow field. Compared are, at a time, the two vectors $(u, v, 1)$ and $(\hat{u}, \hat{v}, 1)$, where (u, v) and $(\hat{u}, \hat{v})$ are the ground truth and estimated image motions, respectively. The length of a flow vector is computed using the Euclidean norm. The relative error between two vectors is defined as the difference of length in percentage between a flow vector in the estimated flow field and the corresponding reference flow field:

$$\frac{\sum ||(\hat{u} - u, \hat{v} - v)||_2}{\sum ||(u, v)||_2} \cdot 100. \tag{7}$$

The angular error between two vectors is defined as the difference in degrees between the direction of the estimated flow vector and the direction of the corresponding reference flow vector.

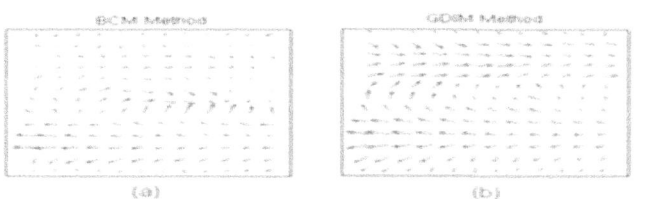

Fig. 3. Computed optical flow in case of both geometric and radiometric transformations. (a) and (b) represents the computed flow field using BCM and GDIM methods, respectively

We first tested on synthetically generated images with ground truth to show both the qualitative and the quantitative performance of the proposed algorithm. Figure 2(b) shows the true log-mapped image motion field which has been used to transform the image sequence $1 - 2$. Figures 3(a) and 3(b) shows the computed image motion as Quiver diagram for the sequence $1 - 2$ using the BCM and GDIM, respectively. A visual comparison of Fig. 2(b) with Figs. 3(a) and 3(b) reveals that the image motion field estimated using GDIM method is similar to that of the true image motion field, unlike the BCM method.

We provide a quantitative error measure to compare the performance of the proposed algorithm with the traditional method. First, we consider the the average relative error, which in some sense reflects the error in estimating the magnitude of the flow field. We obtain the average relative error 7.68 and 6.12 percent for the BCM and GDIM, respectively. To provide a more meaningful information about the error statistics we now look at the average angular error, which in some sense reflects the error in estimating the phase of the flow field. The average angular error was found to be 25.23 and 5.02 degree for the BCM and GDIM, respectively. We also compute RMS error, which was found to be 0.5346 and 0.1732 for the BCM and the GDIM method, respectively. The above error statistics clearly indicates that the performance of the proposed GDIM-based method is superior to the BCM method. Figures 4(a) and 4(b) display the computed optical flow using sequence $1 - 3$, where there is no motion (only

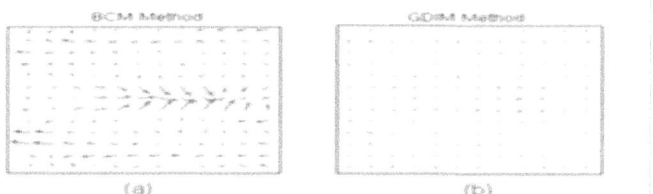

Fig. 4. Computed optical flow in case of radiometric transformation. (a) and (b) represents the computed flow using BCM and GDIM methods, respectively

the radiometric transformation has been considered to transform the image). It is clear from the Fig. 4(a), when employing BCM, we obtain the erroneous interpretation of geometric transformation due to the presence of radiometric variation. On the contrary, the proposed GDIM-based method shows no image motion (see Fig. 4(b)), which is consistent with ground truth.

4.2 Real Image Sequences

To further exemplify the robustness and accuracy of the proposed method, we experimented with real sequence of images captured under a wide range of veiling condition. We capture the real images (both the indoor and the outdoor) using an Olympus digital camera and use a special underwater camera to capture under water images by fixing the camera parameters. The motion for the under water and the outdoor sequence of images were dominantly horizontal motion, while the motion for the indoor laboratory was chosen to be the combination of rotation and horizontal translational motion.

In all our experiments we use the peripheral portion of images i.e., right side to the white vertical line (see Figs. 5(b), 6(b) and 7(b)) for the computation of optical flow. Figures 5(a)-(c), 6(a)-(c) and 7(a)-(c) show a sample frame, log-polar transformed image and the computed image motion for under water, indoor, and outdoor scenery images, respectively. We use the GDIM method only to show the results. We do not present results using BCM method as we have already shown that the performance of the BCM method is inferior to that of modified GDIM-based method.

From Figs. 5(c) and 6(c) it is clear that the flow distribution for the underwater and outdoor scenery images are similar to that of the Fig. 2(b), as expected. But, the flow distribution of the indoor laboratory sequence (see Fig. 7(c)) is different from that of the Fig. 2(b), which is due to that we have chosen different motion profile. As mentioned earlier, the rotation in the image plane produces a constant flow along the radial direction. Hence, the flow distribution of Fig. 7(c) can be seen as the superposition of the flow distribution of the translational flow and that of the constant angular flow.

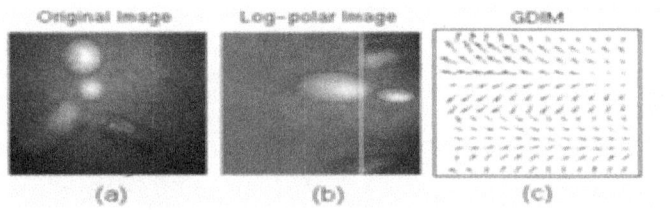

Fig. 5. Optical flow computation using an under water scene. (a) sample image from the under water scene; (b) the log-polar transformed image and, (c) the computed image motion

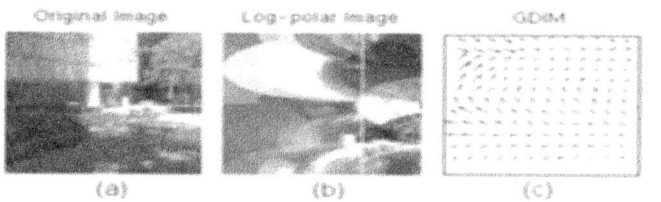

Fig. 6. Similar results shown in Fig. 5 using outdoor scene. (a) sample image from the outdoor scenery image; (b) the log-polar transformed image and (c) the computed image motion

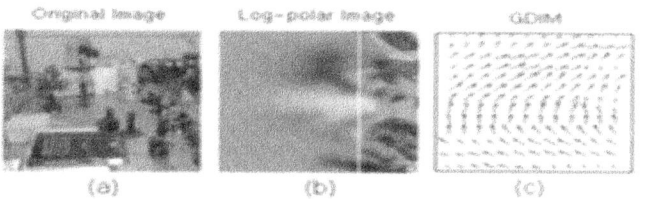

Fig. 7. Similar results shown in Fig. 5 using an indoor laboratory scene. (a) sample image from the outdoor scene; (b) the log-polar transformed image and (c) the computed image motion

These results show the importance of taking into account the radiometric variation as well as the space-variant form of the derivative operator for log-mapped images by providing a accurate image motion estimation and unambiguous interpretation of image motion. It is clear from the results that the proposed method is numerically accurate, robust and provide consistent interpretation. It is important to note that the proposed method has error in computing optical flow. The main source of error is due to the non-uniform sampling as we have transformed the standard images using the logarithmic transformation. The error could be reduced further by filtering the images after sampling.

5   Conclusions

We proposed a GDIM-based method to compute the optical flow which allows image intensity to be vary in the subsequent images as opposed to the traditional BCM-based method. We use the *space-variant form of the derivative operator* to calculate the spatio-temporal gradients as opposed to the Cartesian domain derivative operator, which fails to take into account the scaling factor inherent in logarithmic mapping. We have also introduced a notion called *variable window* while computing the spatio-temporal gradient to ensure that the local neighborhood is preserved. Experimental results on both the synthetic and real images clearly indicates that the proposed method provides a consistent interpretation and yields an accurate result.

References

1. K. Daniilidis, C. Krauss, M. Hansen, G. Sommer: Real-time tracking with moving objects with an active camera. *J. of Real-time Imaging*, Academic Press, (1997). 252, 254

2. M. Tistarelli, G. Sandini: On the advantage of log-polar mapping for estimation of time to impact from the optical flow. *IEEE Trans. on Patt. Analysis and Mach. Intl.*, **15** (1993) 401–410. 252

3. K. Daniilidis, V. Krüger: Optical flow computation in the log-polar plane. Technical report, CAIP, (1995). 252

4. K. Daniilidis: Computation of 3d-motion parameters using the log-polar transform. Technical report, CAIP, (1995). 252

5. B. Fischl, A. Cohen, E. L. Schwartz: Rapid anisotropic diffusion using space-variant vision. *Internat. J. of Comp. Vision*, **28** (1998) 199–212. 253

6. S. Negadharipour: Revised definition of optical flow: Integration of radio-metric and geometric cues for dynamic scene analysis. *IEEE Trans. on Pat. Analysis and Mach. Intl.*, **20** (1998) 961–979. 253, 255

7. R. J. Woodham: Multiple light source optical flow. In: *the Proceed. of Internat. Conf. on Computer Vision*, Osaka, Japan, (Dec. 1990).

8. B. K. P. Horn, B. G. Schunk: Determining optical flow. *Artificial Intelligence*, **17** (1981) 185–203. 254

9. J. K. Kearney, W. R. Thompson, D.L. Bolly: Optical flow estimation, an error analysis of gradient based methods with local optimization. *IEEE Trans. on Pattern Anal. and Mach. Intell.*, **14** (1987) 229–244.

10. B. Horn, B. Schunck: Determining optical flow, *Artificial Intelligence*, **17** (1981) 185–204. 253

11. P. Anandan: *Measuring visual motion from image sequences*, PhD thesis, University of Massachussetts, Amherst, MA, (1987). 253

12. A. B. Watson, A. J. Ahumada: A look at motion in the frequency domain. In: *Motion: perception and representation*, J. K. Tsotsos (ed.), (1983) 1–10. 253

13. J. L. Barron, D. J. Fleet, S. S. Beauchemin: Performance of optical flow techniques. *Internat. J. of Computer Vision*, **12** (1994) 43–77. 253

14. E. L. Schwartz: Computational studies of spatial architecture of primate visual cortex. Vol 10, Chap. 9, Plenum, New York, (1994) 359–411. 253

Hypothetically Modeled Perceptual Sensory Modality of Human Visual Selective Attention Scheme by PFC-Based Network

Takamasa Koshizen and Hiroshi Tsujino

Wako Research Center, Honda R & D Co. Ltd.,
1-4-1 Chuo Wako-Shi, Saitama 351-0193, Japan
koshiz@f.w.rd.honda.co.jp

Abstract. The Selective Attention Scheme has attracted renowned interest in the field of sensorimotor control and visual recognition problems. Especially, selective attention is crucial in terms of saving computational cost for constructing a sensorimotor control system, as the amount of sensory inputs over the system far exceeds its information processing capacity. In fact, selective attention plays an integral role in sensory information processing, enhancing neuronal responses to important or task-relevant stimuli at the expense of the neuronal responses to irrelevant stimuli. To compute human selective attention scheme, we assume that each attention modeled as a probabilistic class must correctly be learned to yield the relationship with different sensory inputs by learning schemes in the first place (sensory modality). Afterwards, their learned probabilistic attention classes can straightforwardly be used for the control property of selecting attention (shifting attention). In this paper, the soundness of proposed human selective attention scheme has been shown in particular with perceptual sensory modality. The scheme is actually realized by a neural network, namely PFC based network.

1 Introduction

Recently, researchers in different fields such as cognitive science and computer science have explored the development of brain-like computers. Especially, for the case of motor learning, the brain processes sensory and motor information in multiple stages. At each stage, neural representations of neural stimulus features or motor commands are manipulated. Gomi (1996) has proposed a hypothesis that the brain possesses and utilizes an "internal model" of arm for determining the motor command, and then a *feedback* process is performed based on visual information about the location of the target to manually figure out the final angle joints and muscle lengths in a 'supervised' fashion. They however haven't described on how the location of a target can computationally be obtained in 'self-organized' fashion.

Koshizen (2000) has recently proposed a hypothesis for controlling (or sifting) the location of a target by revealing the relationship between minimum

R. Klette, S. Peleg, G. Sommer (Eds.): Robot Vision 2001, LNCS 1998, pp. 261–269, 2001.
© Springer-Verlag Berlin Heidelberg 2001

variance theory initiated by (Harris, 1998) and human selective attention. According to the minimum variance theory there is an *infinite* number of possible motor control outputs in time-varying, whereas ones tend to learn stereotyped actions concerning to minimize the control variance. We then posited that the minimum variance theory can strongly be tied with human selective attention scheme which involves a neural network engendered by the PreFrontal Cortex (PFC) in particular.

Selective attention is also important in the view of saving computational cost about the sensorimotor control system, because the amount of sensory information flowing into the system far exceeds its information processing capacity. In fact, neurobiologically it has already been recognized that the PFC could undertake the position of a governor between archicortex and neocortex because of its high cognitive capability. In our framework, PFC-based network allows initially each attention to be modeled as a probabilistic class. Afterwards, each probabilistic class of attention can be correlated to different kinds of modality over the sensory input of sensorimotor control system.

In this paper, a selective attention scheme is considered in particular with visual sensor. Selective attention in vision involves dynamic interplay between attentional control systems and sensory brain structures. Several works in relation to understand the neural mechanism of human selective attention in vision has been investigated so far. Especially, the perceptual ability of the pattern of attentional modulations of sensory processing, visual sensory modality, which motivates toward shifting (controlling) attention, has been observed in many visual sustained selective attention tasks, using PET (e.g., Heinze, 1994), event-related potentials (ERPs) (e.g., Mangun, 1993) and functional magnetic resonance (fMRI) (e.g., Hillyard et al., 1997). A saliency-based selective attention in vision has been computed by Itti (1999). It is actually designed based on underlying neuronal architecture of the *early* primate visual selective attention, whereas our selective attention scheme is computed based on human selective attention involving statistical nature in neocortex where PFC enables the perceptual ability of sensor modality.

Our experimental results relevant to the control of a radio-based helicopter indicate the soundness of hypothetical modeled human selective attention by a PFC-based network with respect to the perceptual sensory modality.

2   Proposed Selective Attention Mechanism

2.1   Probabilistic Class of Attention

In our scheme, each class of attention Ω_l is a (hidden) parameter modeled by a *normal* probability distribution function (*pdf*) as $\mathcal{N}(\mu, \varsigma)$, where μ and ς are the mean and variance of the probability distribution $p(\Omega_l)$ because a parameter of the class Ω_l implicates a statistical column over neurons belonging to the frontal cortex (LaBerge, 1995).

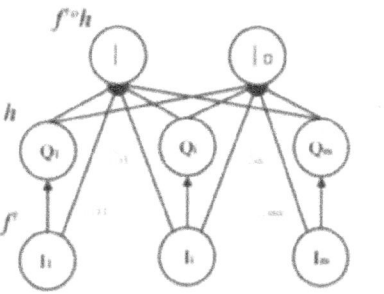

Fig. 1. PFC-Based Network

Suppose that the following two mappings are given:

$$f : I_i(t) \mapsto Q_i(t), \qquad f \in \mathcal{F}, \quad \text{and} \tag{1}$$

$$h : Q_i(t) \mapsto \Omega_l(t), \qquad h \in \mathcal{H} \tag{2}$$

where Ω_l is the l-th class of attention. In addition, I_i denotes sensory inputs, whereas Q_i is motor control outputs.

2.2   PFC-Based Network

We hereby describe the architecture of a PFC-based network that has a combined structure of an *unsupervised* and *supervised* scheme that is somehow a reminiscent of the Radial Basis Function Network, initiated by Poggio (1990). The network basically is composed of three layers of nodes as shown in Fig. 1. Fig. 2 illustrates an overview of the proposed selective attention scheme considering with its perceptual ability of sensory modality.

The algorithm of our PFC-based network is described as follows.

<u>**The algorithm**</u>:

1. Suppose we have a dataset over motor control outputs $\mathbf{Q}(t) := \{Q_i(t)|i = 1, ..., m\}$ given each sensory input, which can be obtained by mapping $f \in \mathcal{F}$.

2. The obtained motor control outputs where $\mathbf{Q}(t)$ is categorized by the number of the column l as shown in Fig. 2 in accordance with a value function in the basal ganglia that which qualifies $\mathbf{Q}^l(t)$ as either *positive* or *negative* with the reinforcement value task. In this sense, the form of the value function must satisfy the minimum variance theory.

3. An *unsupervised* EM algorithm is used to compute the expected value of attention Ω_l as *hidden* parameters using each motor control output $Q_i^l(t)$. The

computation actually corresponds to the mapping $h \in \mathcal{H}$ given by Equ. (2). The EM algorithm (Dempster, 1977) is to find θ to maximize the expectation of $log(p(Q_i^{l}(t)|\Omega_l, \theta))$ given the observed data $Q_i^{l}(t)$ as well as current estimate of θ, by iterating the two following steps - **E** step of the following expectation of the complete (observed and hidden) data:

$$\varphi(\theta|\theta^{(k)}) \quad = \quad \Sigma_i \Sigma_l p(Q_i^{l}(t)|\Omega_l; \theta^{(k)})log \quad \times \quad (p(Q_i^{l}(t), \Omega_l; \theta^{(k)})) \quad (3)$$

and **M** step on updating of the parameter $\theta^{(k)}$ at each iteration k as follows,

$$\theta^{(k+1)} = argmax_\theta \varphi(\theta, \theta^{(k)}) \quad (4)$$

As a consequence, each class of selective attention Ω_l can be assigned to each component of the density $p(\mathbf{Q}_i^{l}(t)|\Omega_l; \theta^{(k)})$ using by *a maximum posterior*. Note that $\hat{l}$ denotes the l-th *predictive* class of the probabilistic attention.

4. PFC-based network enable to learn the relationship between each class of attention Ω_l and sensory data $\mathbf{I}(t): = \{I_i(t), i = 1, ..., m\}$ to estimate the probability density $p_\lambda(I_i(t)|\Omega_l; \theta^{(k)})$. This is also resulted by the synthetic function in $f \circ h$ to calculate $p_\lambda(I_i(t)|\Omega_l)$ as implicated in Fig. 1. Symbol λ stands for the *synaptic* weight matrices of the PFC-based network that is especially learned by an error correction in a supervised fashion.

5. After computing each $p_\lambda(I_i(t)|\Omega_l)$, the Bayes' rule is applied with the following manner:

$$\hat{l} = argmax\{\Omega_l\}_l, \qquad l = 1, ..., n \quad (5)$$

subject to *maximum a posterior probability* (MAP). Proposed human selective attention scheme (see Fig. 2) enables sensory information to distinctly discriminated with each probabilistic class of attention using the Bayes' rule:

$$p(\Omega_{\hat{l}}(t)) = \frac{\bar{p}(\Omega_{\hat{l}}(t))p(I_i(t)|\Omega_{\hat{l}}(t))}{\sum_k p(\Omega_k(t))p(I_i(t)|\Omega_k(t))} \quad (6)$$

Where $\bar{p}$ denotes the *prior* probability, and $p(\Omega_{\hat{l}}(t))$ may be called the belief of $\Omega_{\hat{l}}(t)$. Note that Fig. 2 also shows that the relationship between the statistical attention classes and the columns is identified by *1:1*.

3   Example

In this section, we implement a PFC-based network as elucidated in the previous section with respect to two probabilistic attention classes for controlling the radio-based helicopter shown in Fig. 3 with the use of visual sensor. Selective visual attention mechanism in playing a rule of a control processing, which is also essential to visual perception. Fig. 4 shows visual sensory input (x-axis only) derived from Laplacian filter of raw image taken at each time and the obtained

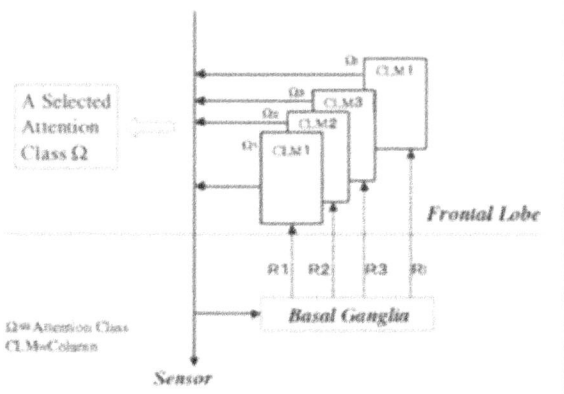

Fig. 2. Overview of the proposed Human Selective Attention Scheme

motor control outputs. Fig. 4 also illustrates a lineament of 100 data points for the visual sensory input represented by (o) and for the motor control output (∗) . It is noted that motor control outputs were collected under the manual operation which was alternatively repeated by stability ($l=1$) and instability ($l=2$) during this experiment. In this implementation, 360 data points were used for training for each probabilistic class of attention as well as 150 test data points. Moreover, each motor control output was divided into $l=1$ or $l=2$ by the following value function:

$$\text{if } |\check{Q}_i - Q_i| \geq \delta \text{ then } Q_i{}^1 \Leftarrow Q_i \ (Positive) \text{ else } Q_i{}^2 \Leftarrow Q_i \ (Negative) \quad (7)$$

In this case, the 'positive' reward corresponds to $l = 1$ but the 'negative' reward is $l = 2$. In addition, $\check{Q}_i$ means the motor control values that maintain under the stability. $\check{Q}_i$ also sets to 82 as well as $\delta = 1.0$ being the threshold that represents a tolerance of stability.

Note that $\check{Q}_i$ denotes the mean value of the probability distribution $p(\mathbf{Q})$. Basically, the value function Equ.(7) plays a role of *reinforcement* task sufficing $\sigma(\mathbf{Q}^1) < \sigma(\mathbf{Q}^2)$ (the minimum variance theory). Such reinforcement-type value

Fig. 3. Radio-Based Helicopter

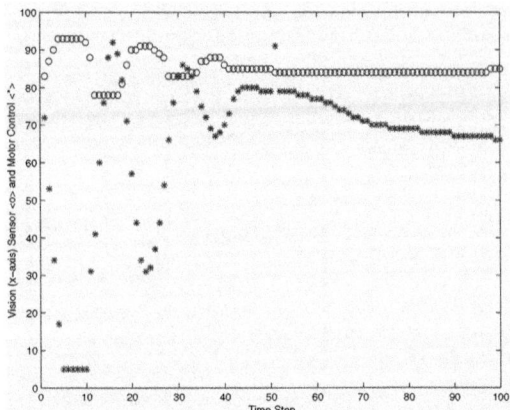

Fig. 4. Visual Sensor (*) and Motor Control (o)

function neurobiologically known to exist in the basal ganglia, including ventral tegmental area in the brain. That is, a main feature of the value function in the basal ganglia is to predict the reward (Schultz, 1990) for each motor control output Q_i^l.

Fig. 6 shows the network in the case of two attention classes. Initially, each sensory input vector $\mathbf{I}_i$ is transformed into each motor control output $\mathbf{Q}_i$ by $f^\dagger$ in Equ.(1). Each motor control output $\mathbf{Q}$ then is able to be divided into either $\mathbf{Q}^1$ or $\mathbf{Q}^2$ by the value function Equ.(7). Since PFC-based network enables such probabilistic attention classes to be regarded as hidden parameters that can be obtained by the EM algorithm in accordance with Equ.(3) and (4), the relationship between the two attention classes (Ω_1, Ω_2) and the given sensory information can be yield as well as training the synaptic weights λ. Importantly, the two attention classes Ω_1, Ω_2 could be identical to excitatory neurons *recurrently* connected via excitatory synapses $S1$ and $S2$ existed in neocortex, with the input space $\mathbf{I}$. In this sense, the excitatory neurons inhabit the input space $\mathbf{I}$ via inhibitory neurons endowed with neuronal signals (Rao, 2000) from the basal ganglia.

In addition, each (actual) variance of the motor control outputs $\mathbf{Q}^1, \mathbf{Q}^2$ over the two attention classes is 0.6542 (positive reward) and 6.5912 (negative reward) where the outputs are collected in the experiment. Fig. 6 shows that the probability distribution of two attention classes which are calculated by the EM after each motor control output is reinforced discriminated by the Equ.(7). The dashed line denotes the predicted parameters by the EM algorithm, while the solid line is the actual ones. In addition, Fig. 6 represents the result of the predicted distributions in which EM is required 6 iterations for its convergence. The predicted variances of the motor control outputs are 0.5288 to $\mathbf{Q}^1$(Left) and 8.6246 to $\mathbf{Q}^2$(Right) are illustrated in Fig. 6.

The belief error of probabilistic attention class is also illustrated in Fig. 7 where each result is obtained based on the predicted distribution shown in Fig. 6. In particular, the lower result in Fig. 7 is obtained based on the predicted distribution incorporated with Bayes' rule shown in Equ.(6) whereas the upper one does not consolidate with the rule. Apparently, Fig. 7 indicates that the lower one in Fig. 7 is more accurate and smoothed estimates than the upper one because of the Bayesian which ensures to be able to further reduce uncertainty of the predicted distribution shown in Fig. 6.

4 Conclusion

In this paper, human selective attention scheme in vision has computationally been proposed. Selective attention, in fact, plays an integral role in sensory information processing, enhancing neuronal responses to important or task-relevant stimuli at the expense of the neuronal responses to irrelevant stimuli. In this paper, prefrontal cortex (PFC) was hypothetically modeled in terms of its perceptual ability of the visual sensory modality over different motions. In this sense, the PFC can also be true to regulate inhibition and excitation in distributed (recurrent) networks existing in visual cortex. Moreover, proposed selective attention scheme requires Bayesian property which can favorably be coped with spatiotemporal data such as image by the effective use of its prior information as suggested by Weiss (1998). Consequently, the experiment result has shown the soundness of the proposed scheme in particular for the perceptual ability of sensory modality which was actually computed by PFC-based network. Further work may allow the PFC-based network to consolidate with underlying neuronal property of posterior parietal cortex (PPC), which is concurrently known as shifting attention based on the perceptual ability of PFC. In other worlds, modeling the flotal lobe including PFC, premotor and the PPC computationally would be attractive, in order to reveal human visual selective attention mechanism.

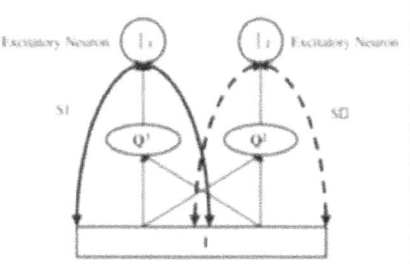

Fig. 5. Network in Two Excitatory Neurons on Attention

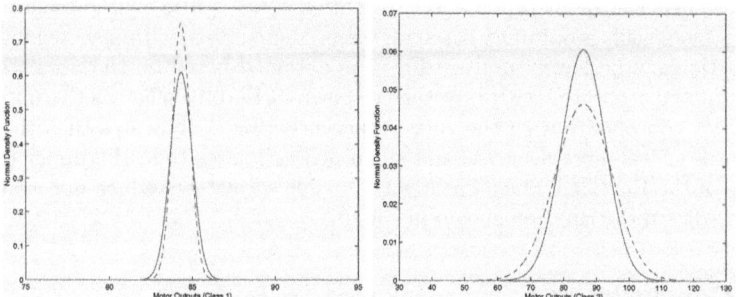

Fig. 6. Probability Distributions over Two Attention Classes (at Mature EM Training)

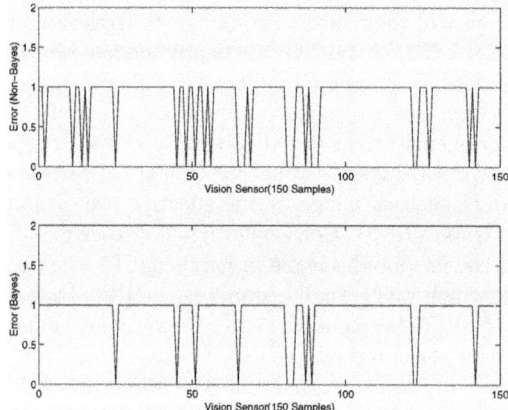

Fig. 7. Error on Two Attention Classes (0:Incorrect; 1:Correct)

References

1. A. P. Dempster, N. M. Laird, D. B. Rubin: Maximum likelihood from incomplete data via the EM algorithm. *J. of Roy. Statist. Soc. Ser. B 39*, **13** (1977) 1–38.
2. H. Gomi, M. Kawato, M.: Equilibrium-point control hypothesis examined by measured arm stiffness during multijoint movement. *Science* **272** (1996) 117–120.
3. C. M. Harris: Signal-dependent noise determine motor planning. *Nature* **394** (1998) 780–784.
4. H. J. Heinze, G. R. Mangun, W. Burchert, H. Hinrichs, M. Scholz, T. F. Munte, A. Gos, M. Scherg, S. Jahannes, H. Hundeshagen, M. S. Gazzaniga, S. A. Hillyard: Combined spatial and temporal imaging of brain activity during visual selective attention in humans. *Nature* **372** (1994) 543–546.
5. S. A. Hillyard, H. Hinrichs, C. Tempelmann, S. Morgan, J. Hansen, H. Scheich, H. J. Heinze: Combining steady-state visual evoked potentials and fMRI to localize brain activity during selective attention. *Human Brain Mapping* **5** (1997) 287–292.

6. L. Itti, C. Koch, E. Neibur: A model of saliency-based visual attention for rapid scene analysis. *Proceed. of Image Understanding Workshop* **11** (1999) 1254–1259.
7. T. Koshizen, Y. Ueda H. Tsujino: New conscious sensorimotor control system induced by human selective attention mechanism with minimum variance theory. Technical Report of Honda R&D Co. Ltd. (In preparation).
8. D. LaBerge M. S. Buchsbaum:Attentional processing: the brain's art of mindfulness. (1995) MA:Harvard University Press.
9. G. R. Mangun, S. A. Hillyard, S. J. Luck: Attention and performance XIV:synergies in experimental psychology, artificial intelligence, and cognitive neuroscience. Cambridge MA: MIT Press (1993) 219–243.
10. T. Poggio, F. Girosi: Networks for approximation and learning. *Proceed. of IEEE* **78** (1990) 1481–1497.
11. W. Schultz: Predictive reward signal of dopamine neurons. *J. of Neurophysiology* **80** (1990) 1–27.
12. R. P. N. Rao: Predictive sequence learning in recurrent neocortical circuits. *Advances in neural information processing systems* **12** (2000) 164-170.
13. Y. Weiss: Slow and Smooth: a Bayesian theory for the combination of local motion signals in human vision. Technical report of Massachusetts Institute of Technology A. I.Memo **1624** (1998).

Results of Test Flights with the Airborne Digital Sensor ADS40

Anko Börner and Ralf Reulke

German Aerospace Center (DLR)
Institute of Space Sensor Technology and Planetary Exploration
Rutherfordstrasse 2, D-12489 Berlin, Germany
Tel: ++49-30-67055509; Fax: ++49-30-67055529
{anko.boerner}@dlr.de

Abstract. During the past two years the company LH Systems and the German Aerospace Center (DLR) have developed the commercial airborne digital sensor ADS40 based on the three-line principle. By assembling additional CCD lines into the same focal plane, the sensor is capable of generating a number of color images. In the first part, the sensor system itself is introduced shortly. The main concept and the key features are described and an overview of the data processing scheme is given. After that, we will focus on the results of test flights. The emphasis is placed on the properties of the overall system including the sensor itself, platform, airplane, and inertial measurement unit. The effect of using staggered CCD lines is discussed. Flights over well known test areas are used to prove the accuracy of derived data products. Differences in data processing methods are pointed out in comparison to sensor systems based on CCD matrices or film.

1 Introduction

LH Systems, in co-operation with the German Aerospace Center (DLR), has developed an airborne three-line digital imaging sensor, the ADS40 [1]. With the help of this sensor a completely digital chain is realized, starting with image acquisition, continuing with image processing, and finally leading to digital products. The new sensor provides imagery suitable for both high precision photogrammetric mapping and thematic data interpretation.

During the development phase certain test models were built and flown for a stepwise development and in order to provide continuous tests. The most important goal of the test flights was to prove the functionality of the entire ADS system including the sensor, the platform and the inertial measurement unit (IMU). A check of the system under real flight conditions was necessary. The main objectives of the ADS test flights were defined as: test of the entire sensor system, verification of the hardware noise behavior, test of the synchronization between image and attitude data, test of the pre-processing facilities, evaluation of the staggered array approach, proof of the accuracy demands.

This paper describes essential parameters of the ADS40 system, the necessary data pre-processing and the results obtained from the ADS tests.

R. Klette, S. Peleg, G. Sommer (Eds.): Robot Vision 2001, LNCS 1998, pp. 270–277, 2001.
© Springer-Verlag Berlin Heidelberg 2001

Table 1. Parameters of the ADS40 sensor

Focal length	62.5 mm
Pixel size	6.5 μm
Panchromatic line	2 X 12.000 pixels
Color lines	12.000 pixels
Field of View (across track)	64^0
Stereo angles	$14^0, 28^0, 42^0$
Dynamic range	12 bit
Radiometric resolution	8 bit
Ground sampling distance (3000 m altitude)	16 cm
Swath width (3000 m altitude)	3.75 km
Read out frequency per CCD line	200 - 800 Hz
In flight storage capacity	200 - 500 GByte

2   The Sensor System ADS40

The ADS40 is based on the tree-line principle. Although the application of CCD matrices would have a number of advantages over linear arrays with respect to data processing, CCD lines are the only sensors that provide the ground resolution combined with a swath width which fulfil the specifications given by photogrammetry at present and in the near future. The stereo capability of the ADS40 is achieved by using three lines with different viewing directions (forward, backward, nadir). Since line scanners are not able to vary the base-to-height ratio the stereo angles between the three lines are set to different values (Table 1) in order to have a certain flexibility.

Because of the linear sensor structure, the second dimension of the image is generated by the aircraft movement. So the attitude disturbances influence the raw image. To overcome this problem and to correct this effect, exact attitude and position measurements are necessary for each image scan line. This is realized by an Inertial Measurement Unit (IMU) from Applanix mounted directly to the sensor.

In order to assure the desired high number of detector elements per line, so-called staggered arrays are used. These detectors consist of two single 12k CCD lines positioned close to each other with an across-track shift of half a pixel.

In addition, several 12k color lines are assembled allowing multispectral imaging. Therefore, interference filters are placed directly on the CCDs. A telecentric optics provides the optical path required for these filters. The RGB lines are optically superimposed during the flight using a beam splitter consisting of dichroitic mirrors. These mirrors divide the light into different color components. So the loss of energy is small contrary to classical beam splitters or sequentially applied filters. The RGB bands are co-registered without postprocessing. Table 1 shows the most important sensor parameters.

The ADS40 system has to be understood not only as a camera, it also includes additional modules, like IMU, platform, flight control and management system, camera computer, etc. Therefore, test flights must take into consideration all these items.

3   Data Processing

All data retrieved during the flight must be extracted from the mass memory system and converted into a usable format. In order to prove the overall system functionality including the sensor electronics and optics, the platform and the IMU data processing was not stopped after consistency checks, but was extended to more complex tasks, e.g. generating elevation models.

All CCD lines with 24,000 pixels each generate a file with a size of about 1.5 GByte assuming a typical number of scan lines (60,000) and a radiometric resolution requiring a one-byte data format (normalized data). The size of each color line image amounts a quarter of a panchromatic image. Since this procedure is independent of the ADS40 hardware itself, it is the first part in which one is able to detect systematic electronic errors.

The processing of the attitude data was done using the Applanix software POSProc. It determines 50 and 200 Hz solutions including the real-time data and the reference data by applying certain signal processing algorithms, e.g. Kalman filtering. The position data were transformed to a local coordinate system or to UTM. Image and attitude data are synchronized by using the events triggering the sensor. This results in an attitude file containing the parameters of the exterior orientation for each scan line. The parameters of interior orientation are determined using the results of the geometric calibration performed at the DLR calibration labs.

4   Results of the Test Flights

The following system parameters were of special interest during the test flights in order to check the functionality and to evaluate the capabilities of ADS40:

- radiometry and noise behavior of the sensor
- evaluation of the platforms,
- accuracy of the attitude parameters,
- staggered arrays.

4.1   Radiometry

One of the main advantages in favor of digital sensor systems over film-based cameras is the much higher radiometric dynamics, equivalent to a greater range of grey values within an image. This has a significant impact on the subsequent data processing (e.g. matching). The dynamic range is described by the radiometric resolution and is limited by different noise sources. Both parameters, radiometric dynamics and noise, were main points of our investigations.

Fig. 1. Radiometric zoom, depicting grey values of the first (left) and the last (right) 8 bits within the original 12 bit grey level range

Figure 1 gives an impression of having 12 bit data instead of 8 bit data at one's disposal. Depicting either the first 8 or the last 8 bits within the original 12 bit grey level range results in two different images. The first shows the bright parts, but no structure in the dark parts is detectable. The second image inverts the situation: in the dark parts we can see details, the bright pixels look saturated. But again, both images result from the same 12 bit image.

A noise analysis must be performed to ensure the estimated radiometric resolution. In general, it can be differentiated between dynamic and static noise. A number of homogeneous targets in certain test images were selected in order to investigate the dynamic, time dependent noise. A nearly gaussian distribution with a standard deviation is about 2 grey values could be observed.

The pixel response non-uniformity (PRNU) describes the pixel-dependent (and time independent) digital output of a CCD sensor for a constant amount of input radiance. Limb shading caused by the optical parts can be described in the same way. This effect is important for real-time compression algorithms and must be corrected before data compression, because it influences the quality of decompressed image data drastically [2]. The ADS40 system compensates the PRNU in real-time by using look-up tables for correction values. Tests showed that remaining PRNU effects are smaller than 1 grey level.

With these tests we could prove the capability of the ADS40 to generate images with a radiometric resolution of 12 bits.

4.2   Evaluation of the Platforms

An adaptation of the platform control concerning the sensor allows an optimal damping of the system. No special adaptation was done for the test systems.

In all EM raw images an oscillation could be observed (Fig. 2). Even if these motions were recorded correctly by the IMU, the image quality decreases because of the necessary resampling process. The amplitude of this oscillations is about 1-2 pixels in image space and the period is about 15 scan lines, which is related to

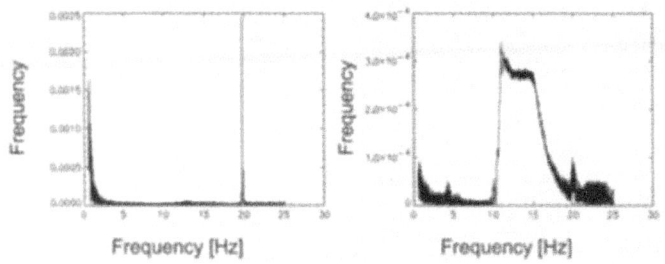

Fig. 2. Power spectrum of the oscillations of ADS40 EM on the non-adapted PAV11 during the flight (left) and on the shaker (right)

a time interval of about 50 ms. This oscillation results from the 20 Hz revolution frequency of the Cessna engine. The applied platforms were not able to suppress these vibrations. Figure 3 (left) shows the power spectrum of the processed IMU data. The 20 Hz peak can be seen clearly.

In order to validate these results the EM and the Applanix system were put together on a shaker. Its stimulating oscillation was swept from 1 Hz to 100 Hz and the system response was recorded by the IMU. Figure 3 (right) shows a part of the test data. For vibrations between 10 and 15 Hz the platform is too inert for a damping of the EM. Damping works for stimulating oscillations beyond 15 Hz. The most interesting detail in this figure is the resonance frequency at 20 Hz which confirms the evaluation of the image and attitude flight data.

In order to minimize the oscillation effects even for test flights in future it was decided to apply additional weights to the sensor until the new, adapted platform is available. The commercial ADS40 uses an adapted platform.

4.3   Accuracy of the Attitude Parameters

To evaluate the quality of the attitude data we performed a rectification of image data to a reference plane (correction of the flight motions) and the generation of a coarse digital elevation model (DEM).

After processing of the IMU data, a simple way to verify the alignment between image scan lines and an appropriate attitude data set can be done by projecting all pixels of the raw image to a reference plane (rectification, Fig. 3). The visual impression of a rectified image is the first, essential indicator of the quality of the attitude data. This method helped us to detect different synchronization problems, e.g. a temporary shift between attitude and image data.

Although the exact photogrammetric data interpretation was not our goal at this stage, we proved the stereo capability of the test systems. With a small number of control points and simple algorithms we already achieved an error (rms) of about one pixel in the x- and y-directions and about three pixels in

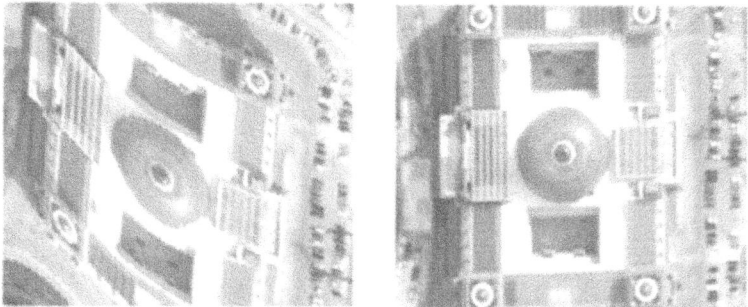

Fig. 3. Raw and rectified image

the z-direction taking into account additional built-in offsets for roll, pitch, and yaw. The sources of error were in our opinion:

– incorrect synchronization between image and attitude data,
– difficulties during the attitude data processing,
– inexact geometric calibration.

With the help of the position and attitude data we generated digital elevation models. This task is very interesting, because it is the last element in a huge, complex chain, starting with the optics and electronics, including aeroplane and platform, and ending with data processing. Some errors can only be detected at this final stage. Only when all elements of this chain work together successfully, high quality photogrammetric products can be obtained.

The first additional criterion for successful processing is the correlation between the stereo images. The matching algorithm worked very well (about 90the 12 bit data.

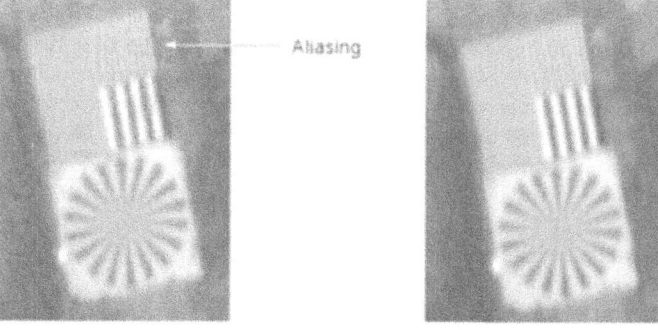

Fig. 4. Test pattern scanned with a 12k sensor (left) and a 24k staggered sensor (right)

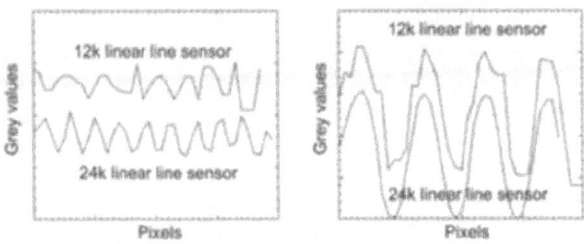

Fig. 5. Cross-section through 20 cm bars (left) within the test pattern and through the 50 cm bars (right)

Two independent DEMs were built automatically with the help of an tool developed at DLR [3] for two different tracks (one south-north, another one north-south) flown over the same target area within one hour. Superimposing both DEMs showed deviations smaller than 1.5 pixels (rms). This relative height error includes all error sources, e.g. attitude data error, matching error, and reconstruction error.

4.4 Staggered Arrays

One of the special points of interest during the test phase was the effect of using staggered arrays instead of linear arrays. The main question was whether or not the theoretical expectations can be fulfilled and the application of such detectors can improve the spatial resolution. Therefore, two special test targets were customized, a Siemens star and different patches with test bars both with a size of $8m \times 8m$.

In order to compare a sensor with a single CCD line (12k) versus a staggered array sensor (24k), a single line sensor was simulated using image data just of one of the two staggered lines with a clock time double as high as the staggered one. The results are shown in Figs. 4 and 5. The following conclusions could be drawn:

- Observing the Siemens star in both images lead to similar results. In both cases the modulation transfer functions (MTF) of the optics and the detector element are evaluated. Please note, that actually the optics are too good for a 12k sensor, because spatial frequencies larger than the Nyquist frequency are transmitted. So in worst-case situations aliasing can occur. If the optics were adapted to the 12k sensor, the blurred region would be larger in the 24k image.
- The comparison of the bar patterns shows the improvement of using a 24k sensor instead of the 12k sensor. The most interesting part is the upper right patch. The 12k image low contrast and aliasing (non-parallel bars) can be observed. Cross sections in the 20 cm and 50 cm patches illustrate the situation. Both, the spatial frequencies and the contrast are better for the staggered arrays.

– The spatial resolution does not only depend on the system MTF (mainly determined by the optics and the pixel size), but as expected, from the sampling distance (sampling theorem) as well.

5   Conclusions

This paper describes the results of the flights of the ADS test systems. Our main goal was to prove the capabilities of the new sensor system under field conditions. These tests must complement laboratory tests and the development of retrieval algorithms. It is the only way to detect error sources at an early stage and can confine the development risks drastically. With the help of the test flights, we were able to identify a number of sources of hardware errors and could acquire a lot of know-how dealing with the sensor system. The resolution-enhancing effect of applying staggered arrays could be demonstrated.

References

1. R. Sandau et al.: Design principles of the LH systems ADS40 airborne digital sensor. *Internat. Archives of Photogrammetry and Remote Sensing*, **XXXIII** Part B1, Commission I, (2000) 258–265. 270
2. G. Schlotzhauer et al.: Effects of pixel faults on the efficiency of WAOSS compression and image quality. *Internat. Archives of Photogrammetry and Remote Sensing, Italy*, **XXX** Part 1, Como, (1994) 12–17. 273
3. A. Börner: Entwicklung und Test von Onboard-Algorithmen für die Landfernerkundung. Dissertation, Technische Universität Berlin, (1999) pages 29–64. 276

Localized Video Compression for Machine Vision

Moshe Porat

Department of Electrical Engineering, Technion-Israel Institute of Technology,
Haifa 32000, Israel,
mp@ee.technion.ac.il

Abstract. A three-dimensional vector quantization system is introduced suitable for video compression. The basic characteristics of slow or repeated scenes in robot vision are used as the basic assumptions of the proposed approach. Accordingly, the localized history of the sequence is used to create localized codebooks, thus representing current visual information as transformed versions of previous details. The results indicate a high compression ratio with high quality of the perceived sequence. The structure of the algorithm is mostly parallel, making it suitable for efficient hardware implementation[1].

1 Introduction

Video compression has been a basic design goal for years [1,2,3]. In many applications, such as robot vision, where the communication channels are of limited bandwidth, this requirement of video compression is further emphasized. Since the distortions of video sequences tend to concentrate mainly at very specific areas like edges and moving details, their contribution to the global SNR is sometimes rather small, giving a good result although the artifacts are quite noticeable. The concept of edge coding has been dealt with successfully in several works (e.g., [4]) on coding the high frequency component of video sequences. A similar approach to coding of oriented edges is reported also by Giunta et al. [5] for the case of 64 Kbit/sec coding. Those methods are based on decomposition of the sequence into several bands [6], [7], with various techniques applied to different bands. This approach of subband decomposition is in accordance with many findings related to the basic structure of the human visual system, where cells and groups of cells are sensitive to limited spatial-frequency bands, and are likely to be parts of different processing mechanisms of the human pattern recognition systems [8].

Naturally, there is a trade-off between the complexity of the transmitter/receiver and the compression ratio obtained. However, higher complexity is more acceptable if parallel processing is available. In such a case the speed of the

---

[1] Initial parts of this research were carried out at Bell Labs, Murray Hill NJ. A patent with N.S. Jayant was applied for by Bell Labs. Later work was carried out at the Technion and was supported in part by the Fund for the Promotion of Research at the Technion and by the Ollendorff Research Center.

R. Klette, S. Peleg, G. Sommer (Eds.): Robot Vision 2001, LNCS 1998, pp. 278–283, 2001.
© Springer-Verlag Berlin Heidelberg 2001

whole process is higher, and the implementation in hardware (VLSI) is simpler, as repeated sub-circuits organized systematically. In this paper a new model-based approach is introduced based on the above principles and motivation. The proposed system is based on a model of slow scenes, and provides a very high compression ratio that can be implemented in a parallel manner.

2   The Model and the System

The model used in this study assumes specific properties related to the nature of slow machine vision. In particular, the main assumption is that motions of recent history are likely to re-appear in the same area of the frame where they have previously appeared. If not similarly repeated (up to allowed distortion), they are likely to be encoded as transformed versions of previous details [9]. Furthermore, if a specific motion cannot be matched accurately according to coarse partitioning of the frame, a more refined description is used [10].

According to this model, frames are divided into blocks and sub-block as shown in Fig. 1. In this figure an image is divided into sub-blocks or vectors of 4x4x4 pixels. Sixteen vectors comprise a block. Typically, an image is divided into at least 25 or 36 blocks, each of them plays a localized role in the compression of the sequence.

Based on the above assumptions and structure, the proposed system is presented in Figs. 2 and 4. In the first stage of the process (Fig. 2), the blocks of the image are used for training localized codebooks using the LBG algorithm [11]. These codebooks are based on the recent localized history of the sequence, thus contain significant information which can be readily used for quantizing the next frames of the sequence. An illustration, which indicates the size of the localized codebooks, is shown in Fig. 3. This example relates to a sequence of

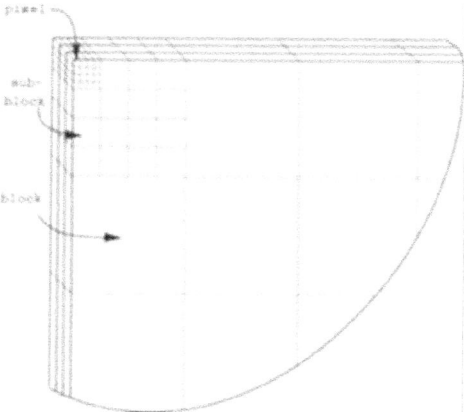

Fig. 1. Blocks and sub-blocks. Each block contains 4 consecutive frames

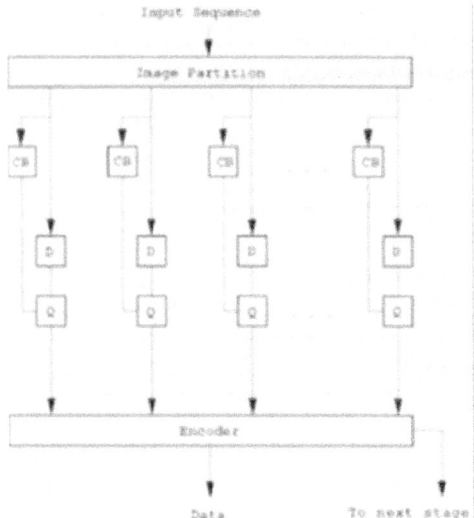

Fig. 2. Block diagram of the first stage. CB=Codebook, D=Delay, Q=Quantizer

Miss America, which represents a slow scene. In this illustration, codebooks of the background are small in size (indicated by dark gray and black) while codebooks of active areas like the head, and in particular the eyes and the mouth, are larger (light gray and white).

Most of the sub-blocks (vectors) of each block are adequately encoded in this stage by relatively sparse codebooks, with up to 512 code-words in the illustrated example. Some of the vectors, however, require additional attention due to distortion above the pre-determined threshold. It is assumed that these vectors refer to motions that have crossed the borders of their block (codebook), thus can be found in one of the adjacent codebooks (Fig. 4). It is also assumed that in such a case translation or other similarity-transform operators might be needed (e.g., rotation, affine).

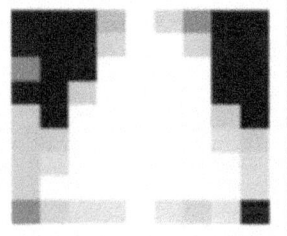

Fig. 3. Size of codebooks, black=0, white=512 (Miss America)

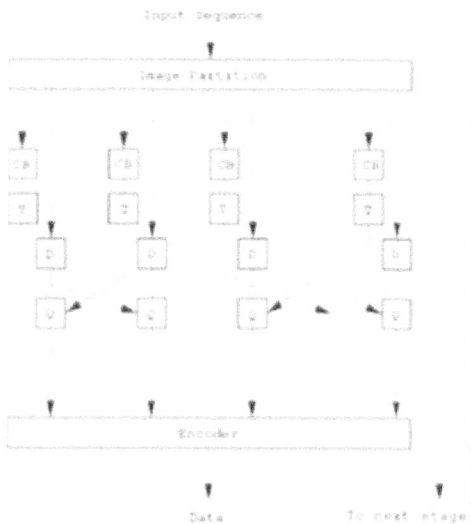

Fig. 4. The second stage of the system. Similar notations as in Fig. 2, here T=Transform

The model has several parameters. In terms of history, two parameters relate to the duration of the history used, and to the typical duration of each movement, respectively. The latter also determines the minimum expected delay of the system. A localization parameter relates to the size of areas considered as

Fig. 5. Original (left) and reconstructed (right) frame from the sequence "Miss America" using localized codebooks with 4x4x4 blocks. The compression is to 0.078 bpp

likely to contain repeated motions during the time period defined as history. As to the transformation process of adjacent codebooks, an additional parameter determines where similar transformed details are searched.

In the next stages of the system, a refined process is carried out with regard to those areas that were not encoded adequately in the first two stages. Usually only very small number of vectors require this additional process. This pyramidal approach introduces additional parameters, mainly the factor by which the size of the blocks is changed between adjacent levels of the pyramidal representation. There are 9x9 blocks in this example (from Miss America), where the gray level of each block represents the number of code-words used. The maximum number of codewords is 512 (white in Fig. 3).

3   Implementation and Results

The system was implemented according to the structure illustrated in Figs. 2 and 4, with three pyramidal levels. The length of the history used for training the codebooks was of 28 frames (approx. one second), with a delay of 4 frames. To avoid the need to transmit the codebooks to the receiving end, the recent history of the transmitted sequence was used for training the codebooks so that the same codebooks could be created at the remote receiver, thus only an index representing each codebook was transmitted via the transmission medium.

A typical result is shown in Fig. 5. In this example a compression ratio of more than 100:1 at 0.078 bpp was obtained, with high quality of the perceived sequence.

4   Summary

This paper has presented a new model-based approach to video compression using localized history of the sequence as a training set for vector quantization. In addition to the quality of the resultant sequences, the implementation of this approach can be systematically organized in parallel by its very nature, since localized codebooks are created for each block and searched for independently. It should be noted that even for communication by serial machines, reduced complexity is achieved by processing of many small codebooks instead of a combined one.

This approach resembles the technique described in [12,13], however, here the codebooks and the VQ process are localized, thus providing both higher quality and more efficient results. Based on its performance and the quality of the compressed sequences, it is suggested that the new localized approach to video compression be further analyzed and integrated in presently available methods.

References

1. IEEE Trans. Image Processing, *Special Issue on Sequence Coding*, (September 1994). 278

2. C. S. Choi, H. Harashima, T. Takebe: Analysis and synthesis of facial expressions in knowledge-based coding of facial image sequences. *IEEE ICASSP* (1991). 278

3. M. Kunt, A. Ikonomopoulos, M. Kocher: Second-generation image-coding techniques. *Proc. of the IEEE*, **73** (1985) 549–573. 278

4. C. I Podilchuk, N. S. Jayant, P. Noll: Sparse codebooks for the quantization of non-dominant sub-bands in image coding. *IEEE ICASSP*, (1990) 2101–2104. 278

5. G. R. Giunta, T. R. Reed, M. Kunt: Image sequence coding using oriented edges. *Image Communication*, **2** (1990) 429–440. 278

6. MC. I. Podilchuk, N. S. Jayant, N. Farvardin: 3-D subband coding of video. *IEEE Trans. on Image Processing*, **4** (1995) 125–139. 278

7. J. W. Woods, S. D. ONeil: Subband coding of images. *IEEE Trans. on Signal Processing*, **ASSP-34** (1986) 1278–1288. 278

8. M. Porat, Y. Y. Zeevi: The generalized gabor scheme in biological and machine vision. *IEEE Trans. on Pattern Analysis and Machine Intelligence*, **PAMI-10** (1988) 452–468. 278

9. N. Katzir, M. Lindenbaum, M. Porat: Curve segmentation under partial occlusion. *IEEE Trans. on Pattern Analysis and Machine Intelligence*, **PAMI-16** (1994) 513–519. 279

10. M. Porat, Y. Y. Zeevi: Localized Texture processing in vision: analysis and synthesis in the gaborian space. *IEEE Trans. on Biomedical Engineering*, **BME-36** (1989) 115–129. 279

11. Y. L. Linde, A. Buzo, R. M. Gray: An algorithm for vector quantizer design. *IEEE Trans. on Communication*, **28** (1980) 84–95. 279

12. S. Panchanathan, M. Goldberg: Adaptive algorithms for image coding using vector quantization. *Signal processing: Image Communication 4*, (1991) 81–92. 282

13. M. Goldberg, H. -F. Sun: Image sequence coding by three-dimensional block vector quantization. *IEEE Proceedings*, **133**, Pt. F, No. 5 (1986) 482–486. 282

Author Index

Lecture Notes in Computer Science

For information about Vols. 1–1913
please contact your bookseller or Springer-Verlag

Vol. 1944: K.R. Dittrich, G. Guerrini, I. Merlo, M. Oliva, M.E. Rodriguez (Eds.), Objects and Databases. Proceedings, 2000. X, 199 pages. 2001.

Vol. 1945: W. Grieskamp, T. Santen, B. Stoddart (Eds.), Integrated Formal Methods. Proceedings, 2000. X, 441 pages. 2000.

Vol. 1946: P. Palanque, F. Paternò (Eds.), Interactive Systems. Proceedings, 2000. X, 251 pages. 2001.

Vol. 1948: T. Tan, Y. Shi, W. Gao (Eds.), Advances in Multimodal Interfaces – ICMI 2000. Proceedings, 2000. XVI, 678 pages. 2000.

Vol. 1949: R. Connor, A. Mendelzon (Eds.), Research Issues in Structured and Semistructured Database Programming. Proceedings, 1999. XII, 325 pages. 2000.

Vol. 1950: D. van Melkebeek, Randomness and Completeness in Computational Complexity. XV, 196 pages. 2000.

Vol. 1951: F. van der Linden (Ed.), Software Architectures for Product Families. Proceedings, 2000. VIII, 255 pages. 2000.

Vol. 1952: M.C. Monard, J. Simão Sichman (Eds.), Advances in Artificial Intelligence. Proceedings, 2000. XV, 498 pages. 2000. (Subseries LNAI).

Vol. 1953: G. Borgefors, I. Nyström, G. Sanniti di Baja (Eds.), Discrete Geometry for Computer Imagery. Proceedings, 2000. XI, 544 pages. 2000.

Vol. 1954: W.A. Hunt, Jr., S.D. Johnson (Eds.), Formal Methods in Computer-Aided Design. Proceedings, 2000. XI, 539 pages. 2000.

Vol. 1955: M. Parigot, A. Voronkov (Eds.), Logic for Programming and Automated Reasoning. Proceedings, 2000. XIII, 487 pages. 2000. (Subseries LNAI).

Vol. 1956: T. Coquand, P. Dybjer, B. Nordström, J. Smith (Eds.), Types for Proofs and Programs. Proceedings, 1999. VII, 195 pages. 2000.

Vol. 1957: P. Ciancarini, M. Wooldridge (Eds.), Agent-Oriented Software Engineering. Proceedings, 2000. X, 323 pages. 2001.

Vol. 1960: A. Ambler, S.B. Calo, G. Kar (Eds.), Services Management in Intelligent Networks. Proceedings, 2000. X, 259 pages. 2000.

Vol. 1961: J. He, M. Sato (Eds.), Advances in Computing Science – ASIAN 2000. Proceedings, 2000. X, 299 pages. 2000.

Vol. 1963: V. Hlaváč, K.G. Jeffery, J. Wiedermann (Eds.), SOFSEM 2000: Theory and Practice of Informatics. Proceedings, 2000. XI, 460 pages. 2000.

Vol. 1964: J. Malenfant, S. Moisan, A. Moreira (Eds.), Object-Oriented Technology. Proceedings, 2000. XI, 309 pages. 2000.

Vol. 1965: Ç. K. Koç, C. Paar (Eds.), Cryptographic Hardware and Embedded Systems – CHES 2000. Proceedings, 2000. XI, 355 pages. 2000.

Vol. 1966: S. Bhalla (Ed.), Databases in Networked Information Systems. Proceedings, 2000. VIII, 247 pages. 2000.

Vol. 1967: S. Arikawa, S. Morishita (Eds.), Discovery Science. Proceedings, 2000. XII, 332 pages. 2000. (Subseries LNAI).

Vol. 1968: H. Arimura, S. Jain, A. Sharma (Eds.), Algorithmic Learning Theory. Proceedings, 2000. XI, 335 pages. 2000. (Subseries LNAI).

Vol. 1969: D.T. Lee, S.-H. Teng (Eds.), Algorithms and Computation. Proceedings, 2000. XIV, 578 pages. 2000.

Vol. 1970: M. Valero, V.K. Prasanna, S. Vajapeyam (Eds.), High Performance Computing – HiPC 2000. Proceedings, 2000. XVIII, 568 pages. 2000.

Vol. 1971: R. Buyya, M. Baker (Eds.), Grid Computing – GRID 2000. Proceedings, 2000. XIV, 229 pages. 2000.

Vol. 1972: A. Omicini, R. Tolksdorf, F. Zambonelli (Eds.), Engineering Societies in the Agents World. Proceedings, 2000. IX, 143 pages. 2000. (Subseries LNAI).

Vol. 1973: J. Van den Bussche, V. Vianu (Eds.), Database Theory – ICDT 2001. Proceedings, 2001. X, 451 pages. 2001.

Vol. 1974: S. Kapoor, S. Prasad (Eds.), FST TCS 2000: Foundations of Software Technology and Theoretical Computer Science. Proceedings, 2000. XIII, 532 pages. 2000.

Vol. 1975: J. Pieprzyk, E. Okamoto, J. Seberry (Eds.), Information Security. Proceedings, 2000. X, 323 pages. 2000.

Vol. 1976: T. Okamoto (Ed.), Advances in Cryptology – ASIACRYPT 2000. Proceedings, 2000. XII, 630 pages. 2000.

Vol. 1977: B. Roy, E. Okamoto (Eds.), Progress in Cryptology – INDOCRYPT 2000. Proceedings, 2000. X, 295 pages. 2000.

Vol. 1979: S. Moss, P. Davidsson (Eds.), Multi-Agent-Based Simulation. Proceedings, 2000. VIII, 267 pages. 2001. (Subseries LNAI).

Vol. 1983: K.S. Leung, L.-W. Chan, H. Meng (Eds.), Intelligent Data Engineering and Automated Learning – IDEAL 2000. Proceedings, 2000. XVI, 573 pages. 2000.

Vol. 1984: J. Marks (Ed.), Graph Drawing. Proceedings, 2001. XII, 419 pages. 2001.

Vol. 1987: K.-L. Tan, M.J. Franklin, J. C.-S. Lui (Eds.), Mobile Data Management. Proceedings, 2001. XIII, 289 pages. 2001.

Vol. 1989: M. Ajmone Marsan, A. Bianco (Eds.), Quality of Service in Multiservice IP Networks. Proceedings, 2001. XII, 440 pages. 2001.

Vol. 1991: F. Dignum, C. Sierra (Eds.), Agent Mediated Electronic Commerce. VIII, 241 pages. 2001. (Subseries LNAI).

Vol. 1992: K. Kim (Ed.), Public Key Cryptography. Proceedings, 2001. XI, 423 pages. 2001.

Vol. 1995: M. Sloman, J. Lobo, E.C. Lupu (Eds.), Policies for Distributed Systems and Networks. Proceedings, 2001. X, 263 pages. 2001.

Vol. 1998: R. Klette, S. Peleg, G. Sommer (Eds.), Robot Vision. Proceedings, 2001. IX, 285 pages. 2001.

Vol. 2004: A. Gelbukh (Ed.), Computational Linguistics and Intelligent Text Processing. Proceedings, 2001. XII, 528 pages. 2001.

Vol. 2010: A. Ferreira, H. Reichel (Eds.), STACS 2001. Proceedings, 2001. XV, 576 pages. 2001.

.